Concrete durability

A practical guide to the design of durable
concrete structures

Concrete durability

A practical guide to the design of durable concrete structures

Edited by
Marios Soutsos
University of Liverpool, UK

Published by Thomas Telford Limited, 40 Marsh Wall, London E14 9TP.
www.thomastelford.com

Distributors for Thomas Telford books are
USA: ASCE Press, 1801 Alexander Bell Drive, Reston, VA 20191-4400
Australia: DA Books and Journals, 648 Whitehorse Road, Mitcham 3132, Victoria

First published 2010
Also available from Thomas Telford Limited
Concrete: Neville's insights and issues. A. Neville. ISBN: 978-0-7277-3468-6
ICE manual of construction materials. Editor M. Forde. ISBN: 978-0-7277-3597-3
Concrete bridge strengthening and repair. I. Kennedy Reid. ISBN: 978-0-7277-3603-1

A catalogue record for this book is available from the British Library

ISBN: 978-0-7277-3517-1

FSC
Mixed Sources
Product group from well-managed
forests and other controlled sources
Cert no. SGS-COC-2953
www.fsc.org
© 1996 Forest Stewardship Council

Typeset by Academic + Technical, Bristol
Printed and bound in Great Britain by Antony Rowe Limited, Chippenham
Index created by Indexing Specialists (UK) Ltd, Hove, East Sussex

Contents

Foreword

Far more concrete is produced than any other manufactured material. It is the basic material for the construction industry that employs 7% of the workforce worldwide and over half in some countries. One tonne of concrete is produced every year per head of population on the planet.

Over a hundred years ago in 1907 Knudson showed in his paper entitled 'Electrolytic Corrosion of Iron and Steel in Concrete' that the passage of a small current through the reinforcement in concrete would cause corrosion. His opinion that stray currents were the cause of the problem was shared by most researchers at the time but within a few years Rosa, McCullom and Peters had concluded that 'the presence of chlorides always facilitated trouble' ('Electrolysis of Concrete', 1912). At the time this was written the chlorides that did the damage were almost always from sea water. It was not until half a century later that winter salting of roads became widespread and soon after that extensive deterioration in highway structures was observed. This corrosion is the most significant of a wide range of different problems that affect modern structures that are discussed in this book.

Clients are now expecting their structures to last longer with service lives over 100 years being demanded in some cases. In order to avoid the liability for long term repairs they are looking for a 'warranty' from the construction company and, in many cases, giving them contractual responsibilities for long term maintenance. The contractors can therefore no longer bid on the cheapest solution. They must consider the long term durability of the structure to avoid spending their profits on repairs.

Durability is a very difficult area for design. A structural engineer can decide on the size and shape of components and carry out the necessary calculations (or get a computer to do them) and determine load bearing capabilities with great accuracy. No such process is possible with considerations of durability. The mix design of concrete can be specified and the exposure conditions determined (often not very accurately) but the only things that these will affect directly are properties of the matrix such as the hydrate structure and the pore size and fluid chemistry. All that these will affect is the transport properties and it is these that finally determine the durability of the structure. The link between what we can control and the result we get is therefore extremely complex and can only be determined by experimental or probabilistic methods.

To add further complications to a complex problem, concrete practice is currently going through a series of fundamental changes. These have been driven partly by costs but primarily by environmental concerns. Cement production is responsible for approximately 5% of the world's carbon dioxide emissions. To put this into context it should be noted that CO_2 emissions for concrete are 20 times less per tonne than that of steel but the quantities are vastly greater. Thus there is enormous pressure to replace as much cement as possible with other materials. The use of these materials and their effects on durability are discussed in detail in this book.

The final issue is that of workmanship. Despite many advances in quality control the plans and specifications which are produced by the designer are often not accurately followed. The depth of cover concrete which protects the steel from the exterior environment often ends up less than that specified. Also the curing which should prevent the concrete from drying out during the days after it is poured is often not effective. On road bridges drainage systems are notorious for failing to keep the damaging salt water away from vulnerable concrete surfaces. These factors combine to make failures of even the best designs if inadequate provision is made to prevent, or allow for, poor construction.

In the past there has been a clear division between the structural engineers who have carried out designs and the concrete technologists who have often only become involved when durability problems have arisen. This has come about due to conflicting pressures on the syllabus in civil engineering courses. But the designers are now finding that durability considerations have become a central part of the process and must be considered at every stage from concept to completion. Those wishing to study the topic have found the relevant information to be fragmented partly due to the massive volume of research that has been published in the numerous journals that cover the area. Much of the information is also presented in terms of complex physical and chemical theories. This book is a compilation of essays by a selection of leading practitioners covering the full range of durability topics at a level that can be understood by engineers without the need to study the science in great detail.

Professor Peter Claisse
Coventry University

xii

List of contributors

Christopher Atkins, *Mott MacDonald, Altrincham, UK*
P. A. Muhammed Basheer, *Queen's University Belfast, Northern Ireland, UK*
Salim Barbhuiya, *Queen's University Belfast, Northern Ireland, UK*
John A. Bickley, *John A. Bickley Associates Ltd., Toronto, Canada*
John P. Broomfield, *Broomfield Consultants, Surrey, UK*
Klaas van Breugel, *Civil Engineering & Geosciences, TU Delft, The Netherlands*
David J. Cochrane, *Consultant to the Nickel Institute, Sidcup, UK*
Michael D. Connell, *Hanson Cement, UK*
Oguzhan Copuroglu, *TU Delft, The Netherlands*
Ken Day , *Independent Concrete Technologist, Australia, www.kenday.id.au*
Odd E. Gjorv, *Norwegian University of Science and Technology, Trondheim, Norway*
Douglas Gremel, *Hughes Brothers, Inc., Seward, NE, USA*
Robert D. Hooton, *University of Toronto, Canada*
Kenneth C. Hover, *Cornell University, Ithaca, New York, USA*
Scott Humphreys, *Concrete Reinforcing Steel Institute, Schaumburg, Illinois, USA*
Desmond Makepeace, *Galvanizers Association, Sutton Coldfield, UK*
Konstantin Kovler, *Technion – Israel Institute of Technology, Haifa, Israel*
Robert C. Lewis, *Elkem Ltd, Sheffield, UK*
Bryan Marsh, *Arup, London, UK*
Alan Poole, *A. B. Poole & Associates, Oxford, UK*
Alan Richardson, *Northumbria University, UK*
Mustafa Sahmaran, *Gaziantep University, Gaziantep, Turkey*
Lindan Sear, *UK Quality Ash Association, Wolverhampton, UK*
Eric Schlangen, *TU Delft, The Netherlands*
Asia Shvarzman, *Sami Shamoon College of Engineering, Beer-Sheva, Israel*

Marios Soutsos, *University of Liverpool, UK*
Robert F. Viles, *Fosroc International Ltd, Tamworth, UK*
Guang Ye, *TU Delft, The Netherlands*
Donald Wimpenny, *Halcrow Pacific Pty, Australia*

1

Introduction

Marios Soutsos, *University of Liverpool, UK*

Concrete is, due to its versatility, comparative cheapness and energy efficiency, of great and increasing importance for all types of construction throughout the world. Concrete structures can be durable and long lasting, but to be so, due consideration needs to be given at the design stage to the effect that the environment to which the structure will be exposed to will have on the concrete. Degradation can result from either the environment to which the concrete is exposed, for example frost damage, or from internal causes within the concrete, as in the alkali–aggregate reaction. It is also necessary to distinguish between degradation of the concrete itself and loss of protection and subsequent corrosion of the steel reinforcement. ACI Committee 201[1] defined concrete durability as: 'its resistance to deteriorating influences which may through inadvertence or ignorance reside in the concrete itself, or which are inherent in the environment to which it is exposed'.

Initially, concrete was regarded as having an inherently high durability, but more recent experiences have shown that this is not necessarily the case unless durability design forms an integral part of the design and construction process. There is a need to consider all potential deterioration mechanisms at the design stage in order to select and specify an appropriate concrete mixture from a durability perspective. Prescriptive specification for concrete based on the permissible maximum water/cement ratio and the minimum cement content has received much criticism in recent years. It may even have inadvertently allowed designers and contractors a way of avoiding to consider or implement all the available information required for a sound design for durable construction. Unexpected maintenance and repairs arising very early in the specified service life of structures have caused enormous financial burdens to clients. The US Department of Defense funded a major research project entitled 'Concrete durability – a multibillion-dollar opportunity' to try and determine whether the problem with

Concrete durability
978-0-7277-3517-1

concrete durability was a technical or an institutional problem.[2] It concluded that:

> Most of the knowledge exists which, if properly applied, would produce durable concrete. The lack of proper application may be attributed to a lack of knowledge by practitioners and the system's failure to make durability the responsibility of the organisation which can most directly provide it – the contractor.

It is not so much the lack of knowledge of the design team but the failure to recognise the magnitude and financial costs of potential durability problems leading to unexpected and very high maintenance and repair costs. Yet, one has only to look through some back issues of the *New Civil Engineer*, the weekly journal of the Institution of Civil Engineers, to realise that it is not only engineering achievements that are reported but also engineering failures, irrespective of whether these are structural or serviceability, i.e. durability failures. Such examples include the following:

1.1   Concrete condemned[3]

A multi-storey car park in Colchester is to be pulled down because concrete corrosion problems are beyond economic repair (Fig. 1.1).

Fig. 1.1 A multi-storey car park in Colchester was pulled down in 1995 because concrete corrosion problems were beyond economic repair

The Queens Street car park was only put up in 1971 under a design and build contract by the now defunct contractor Shears-Neal. By 1985 evidence of corrosion of precast concrete units on the in-situ concrete frame prompted Colchester Borough Council to commission consultant Eastwood & Partners to report on the structure. Eastwood found 3% of units to be affected and made repair recommendations to prolong its life by five years. In 1992 the car park, which sits over a live bus station, was closed for safety reasons. Eastwood further reported that 40% of units were affected. Refurbishment costs were put at £1.5M, rebuild at £3M and demolition at £350,000. The council has opted for demolition which, subject to final approval on 4 April, will go ahead within the next two months.

1.2 Approach shot – repairs to a heavily used bridge in Runcorn have to be carried out without disruption to traffic[4]

Problems on the 34 year old 27 span crossing with its striking 330 m central lattice steel arch (Fig. 1.2), were first identified in 1989. The

Fig. 1.2 Runcorn bridge

three span bridge itself, with its concrete deck, remains in reasonably good condition thanks to owner Cheshire County Council's £1M a year maintenance programme. But the all concrete approaches presented a less favourable report. Each is made up of four in-situ longitudinal deck beams carried generally on a single central pier with integral crosshead. The ever familiar story of road salts seeping down through leaking deck expansion joints to attack beams, crossheads and piers, is all too evident on both sides of the Mersey.

Worst damage lies beneath the Widnes approaches which were widened in 1977 by adding a separately supported 5 m wide strip of deck. Here the main culprit is a 500 mm wide longitudinal infill slab which connects original and extension sections of the deck. Flexible joints supporting both sides of the infill are leaking, allowing chloride rich surface water to run down. Chloride levels in the rectangular beam which is up to 2 m deep, approach 2% by weight of cement. Concrete has spalled and delaminated with the link steel attacked. But the four layers of densely packed main 50 mm rebar are so far relatively unscathed.

Cathodic protection, now being installed in the chloride riddled approach spans to Europe's largest steel arch bridge, is fulfilling two vital functions. It is avoiding widespread demolition and reconstruction of the damaged spans for one of the River Mersey's most strategic crossings. And it is also hosting the first full scale application of micro concrete gunite overlay claimed to have solved the historic package of problems surrounding such specialised sprayed concrete applications.

1.3   Terminal Operation – miracle cure[5]

What to do about Marsh Mills Viaducts has been a major preoccupation of engineers in the West Country ever since the realisation 15 years ago that the elegant concrete structure was condemned to a lingering but terminal decline. Revelation that Marsh Mills viaducts were afflicted by alkali silica reaction (ASR) which would inexorably burst it apart came as a shock. The industry had assumed that ASR was a technically interesting cause of deterioration to concrete overseas but generally of only academic interest in Britain. ASR deterioration of structures such as Charles Cross car park in Plymouth and the foundations of electricity sub-stations in the South West had previously been dismissed as freak incidents. Cracking on the viaduct, which it transpired had already been monitored for 18 months when headlined by *New Civil*

Engineer (NCE), was dismissed at the time by Devon County Council engineers as something 'rather superficial'.

The trouble was caused by alkali rich cement from the nearby Plymstock works used in combination with certain sea dredged aggregates and aggravated by road deicing salt. Moisture is required for the reaction, which produces an expansive gel which bursts the concrete structure apart, the internal expansion causing a characteristic map cracking effect on the surface.

Discovery of the full extent of problems at Marsh Mills prompted a nationwide examination of other highway structures. Many were found to be in trouble to a greater or lesser degree and several were replaced. As well as March Mills, there were several other reinforced concrete bridges on the 1969/70 A38 highway between Exeter and Plymouth. Measures adopted were to observe and contain the problem with remedial works such as weather shields to extend the working life of structure until such time as replacements could be built.

Miracle cure? Bold innovation won Hochtief the contract to replace Plymouth's concrete cancer-crippled Marsh Mills viaducts and has given the Highways Agency a design and build bargain at £12.25M. The idea is simple. Its execution nerve wracking. Traffic diversions could be avoided almost entirely figured Hochtief and its designer Tony Gee & Partners; just assemble the new viaduct decks on temporary supports beside the old structures while building the permanent foundations and piers beneath them; transfer traffic away for a few hours while each new viaduct is slewed sideways on to the permanent supports (Fig. 1.3). Slewing in the viaducts will probably involve the biggest such bridge jacking operation ever attempted. Each sliproad deck will be some 400 m long, weigh about 5250 tonnes and be supported on bearings sliding on tracks set on seven or eight intermediate piers. Just for good measure the viaducts are each set out on a curve with a severe gradient and a crossfall.

Motivation for this extreme solution comes from the lane licence charges imposed by the Highways Agency. Overnight closure of any two lanes of the A38 will cost the contractor £5000, at a weekend £18,000 a day and during the week a thumping £25,000 a day. In effect Hochtief is saving these charges and spending money instead on extensive temporary works. The crippled viaducts carry sliproads for the A38 up to 12 m above the Plym Valley. They are heavily trafficked and tightly constrained by obstacles including the main railway line to Penzance, the River Plym and buried services including most of the trunk gas, water and electricity supplies to Plymouth.

5

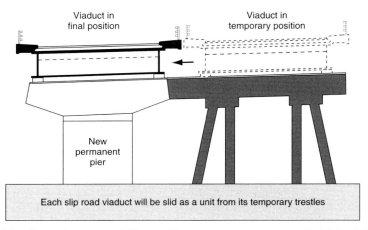

Fig. 1.3 *The replacement of Plymouth's concrete cancer – crippled Marsh Mills viaducts*

1.4 Lasting effect: the collapse of the Ynys-y-Gwas Bridge led to a ban on grouted tendons[6]

Bridge owners have for many years, been concerned about the corrosion of prestressing cables and the difficulty of inspection. These concerns were highlighted in December 1985 with the sudden collapse of a 32-year old 18.3 m span post-tensioned segmental road bridge in South Wales (Fig. 1.4). The failure of the Ynys-y-Gwas Bridge was directly caused by tendons corroded by chlorides from de-icing salts. The salt penetration was eventually attributed to a combination of inadequate tendon protection, poor workmanship and ineffective deck water-proofing. Other key factors identified included the lack of an in-situ top slab and joints opening under load.

Although possibly the most newsworthy, this is by no means the only bridge to have had problems. In September 1992 the Department of Transport's concern as an owner and client led to the announcement of a temporary ban on the commissioning of any new bridges of the 'grout duct post-tensioned type' until specifications had been reviewed. Construction of some bridges, already designed using bonded internal prestress, was allowed to continue. The Department of Trade's decision in effect laid down a challenge to the UK concrete bridge industry to put its house in order and to be able to demonstrate it had done so. The response by the Concrete Society, supported by the Concrete Bridge Development Group, was to set up a working party in June 1992 to

6

Fig. 1.4 The collapse of the Ynys-y-Gwas Bridge led to a ban on grouted tendons

study the problem and prepare recommendations. In May 1994 the working party held a seminar which summarised the position at that time. Detailed discussions started with the Highways Agency in April 1995 with a view to making use of the revised design and construction procedures, to allow a phased re-introduction of bonded post-tensioned bridges. *Concrete Society Technical Report 47: Durable bonded post-tensioned concrete bridges* was published in 1996.[7]

1.5 Thaumasite test for Cotswolds[8]

Fears that thaumasite sulphate attack in concrete could also hit thousands of house foundations have prompted the government to commission a major in-situ research project, it emerged this week (Fig. 1.5). The research will take place at a site in the Cotswold village of Shipston-on-Stour, where cases of thaumasite in house foundations were discovered as long ago as 1990 (*NCE* 23 April). The tests will monitor 176 concrete specimens over periods of three and 10 years. Contractors are expected to finish placing samples by the end of the week. Some 43 mix types will be used with four different aggregates and 10 different binders. A combination of pre-cast and cast in-situ specimens will be placed in two 11 m long by 2.5 m deep trenches.

The research is a clear indication of the Government's increasing fears over the possible scale of the problem. The Highways Agency

Fig. 1.5 Thaumasite attack on bridges on the M5 in Gloucestershire

revealed on Monday that it has discovered 10 more cases of thaumasite on bridges on the M5 in Gloucestershire. A further 27 structures are understood to have been identified as vulnerable to thaumasite attack on Gloucestershire county roads.

The subject of thaumasite attack was in the spotlight for almost a year.[9-12] The government's Expert Group produced, within an eight-month investigation, a guidance document on thaumasite attack.[13] Articles on thaumasite, however, continued to appear months after its publication.[14-17]

1.6   Joint detailing key to flat slab[18]

The Health Safety Executive has this week published its report into the 1997 Pipers Row multistorey car park collapse in Wolverhampton (Fig. 1.6).[19] The research identified the lift slab technique as being susceptible to punching shear when there is concrete deterioration in the slab/column zone.

Pipers Row was a punching shear failure and occurred with only dead load acting, implying an apparent major reduction in strength. The failure was initiated by local deterioration of the concrete, confirmed both by examination of the debris and by subsequent rigorous structural

Fig. 1.6 The collapse of Pipers Row multi-storey car park in Wolverhampton

analysis. The concrete in the collapsed top slab was of lower quality than elsewhere in the structure. It was exposed to the elements for over 30 years and undoubtedly went through many freeze–thaw cycles. Frost action led to the concrete becoming friable, penetrating into the slab near an internal column, as far as the top steel, eventually leading to a complete loss of bond and anchorage. Rigorous analysis of Pipers Row clearly showed that the undeteriorated as-built strength had a safety margin of at least 1.5 against the worst possible in-service load.

The expectation of the owner of a structure is that it will only require very little or no maintenance during its design life. The owners have realised that the cheapest option for constructing a structure may work out to be an expensive option in the long run, and they have therefore sought ways of minimising project risks to themselves. The design–bid–build delivery system was the norm, where the owner contracts separately the design and construction of a project. However, they then adopted design–build delivery systems, where from inception to completion only one organisation is liable to the owner for defects, delays and losses. Streamlining the delivery system reduced the delivery

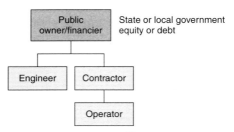

Fig. 1.7 Design–build–operate (Maintain) [20]

time of the completed project, by forcing consultancy/design teams and contractors/construction companies to form collaborations and to complete the separate tasks at the same time, i.e. working in parallel. This system is used to minimise the project risk for an owner and to reduce the delivery schedule by overlapping the design phase and construction phase of a project. However, this approach does not take 'life cycle costing' into account. The benefits of 'life cycle costing' are particularly important, as most infrastructure owners spend more money maintaining their systems than on expansion. In addition, the life cycle approach removes important maintenance issues from the political vagaries affecting many maintenance budgets, with owners often not knowing how much funding will be available to them from year to year. In such cases, they are often forced to spend what money they do have on the most pressing maintenance needs rather than a more rational and cost-effective, preventive approach. Major infrastructure projects have now moved to design–build–operate (maintain) or 'turnkey' procurement, e.g. the US Department of Transportation – Federal Highway Administration defines it as an integrated partnership that combines the design and construction responsibilities of design–build procurements with operations and maintenance (see Fig. 1.7). [20]

The advantage of the design–build–operate (maintain) approach is that it combines responsibility for usually disparate functions – design, construction and maintenance – into a single entity. This allows the private partners to take advantage of a number of efficiencies. The project design can be tailored to the construction equipment and materials that will be used. In addition, the design–build–operate (maintain) team is also required to establish a long-term maintenance programme up front, together with estimates of the associated costs. The team's detailed knowledge of the project design and the materials utilised allows it to develop a tailored maintenance plan that anticipates and addresses needs as they occur, thereby reducing the risk that issues will go unnoticed or unattended and then deteriorate into much more costly problems.

Few structures fall down in the UK, but when they do, the consequences and ramifications are huge. 'Avoid the complacency which leads to tragedy' was the central theme of the Standing Committee on Structural Safety's 12th bi-annual report.[21]

It must be realised that a basic understanding of all the degradation processes is needed by all involved in the design and/or construction of reinforced-concrete structures. Potential problems can be eliminated, or at least minimised, by due consideration to durability criteria in the design and specification of new structures. Ignorance of, or lack of attention to, such criteria in the past has led to a thriving and expanding repair industry in recent years (see Fig. 1.8); it is to be hoped that today's

Concrete repair on the increase

CONCRETE REPAIRS are on the increase. The value of work completed by members of the Concrete Repair Association during the first half of 1995 was a 'marked improvement' on the previous six months.

The latest CRA state of trade survey, published this week, shows a lift in enquires from all sectors – private, public and civil engineering.

A CRA spokesman said: 'This is the fifth half-yearly survey we have conducted. The first three showed a buoyant industry, although the results of the fourth one were bad. The latest shows that things are getting better.'

Encouragingly, the new figures reveal that the interval between enquiries being received and work being let shortened for the first time in three years. However, contractor members report no change with regard to repair work in hand.

The UK concrete repair market is worth £130 million a year. After expansion in 1993, the second half of 1994 brought a large downturn in workload which has left margins under pressure.

Concrete repair is primarily con-

BY JOHN LEITCH

nected with refurbishment work.

Enthusiasm for the improvement in the value of concrete repair work in 1995 is muted by the fact that the number of contracts completed during the peri-

od fell significantly. The main cut back was in larger value projects.

Half of CRA's members are working at less capacity than they expected. They expect no improvement in profit margins over the forthcoming 12 months and are pessimistic about work volumes.

The UK concrete repair market is worth £130 million a year

Fig. 1.8 The thriving and expanding repair industry in the UK[22] (courtesy of Contract Journal)

practitioners will be able to learn from these lessons and reduce the need for such activities in the future.

References

1. ACI Committee 201 (2001) *Guide to durable concrete*, ACI 201.2R-01. American Concrete Institute, Farmington Hills, MI, p. 41.
2. National Research Council Committee (1988) Concrete durability – a multibillion-dollar opportunity. *Concrete International*, Jan., 33–35.
3. New Civil Engineer (1995) Concrete condemned. *New Civil Engineer*, 30 Mar., 6.
4. New Civil Engineer (1995) Approach shot – repairs to a heavily used bridge in Runcorn have to be carried out without disruption to traffic. *New Civil Engineer*, Mar., 26–27.
5. New Civil Engineer (1995) Terminal operation – miracle cure. *New Civil Engineer*, 23 Mar., 18–21.
6. New Civil Engineer (1995) Lasting effect: the collapse of the Ynys-y-Gwas Bridge led to a ban on grouted tendons. NCE concrete supplement, 46–48.
7. Concrete Society Working Party in collaboration with the Concrete Bridge Development Group (1996) *Durable bonded post-tensioned concrete bridges.* Technical Report 47. Concrete Society, Crowthorne.
8. New Civil Engineer (1998) Thaumasite test for Cotswolds. *New Civil Engineer*, 21 May, 3.
9. New Civil Engineer (1998) Sulphate attack hits M5 bridges. *New Civil Engineer*, 2 Apr., 3.
10. New Civil Engineer (1998) Industry ignored warnings on thaumasite threat. *New Civil Engineer*, 9/16 Apr., 3.
11. New Civil Engineer (1998) More sulphate casualties come to light. *New Civil Engineer*, 23 Apr., 8.
12. New Civil Engineer (1998) Crash damage clouds M5 thaumasite case. *New Civil Engineer*, 11 June, 8.
13. Thaumasite Expert Group (1999) *The thaumasite form of sulphate attack: risks, diagnosis, remedial works and guidance on new construction.* Department of the Environment, Transport and the Regions, London.
14. New Civil Engineer (1999) Experts cast doubt on Thaumasite Report. *New Civil Engineer*, 28 Jan., 3–4.
15. New Civil Engineer (1999) No need to panic. *New Civil Engineer*, 11 Feb., 12–13.
16. New Civil Engineer (1999) New thaumasite evidence. *New Civil Engineer*, 4 Mar., 6.
17. New Civil Engineer (1999) Contractors bear thaumasite responsibility. *New Civil Engineer*, 15 Apr., 6.

18. Somerville, G.S. (2004) Joint detailing key to flat slab. *New Civil Engineer*, 8 Jan.

19. Wood, J.G.M. (1997) *Pipers Row Car Park, Wolverhampton – quantitative study of the causes of the partial collapse on 20 March 1997*. Health and Safety Executive, London. www.hse.gov.uk/research/misc/pipersrow.htm.

20. US Federal Highway Administration. *PPPs defined – design build operate (maintain)*. FHA, Washington, DC. www.fhwa.dot.gov/PPP/defined_dbom.htm.

21. Standing Committee on Structural Safety (1999) *Structural Safety 1997–99: review and recommendations. 12th report of SCOSS*. SCOSS, London. www.scoss.org.uk/publications/rtf/12Report.pdf.

22. Leitch, J. (1995) Concrete repair on the increase. *Contract Journal*.

2

Pore structure and transport processes

P. A. Muhammed Basheer and Salim Barbhuiya, Queen's University Belfast, Northern Ireland, UK

2.1   Introduction

Concrete durability depends largely on the ease (or difficulty) with which fluids in the form of liquid (water), gas (carbon dioxide, oxygen) or ions (chlorides, sulfates) can migrate through the hardened concrete mass. Concrete is a porous material and, hence, moisture movement can occur by flow, diffusion or sorption. Generally, the overall potential for moisture, gas and ion ingress in concrete by these three modes is referred to as its *permeability*, but permeability does not strictly represent all the three mechanisms of transport. The durability of concrete is also influenced or controlled by the number, type and size of pores present. It is believed that capillary voids of 50 μm in diameter, referred to as macro-pores, are detrimental to strength and impermeability, whereas voids smaller than 50 μm, referred to as micro-pores, are more related to drying shrinkage and creep.[1] The water existing in pores bigger than 50 nm behaves as free water and plays an important role in the durability of concrete. Therefore, to improve durability a clear understanding of the characteristics of the pore structure and permeation properties of concrete is very much essential.

2.2   Pore structure of concrete

Properties of concrete are strongly affected by the manner in which the pores of various sizes are distributed within it. Concrete with the same total pore volume can exhibit entirely different properties, depending on whether it contains a small number of large pores or a large number of small pores. Concrete is a porous material, and its pore structure is extremely tortuous, complex and spatially inhomogeneous. For example, the cement paste close to aggregates, i.e. in the interfacial transition zone (ITZ), has a different pore structure than the bulk cement paste. In addition, fine and coarse aggregate particles in concrete may have their own pore systems, which might be completely different from the pore system of the cement paste. The pores present in concrete may also vary in size, shape and origin. The various types of pores which are present in concrete include:

- pores in the hydrated cement paste (HCP)
- pores in the aggregates
- pores associated with the ITZ
- water-filled capillary voids
- voids due to construction deficiencies, e.g. honeycombing due to poor compaction
- internal discontinuities in the HCP.

Figure 2.1 shows schematically the sequence of pore structure formation in HCP as the hydration proceeds. This involves the replacement of water that separates individual cement grains in the fluid paste (Fig. 2.1a) with solid hydration products. Owing to their lower specific gravity ($\sim$2.0 compared with 3.2 of cement), the hydration products

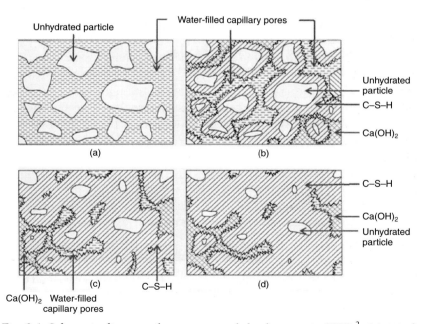

Fig. 2.1 Schematic diagram of pore structural development in HCP:[2] *(a) initial mix, (b) 7 days, (c) 28 days and (d) 90 days. (Reproduced from J. Bensted and P. Barnes, Structure and Performancer of Cement 2e, © 2002 Taylor and Francis)*

occupy a greater volume than the original cement compounds and, thus, form a dense matrix. The hydration products also bind the residual cement grains together over a period of time, as illustrated in Fig. 2.1(b)–(d). Although calcium sulfo-aluminates crystallise as separate phases, for simplification these are included as part of calcium silicate hydrate (C–S–H) in Fig. 2.1.

HCP is produced when cement chemically reacts with water. The HCP contains C–S–H gel, crystals of calcium hydroxide (Ca(OH)$_2$), calcium sulfoaluminate (ettringite), minor residues of original un-hydrated cement and residues of original water-filled spaces in the fresh paste. There are four main types of pores in HCP:

(*a*) gel pores
(*b*) capillary pores
(*c*) hollow-shell pores
(*d*) air voids.

The features and significance of each type of pore in the transport processes in concrete are given below.

2.2.1 Gel pores

Gel pores are the inherent pores in the C–S–H gel. Although the work by Powers[3] assumed that the gel pores have a width of 18 Å (1 Å = 10^{-10} m), work by Feldman and Sereda[4] suggested that these spaces may vary from 5 to 25 Å. These pores do not play any significant role in the flow of water through concrete because it has been shown[5–9] that pores with radius less than 500 Å do not contribute to the water permeability. However, they may contribute to the drying shrinkage and creep.[4]

2.2.2 Capillary pores

Capillary pores represent the spaces not filled by solid components of the HCP. Therefore, the volume and size of the capillary voids depend on the distance between unhydrated cement particles in the freshly mixed cement paste and the degree of hydration.[10,11] Unlike the gel pores, these pores are assumed to have a major effect on transport processes, but only a minor effect on the hydration rates. In well-hydrated and low water/cement ratio pastes, the capillary voids may range from 100 to 500 Å, whereas in high water/cement ratio pastes, at early ages of hydration the capillary voids may be as large as 3000 to 5000 Å.[6]

2.2.3 *Hollow-shell pores*

Hollow-shell pores are closed distinct pores, which are formed from the original cement grain boundaries, and their size ranges from 1 to 15 μm. At low water/cement ratios, they can be larger than the capillary pores by more than two orders of magnitude.[12]

2.2.4 *Air voids*

Air voids in concrete are due to either entrapped air during casting or intentional entraining by using an air-entraining admixture. The entrapped air cavities may be as large as 3 mm and the entrained air voids may range from 50 μm (500 000 Å) to 200 μm (2 000 000 Å). They are much bigger than the capillary voids, and have a significant role in the permeability of concrete.

2.3 Pore structure of the transition zone

A preferential orientation of large crystals of calcium hydroxide and calcium sulfo-aluminate results around the aggregates due to a local increase in the water/cement ratio close to them. A water film formed around the aggregates in the fresh compacted concrete and bleedwater collected underneath flaky coarse aggregates account for this variation in the water/cement ratio. As a result, concrete becomes more porous close to the aggregates, with larger sized capillary voids closer to them. Therefore, the properties of the transition zone significantly affect the permeability of concrete.

2.4 Techniques used to study pore structure

A wide variety of techniques have been used to characterise the pore structure of concrete. Some of the methods have access only to the open pores, whilst others have access to both open and closed pores. Each of these techniques is most suitable for a specific size range. The various techniques used in characterising the pore structure of concrete are listed below:

- mercury intrusion porosimetry
- gas adsorption
- displacement methods
- thermoporometry
- nuclear magnetic resonance

Table 2.1 Classification of pore sizes according to the general classification by IUPAC and concrete science terminology

According to IUPAC[12]		According to Mehta[1]		According to Mindess et al.[13]		
Name	Diameter	Pore type	Size range	Name	Name	Diameter
Micropores	<2 nm	Inter-particle space between C–S–H sheets	1–3 nm	Micropores inter layer	Gel pores	<0.5 nm
				Micropores		0.5–2.5 nm
				Small capillaries		
Mesopores	2–50 nm	Capillary pores (low w/c)	10–50 nm	Medium capillaries	Hollow-shell pores	2.5–10 nm
					Capillary pores	10–50 nm
Macropores	>50 nm	Capillary pores (high w/c)	3–5 µm	Large capillaries		50 nm–10 µm
		Entrained air voids	50 µm–1 mm	Entrained air		0.1–1 mm

w/c, water/cement ratio.

- small-angle scattering
- optical microscopy
- scanning electron microscopy.

Details on the suitability of these test techniques along with the range of pore sizes where each technique is applicable can be found in Aligizaki.[12]

2.5 Factors influencing pore structure

The main factors which influence the pore structure of concrete include the water/cementitious materials (w/c-m) ratio, degree of hydration, use of supplementary cementitious materials, the presence of chemical admixtures and curing conditions. In addition to these, the type of cement used and its age may also influence the pore structure of concrete. The pore structure of HCP and, hence, the concrete largely depends on the w/c-m ratio and the degree of hydration. In general, the higher the w/c-m ratio for a given degree of hydration, the higher will be the volume of larger pores in HCP.[14]

The use of various supplementary cementitious materials, such as pulverised-fuel ash (PFA) (also known as fly ash), ground granulated blast-furnace slag (GGBS), microsilica (MS) (also known as silica fume) and metakaolin (MK), refine the pore structure of concrete mainly by producing secondary C–S–H. In the case of MS and MK, the fine nature of the particles also contributes to the pore refinement in concrete. The pore structure of HCP also changes greatly when different chemical admixtures are used. This is mainly due to the change in the degree of hydration of cement. At a constant w/c-m ratio, the addition of superplasticiser is variously reported to refine the pore structure.[15–17] In particular, Khatib and Mangat[17] demonstrated that the inclusion of superplasticiser decreased the total intruded pore volume of HCP.

Proper curing of concrete is widely recognised as a necessity for assuring adequate field performance of concrete structures. Not only is it recognised as important to minimise evaporation of the concrete mix water, but it is equally emphasised to provide a source of external or internal curing water to replace that consumed by chemical shrinkage during the hydration of the cement. A cement paste, mortar or concrete cured under sealed conditions will self-desiccate, resulting in the creation of coarse capillary pores within the microstructure. For a w/c-m ratio greater than approximately 0.42, there is sufficient water in the mix such that complete hydration of the cement can be achieved theoretically without supplying additional water to the cement paste. Even if complete hydration was achievable, the lack of additional curing water could still result in the creation of relatively large pores within the final microstructure. In this case, the addition of curing water would assure that all pores remain water-filled and eligible as locations for the precipitation and growth of hydration products during curing.

2.6   Transport processes in concrete

The concrete durability depends largely on the resistance it offers to fluid penetrations (such as water, carbon dioxide, oxygen and salts) from its service environment into and through its matrix. As concrete is a porous material, the transport of these substances can occur by capillary-suction-driven flow (absorption), concentration-driven flow (diffusion) or pressure-driven flow (permeation). Generally, the overall potential for the ingress of the substances into concrete by these four modes is referred to as its permeability.

- *Absorption* is the process by which concrete takes in a fluid due to capillary suction in pores in order to fill the space within the material.[18,19]
- *Diffusion* is the random motion of free molecules or ions in a saturated pore system under the influence of a concentration gradient.[18,20] The diffusion can be in either gaseous or ionic form. A typical example of gaseous diffusion is the diffusion of carbon dioxide into concrete, which can lead to the carbonation of concrete; whereas a typical example for ionic diffusion is chloride diffusion, which can cause corrosion of the embedded reinforcing bar in concrete. The diffusion can be either steady or non-steady. More details on this can be found elsewhere.[21]
- *Permeation* is that property which characterises the ease with which a fluid passes into and through the body of the concrete under a pressure difference.[18,20,22] A typical example of this mechanism is that occurring in deep-water marine structures or the bases of retaining structures.

2.6.1 *Fundamentals of transport processes*

Rose[19] distinguished six stages in the transfer of fluid flux through a porous medium. These are presented in Fig. 2.2 from (a) to (f). In

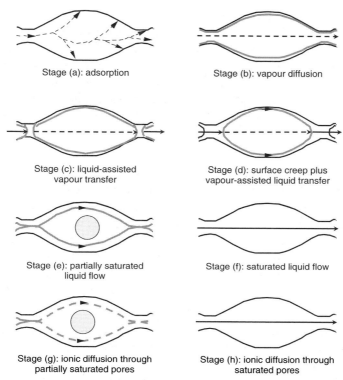

Fig. 2.2 *Idealised model of the transfer of liquids and ions within concrete*[18,19]

addition to these stages, Report No. 31 of the Concrete Society[18] outlined two more stages ((g) and (h)) for the ionic diffusion through a partially saturated and fully saturated porous system. For the convenience of discussion, a single pore with a neck at each end only is presented in Fig. 2.2. A vapour flux is incident from the left in (a) to (f), and an ionic flux is incident from the left in (g) and (h). The various stages presented in Fig. 2.2 are explained below.

The first stage is one of adsorption (stage (a)), and, until this is complete, a vapour flux cannot be transmitted and 'conductivity' has no meaning, even though flux will move to adsorption sites as vapour. This does not prevent surface diffusion in the adsorbed phase. After the initial adsorption, the second stage is one of unimpeded vapour movement (stage (b)), where the vapour behaves like an ideal gas. This is expressed by Fick's first law for diffusion:

$$J_x = -D_x \frac{\delta c}{\delta x} \tag{2.1}$$

21

where

J_x = flux of fluid at a position x in a direction x from the origin
D_x = diffusion coefficient at position x
$\delta c/\delta x$ = concentration gradient at position x

(Fick's second law for diffusion relates the rate of change of concentration gradients by assuming that the diffusion coefficient is independent of position).

The third and fourth stages occur when the necks contain liquid, either without or with a thin film of significant thickness on the walls of the cavity (stages (c) and (d), respectively). In stage (c) the system is impervious to an inert gas and pervious to liquid only by a process of distillation in which the necks act as short-circuits for vapour movement. The process is described as liquid-assisted vapour transfer, the assistance arising because liquids shorten the effective path length for vapour diffusion.

In stage (d) the condition is one of surface creep, i.e. flow in thin liquid films, in which there is vapour-assisted liquid transfer. Eventually, there is a transition to the fifth and sixth stages, where there is liquid flow, with stage (e) representing the unsaturated condition and stage (f) representing the saturated condition. There is an important difference between (d) and (e), best revealed by ignoring the arrows and treating the diagram as depicting a dynamic equilibrium. In (e) the air–water boundary will have the same curvature everywhere, but in (d) the curvature away from the necks is largely determined by the shape of the solid surface.

As the flow occurs in the pore due to the pressure differential across the boundary (meniscus) in stage (e), the rate of flow is given by Washburn's equation:[23]

$$v = \frac{r\gamma}{4d\mu}\cos\theta \tag{2.2}$$

where

v = rate of flow
r = capillary radius
γ = surface tension
d = depth of penetration of fluid
μ = fluid viscosity
θ = contact angle

The flow in stage (f), i.e. in the saturated condition, is due to a high-pressure head existing across the pore. Therefore, the rate of flow is governed by Darcy's law. For a non-compressible fluid and a saturated porous medium, Darcy's law states that the steady state rate of flow is directly proportional to the hydraulic gradient, i.e.

$$v = \frac{Q}{A} = -Ki = -K\frac{\delta h}{\delta l} \tag{2.3}$$

where

v = apparent velocity of flow
Q = flow rate
A = cross-sectional area
i = hydraulic gradient
δh = head loss over a flow path of length δl
K = coefficient of permeability or hydraulic conductivity

It must be emphasised that the application of Darcy's equation to the flow through porous media is based on the following assumptions:

- complete saturation has been achieved
- flow is laminar and viscous
- equilibrium flow conditions have been established.

A more rational concept of permeability, which is independent of the fluid properties and dependent purely on the characteristics of the porous body, is intrinsic permeability. This is expressed as

$$v = \frac{Q}{A} = -\frac{k}{\mu}\frac{\delta p}{\delta l} \tag{2.4}$$

where

v = velocity of flow
Q = volume rate of flow
A = area of cross section
δp = pressure loss over the flow path of length δl
μ = viscosity of the fluid
k = intrinsic permeability of the porous medium

In addition to the stages from (a) to (f), ionic diffusion may takes place in stages (e) and (f), as shown in stages (g) and (h). This is superimposed over the other transport processes already explained with reference to these figures. The ionic diffusion is also governed by Fick's first law. It can be seen in Fig. 2.2 that the moisture

23

condition of the concrete influences various stages of the transfer of fluids through it.

2.6.2 *Methods used to study the transport processes*

The primary test methods used to study the transport processes in concrete can be broadly classified into three categories, namely absorption tests, diffusion tests and permeation tests. The various methods under these categories are listed in Fig. 2.3. Detailed description of these test methods can be found in the literature.[23,24]

2.6.3 *Factors influencing transport processes*

The factors influencing various transport processes in concrete are presented in Table 2.2. It is worthwhile to mention here that the interaction amongst the factors identified in Table 2.2 is complex, as some of the variables are interdependent. For instance, an increase in the cement content normally means a reduction in aggregate content. A detailed discussion on how all these factors influence the transport processes in concrete is beyond the scope of this chapter. More information on these can be found out elsewhere.[24,25]

It can be seen in Table 2.2 that both the constituent materials and treatment given subsequent to concreting accounted for majority of the variations in transport properties reported in publications. Both the ambient temperature and moisture condition were found to be factors influencing transport processes. Therefore, tests for measuring transport properties of concrete should take account of environmental parameters which are known to influence test results. This is applicable to all three transport processes.

2.7 Influence of pore structure and transport processes on the durability of concrete

The overall effect of pore structure and permeability influencing the penetration of various aggressive substances, and thereby their effect on the durability of concrete, can be visualised in Fig. 2.4.[26] The parameter at the start of an arrow influences that at its head. The constituent materials of concrete, the manufacturing method and the subsequent treatment given to the hardened concrete all have a direct effect on the properties of concrete in a structure. These factors also govern the pore structure and the formation of various microcracks

Absorption tests	Tests for water absorption capacity			
	Sorptivity tests	Sorptivity from water absorbed		
		Sorptivity from water penetration depth		
	Absorptivity tests	Surface absorptivity tests	Initial surface absorption test	
			Autoclam sorptivity test	
			Stand pipe sorptivity test	
		Drill-hole absorptivity tests	Figg water absorption test	
			Covercrete absorption test	
Diffusion tests	Gas diffusion tests	Water vapour transmission test		
		Water vapour transpiration test		
		Oxygen diffusion test		
	Ionic diffusion tests	Steady state diffusion test		
		Non-steady state diffusion tests	Immersion test	
			Ponding test	
		Electric field migration tests	Steady state test	
			Non-steady state tests	Rapid chloride permeability test
				CTH method
Permeation tests	Liquid permeability tests	Steady state water flow test		
		Non-steady state water flow test	Steinart guard ring test	
			Autoclam water permeability test	
		Water penetration test		
	Gas permeability tests	Constant head gas permeability test		
		Falling head gas permeability tests	Drill hole suction test	
			Drill hole over-pressure test	
			Surface suction test (Torrent Test)	
			Surface over-pressure test (Autoclam air permeability test)	

Fig. 2.3 Classification of test methods used to study the transport processes in concrete

in concrete. The pore structure characteristics and the presence of various microcracks influence the transport processes. Therefore, these are represented in the top three blocks in Fig. 2.4.

The next set of blocks in Fig. 2.4 represents the penetration of various aggressive substances, which govern the different deterioration

Table 2.2 Factors influencing transport processes in concrete

Factors		Percentage reference made in nearly 150 publications
Constituent materials	Water content	16.55
	Cementitious materials content	7.20
	Aggregate size, type and content	5.76
	Admixtures and cement replacements	7.20
Method of preparation	Mixing	0.72
	Method of casting	0.72
	Compaction	0.72
	Trowelling	0.72
Subsequent treatment	Curing	16.55
	Protective treatments	0.72
	Age	9.32
Interdependence	Porosity	7.19
	Pore size distribution	4.32
	Strength	4.32
Test parameters	Test pressure and duration of test	4.32
	Ambient environmental conditions	3.60
	Moisture content	5.04

mechanisms. The mechanisms of deterioration are shown in the next in a set of blocks, below the aggressive substances. The arrows between these two sets of blocks indicate the dependence between the penetration of aggressive substances and the deterioration mechanisms. As a result of all the mechanisms presented in Fig. 2.4, concrete cracks, and this in turn accelerates the transport processes. In sulfate attack, chloride attack and carbonation, slight modification of the pore structure may occur, which in turn may diminish the transport processes. However, under prolonged action of the mechanisms, concrete will eventually crack.

Experimental evidence exists illustrating the correlation between the relevant transport properties and either the penetration of different aggressive substances or the mechanisms of deterioration. Typical correlations are presented in Figs 2.5–2.8. The dependence of carbonation on the gas permeability of concrete is illustrated in Figs 2.5[27] and 2.6.[26] In Fig. 2.5 the depth of carbonation after 20 weeks in an accelerated carbonation test is presented against the intrinsic permeability in a logarithmic scale. Figure 2.6 illustrates the variation of the depth of

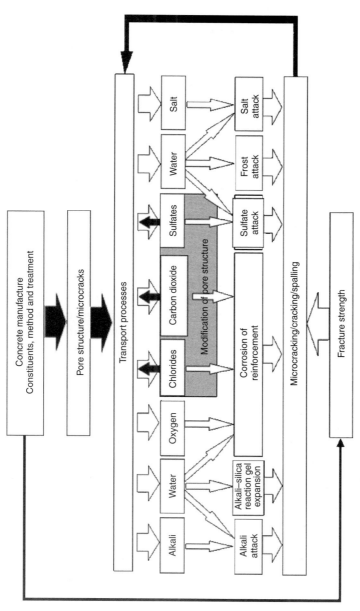

Fig. 2.4 Interdependence of transport processes and the durability of concrete[26]

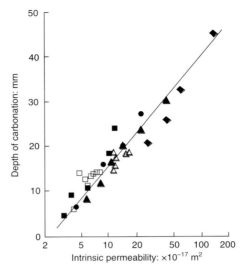

Fig. 2.5 Relationship between permeability and depth of carbonation[27]

carbonation of 27 different mixes after 14 days in an accelerated test with the Autoclam air permeability results. In both sets of data, good correlation can be found between the depth of carbonation and the gas permeability of concrete. The variation between the two sets of data is considered to be due to the difference in sample preparation, testing conditions and different types of cementitious materials used in the concrete.

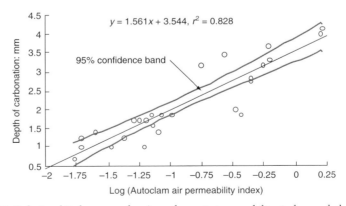

Fig. 2.6 Relationship between the Autoclam air permeability index and the depth of carbonation[24]

28

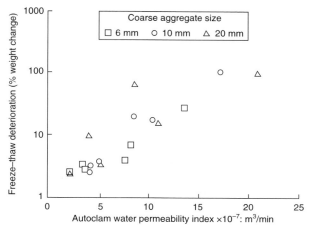

Fig. 2.7 Relationship between the Autoclam water permeability index and freezing and thawing deterioration[24]

Long *et al.*[28] reported the relationship between Autoclam water permeability results and the mass loss (expressed as a percentage of the initial saturated mass) due to 300 cycles of the ASTM C666 Procedure A freezing and thawing test (Fig. 2.7). The data presented in Fig. 2.7 are based on tests carried out on 27 different concrete mixes. As can be seen, there exists a good correlation between the two.

The dependence of chloride penetration on the sorptivity of concrete can be seen in Fig. 2.8.[29] The data in this figure are based on ponding tests carried out on both surface-treated and untreated concrete specimens. The three untreated concrete specimens resulted in a linear correlation between the chloride ingress and the sorptivity

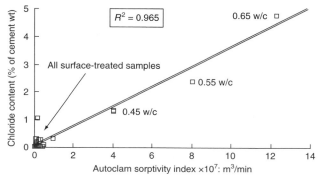

Fig. 2.8 Relationship between the Autoclam sorptivity index and chloride ingress[29]

29

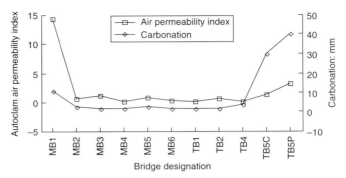

Fig. 2.9 Dependence of carbonation on the air permeability of concrete bridges[31]

values, and the reduction in sorptivity due to surface treatments correspondingly resulted in a decrease in the chloride content. Basheer and Cleland[30] reported that the reduction in both the sorptivity and the chloride ingress was accompanied by a corresponding reduction in the rate of corrosion of steel bars embedded in the test specimens.

Whereas the dependence of various mechanisms of deterioration on permeation properties is well documented for laboratory investigations, not much data exist for in-situ investigations. Two graphs are presented in Figs 2.9 and 2.10 which report the dependence of both carbonation and chloride ingress on the air permeability of concrete, based on tests carried out with the Autoclam Permeability System on 11 concrete motorway bridges in Northern Ireland.[24,31] Although the effect of moisture was not completely accounted for in the air permeability results, large variations in both the depth of carbonation and chloride

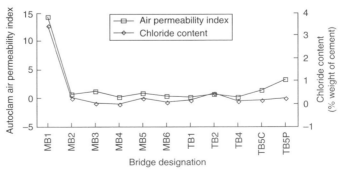

Fig. 2.10 Dependence of chloride ingress on the permeability of concrete bridges[31]

ingress corresponded with a similarly large variation in the air permeability of concrete.

It may be recalled that the conceptual model relating permeation properties to both the penetration of aggressive substances and the deterioration mechanisms (Fig. 2.4) can be validated by data in so far as the mechanisms of deterioration involve the transport of aggressive substances, and these initiate the deterioration. Also, it is relatively easy to show that there exists a good correlation between a specific mechanism of deterioration and a related transport property. However, in reality, one mechanism may act as the catalyst for a number of other mechanisms, and the permeation properties alone may not be sufficient to estimate both the rate and the extent of deterioration. Therefore, permeation properties allied to the knowledge of both the physical and chemical characteristics of concrete should be made use of in assessing the durability of concrete.

2.8 Summary

The pore structure characteristics of cement-based materials play a fundamental role in governing the transport processes in concrete and, hence, its durability. Variations in size of pores in concrete and their influence on both transport properties and durability are highlighted in this chapter. It has been observed that numerous test methods are available to characterise the pore structure and permeability of concrete. Environmental factors, such as temperature and moisture variations, affect all transport processes, and hence there is a need to consider their effects on the measurement of transport properties. Both laboratory and field data highlight the dependence of durability on transport properties. The properties of the constituent materials, the methods of manufacturing concrete and the subsequent treatment of concrete affect both the pore structure and the transport processes, and hence, to design durable structures, attention should be given to these factors.

Further reading

Aligizaki, K.K. (2005) *Pore structure of cement-based materials: testing, interpretation and requirements. Modern concrete technology.* Taylor and Francis, London.

Kropp, J. and Hilsdorf, H.K. (1995) *Performance criteria for concrete durability. State-of-the-art-report prepared by RILEM Technical Committee TC 116-PCD.* RILEM Report 12. RILEM, Bagneux.

Mehta, P.K. and Monteiro, P.J. (2005) *Concrete: structures, properties and materials*. McGraw-Hill, New York, 3rd edn.

Neville, A.M. (1996) *Properties of concrete*. John Wiley, London, 4th edn.

Ramachandran, V.S. and Beaudoin, J.J. (eds) (2000) *Handbook of analytical techniques in concrete science and technology: principles, techniques and applications*. Barnes and Noble, Lyndhurst, NJ.

References

1. Mehta, P.K. and Monteiro, P.J. (2005) *Concrete: structures, properties and materials*. McGraw-Hill, New York, 3rd edn.

2. Bensted, J. and Barnes, P. (2002) *Structure and performance of cements*. Spon, London, 2nd edn.

3. Powers, T.C. (1958) Structure and physical properties of hydrated Portland cement paste. *Journal of the American Ceramic Society*, 41(1), 1–6.

4. Feldman, R.F. and Sereda, P.J. (1968) A model for hydrated Portland cement paste as deduced from sorption-length change and mechanical properties. *RILEM Bulletin*, 1(6), 509–520.

5. Diem, P. (1986) Determination of capillary absorptiveness of very dense concrete. *Betonwerk und Fertigteil-technik*, 52(11), 719–724.

6. Mehta, P.K. and Manmohan, D. (1980) Pore size distribution and permeability of hardened cement pastes. In: *7th International Symposium of the Chemistry of Cement*. Paris, vol. 3, 181–185.

7. Midgley, H.G. and Illuston, J.M. (1983) Some comments on the microstructure of hardened cement pastes. *Cement and Concrete Research*, 13, 197–206.

8. Nieminen, P. and Romu, M.I. (1988) Porosity and frost resistance of clay bricks. *Proceedings of the 8th International Brick/Block Masonry Conference*. Dublin, 103–109

9. Nyame, B.K. (1985) Permeability of normal and lightweight mortars. *Magazine of Concrete Research*, 37(130), 44–48.

10. Parrott, L.J. and Killoh, D.C. (1984) Prediction of cement hydration. *The chemistry and chemically related properties of cement. British Ceramic Proceedings*, No. 35, 41–53.

11. Parrott, L.J. (1985) Effect of changes in U.K. cements upon strength and recommended curing times. *Concrete: Journal of the Concrete Society*, 22–24.

12. Aligizaki, K.K. (2005) Pore structure of cement-based materials: Testing, interpretation and requirements. In: M. Alexander and S. Mindess (eds) *Modern Concrete Technology*. Taylor & Francis, London, ch. 12.

13. Mindess, S., Young, J.F. and Darwin, D. (2002) *Concrete*. Prentice Hall, Englewood Cliff, NJ, 2nd edn.

14. Klieger, P. and Lamond, J.F. (1994) *Significance of tests and properties of concrete and concrete-making materials*. ASTM, Philadelphia, PA.

15. Mor, A. and Mehta, P.K. (1984) Effect of superplasticising admixtures on cement hydration. *Cement and Concrete Research*, 14(5), 754–756.

16. Syal, S.K. and Kataria, S.S. (1982) Development and interaction of a concrete additive for improved performance and durability. *Cement Concrete and Aggregates*, 4(2), 110–114.

17. Khatib, J.M. and Mangat, P.S. (2002) Influence of high-temperature and low-humidity curing on chloride penetration in blended cement concrete. *Cement and Concrete Research*, 32(11), 1743–1753.

18. Report of a Concrete Society Working Party (1988) *Permeability testing of site concrete – a review of methods and experience*. Technical Report No. 31. Concrete Society, London.

19. Rose, D.A. (1965) Water movement in unsaturated porous materials. *RILEM Bulletin*, No. 29, 119–124.

20. Feldman, R.F. and Sereda, P.J. (1968) A model for hydrated Portland cement paste as deduced from sorption-length change and mechanical properties. *RILEM Bulletin*, 1(6), 509–520.

21. Nanukuttan, S.V. (2007) *Development of a new test protocol for the PERMIT ion migration test*. PhD thesis, Queen's University Belfast.

22. Mercer, L.B. (1945) Permeability of concrete – 1, theoretical considerations, laboratory test methods, details of experimental work with new apparatus. *The Commonwealth Engineer*, 349–357.

23. Ramachandran, V.S. and Beaudoin, J.J. (eds) (2000) *Handbook of analytical techniques in concrete science and technology: principles, techniques and applications*. Barnes and Noble, Lyndhurst, NJ.

24. Basheer, P.A.M. (1991) *Clam permeability tests for assessing the durability of concrete*. PhD Thesis, Queen's University Belfast.

25. Kropp, J. and Hilsdorf, H.K. (1995) *Performance criteria for concrete durability, State-of-the-Art-Report prepared by RILEM Technical Committee TC 116-PCD*. RILEM Report 12. RILEM, Bagneux.

26. Basheer, P.A.M., Long, A.E. and Montgomery, F.R. (1994) *An interaction model for causes of deterioration of concrete, Malhotra symposium on concrete technology, San Francisco, SP 144*. American Concrete Institute, Farmington Hills, MI, 213–233.

27. Dhir, R.K., Hewlett, P.C. and Chan, Y.N. (1989) Near-surface characteristics of concrete: prediction of carbonation resistance. *Magazine of Concrete Research*, 41(148), 137–143.

28. Long, A.E., Basheer, P.A.M. and Montgomery, F.R. (1997) In-situ permeability testing – a basis for service life prediction. In: *Advances in Concrete Technology. Proceedings of the CANMET/ACI International Conference, SP 171*. New Zealand, 651–670.

29. Basheer, L. (1994) *Assessment of the durability characteristics of surface treated concrete*. PhD Thesis, The Queen's University of Belfast.

33

30. Basheer, L. and Cleland, D.J. (1995) Surface treatments for concrete quantifying the improvement. *Proceedings of the IABSE Symposium.* San Francisco, 275–280.
31. Basheer, P.A.M., Montgomery, F.R. and Long, A.E. (1990) In-situ assessment of durability of concrete motorway bridges. In: *International Symposium on the Durability of Concrete, Toronto*, SP 131. American Concrete Institute, Farmington Hills, MI, 305–311.

3

Physical deterioration mechanisms

Christopher Atkins, Mott MacDonald, UK

3.1   Introduction

Concrete is a porous material. Generally, the pore structure is produced as a result of excess water added when it was first placed. For a typical structure, the amount of water required to make the cement fully hydrated is approximately 30% of the total mass of cement. If a cement paste is mixed with this amount of water, it is quite stiff. If aggregate is added to make concrete, it would not be possible to adequately mix or place the concrete. The easiest solution is to add more water. This excess water is not chemically combined into the hydrate structure, and causes the formation of voids. More water results in more voids, which in turn results in lower-strength concrete.

There are a range of mechanisms that result in the physical degradation of concrete:

- Conventional abrasion and cavitation will occur.
- Where concrete surfaces are subject to traffic, large hydrostatic pressures can arise when moisture is compressed in defects, and this can result in accelerated abrasion.
- Some physical degradation mechanisms involve material within the pores nearest the surface expanding. This imparts an indirect tensile stress on the concrete, which results in the surface scaling off.
- Finally, fire, and subsequent fire fighting, can produce damage to concrete.

This chapter reviews these mechanisms.

It should be noted that concrete often contains joints. These are typically made up of a softer, less durable material, and so can be subject to more damage than the surrounding concrete. However, once these have degraded, the result is an exposed edge which will always be more vulnerable to damage. If the concrete in question is founded on

Concrete durability
978-0-7277-3517-1

a granular sub-base, the passage of water through failed joints can lead to erosion of material under the concrete with little or no indication that this is occurring until the concrete itself is sufficiently undermined to result in failure. Any exposed joints and edges should always be given special consideration in areas prone to physical damage.

3.2   Abrasion/erosion

Abrasion or erosion can be defined as the wearing away of surfaces. It is commonly considered that it requires the action of a hard material on a softer one, but this is not the case. Over a significant period of time, a soft material can slowly wear away a harder one. For example, at canal locks, cast iron posts are often provided to hold a barge in place while a lock is emptied or filled: the barge is held close to the wall of the lock by looping a rope around the post. Over time these posts can become abraded by the rope. An example is shown in Fig. 3.1.

Similarly, stone steps in very old structures are often worn into an inverted bell curve shape by the passage of hundreds of thousands of feet. Concrete is no exception to this wear, and will be subject to abrasion with time in any environment where other materials pass

Fig. 3.1 Rope abrading cast iron

Fig. 3.2 Abrasion of concrete in a stilling basin

over it. Where it is subjected to a flowing environment, which may contain suspended solids, the resulting damage can be extensive. This is shown in Fig. 3.2.

Typically, it is a very slow process, unless it is subject to other mechanisms discussed later. For conventional reinforced concrete, the abrasion is likely to be of most of a concern if it results in an uneven surface for traffic, or if it significantly reduces cover to the rein-forcement. Uneven surfaces often allow puddles of water to form, which can dramatically accelerate the abrasion process. When water is trapped while being subject to wheel loading, a significant pressure is exerted in the horizontal plane. This can rapidly lead to undermining of the surrounding surfaces, or opening up of existing defects, in turn leading to increasingly more widespread surface degradation. The ability of water to seek out cracks and disrupt concrete has led to the use of ultra-high water pressure as an effective means of removing concrete in a technique commonly referred to as hydro demolition.

The act of repair itself may lead to physical degradation of the surrounding parent concrete. For areas subject to traffic loading or that require a rapid return to service due to operation constraints, it is tempting to use a high-strength material, on the basis that this will achieve the minimum strength required for a return to service rapidly. If this results in the use of a material with a higher Young's modulus to the original parent concrete, this can result in abrasion of the concrete surrounding the repair. The repair material will act as a hard spot, and so the surrounding concrete may be subject to abrasion.

3.2.1 *Avoiding abrasion*

As with most concrete problems, a high-quality concrete properly placed and cured can minimise the risk of abrasion. Higher strengths will typically provide higher abrasion resistance, but the curing and placement of concrete is of equal importance. Anything that results in a soft or porous surface containing cracks, such as plastic settlement or shrinkage, will result in a concrete more susceptible to abrasion. Typically, industrial floors require a high-quality smooth finish. In order to achieve this, the surface is often heavily trowelled or power floated. This draws water to the surface, and can result in a smooth but weak surface, and so in attempting to achieve a smooth surface, the abrasion resistance may in fact have been compromised. This becomes apparent shortly after the floor begins to be subject to wheel loading, when it begins to degrade and the surface breaks up. Increasing requirements for greater joint spacing have improved the technology available for casting floors, but concrete will still shrink and be subject to thermal movements. In addition, clients are more stringent with regard to acceptable tolerances on flooring, resulting in more queries at hand-over as to what is and is not an acceptable floor.

Even high-performance concretes can be subject to abrasion if the environment is severe enough. In this case there are a number of commercial products known as 'dry shake' materials that can provide significant enhancement in surface properties. As the name suggests, these are dry products shaken onto the surface of wet concrete prior to finishing. They typically comprise a cement-based product and carefully selected aggregate material. The most hard-wearing materials are typically metallic and can provide resistance to tracked military vehicles.

Fibre reinforcement can be used to improve the integrity of exposed concrete surfaces. Fibres are typically either polypropylene or steel, and are included in the original mix. After placement, the resultant concrete contains randomly oriented fibres that provide an enhanced tensile strength. The presence of fibres in a mix often means that the concrete has to be placed with greater care, which in turn can lead to an improvement in the abrasion resistance of concrete. Additionally, the fibres provide enhanced tensile properties in the cover, which can also improve the abrasion resistance of concrete. The fibres also serve to hold damaged concrete in place to some extent.

Once the surface begins to break down, achieving an acceptable repair is usually complicated by the fact that areas subject to heavy abrasion are usually in constant use with no alternatives available. As a result, a number of rapid-set materials have been developed that can be used to reinstate lost material. When using such materials, the rapidity of the set requires a contractor experienced with the specific product. Trials should be performed before commencement of the formal repair project, to confirm that the contractor is familiar with the material. As the main purpose of the repair is to fill a hole, corrosion protection requirements can be secondary, and a number of rapid-set materials do not have the level of alkalinity required to prevent reinforcement from corroding.

The nature of abrasion often leaves a surface with a good mechanical key due to exposed aggregate, but it can require shallow repairs that need to be feather edged. Ideally, to achieve a successful repair, all edges should be saw cut to a depth of 10 mm, to provide a sharp edge to form the repair against.

It should also be borne in mind that in order to provide a repair that is structurally compatible with the existing structure, the material to be used should have a similar Young's modulus to the concrete being repaired, and should exhibit little or no shrinkage, and have similar thermal properties. In areas where access times are not limited, there are a range of repair products that can be used to provide an effective repair.

3.3   Cavitation

Cavitation occurs when liquids undergo a change in pressure to a point where the pressure in the liquid is below its vapour pressure. This produces bubbles in the liquid which subsequently collapse and produce shock waves. This can result in a very aggressive environment and extremely high rates of abrasion. This process is accepted as a mechanism that results in coastal erosion whereby air pockets within water are forced into fissures. The continuing inflow of water results in the pockets collapsing and a range of shock waves being produced that result in degradation.

3.4   Frost

Frost damage is caused by a simple mechanism. Moisture within the concrete expands as it freezes, and as a result tensile stresses form near to the surface. Typically, water expands by approximately 9% when it approaches freezing. If this expansion exceeds the available volume of the pores, then the concrete will be damaged. This can result in the surface crazing then scaling off. Repeated cycles result in further damage, as the surface opens more and more, allowing a greater volume of water to penetrate the concrete.

There are two critical factors. The water in the large capillary pores is free to freeze, whereas the water in the gel pores is adsorbed on the cement hydrates and so has limited effect, although it will migrate into the capillary pores and contribute to the damage. This is balanced against the entrapped air voids that are compressible and allow expansion of the frozen water. Controlling frost attack is therefore normally achieved by using low-permeability concrete to keep the water out, and by using air entrainment to increase the compressible volume of voids.

There are a range of products available to achieve air entrainment. Typically, they are organic materials that produce a series of evenly dispersed discontinuous pores or bubbles within the concrete that are relatively uniform in diameter. Generally, they are less than 0.1 mm in diameter, and the volume that is typically required for protection is of the order of 4–7% by volume of the concrete. The size of bubbles and fact that they are discontinuous makes the distinction between the normal entrapped air that occurs in concrete. This provides two benefits. The first is that because the pores are discontinuous, the surface absorbancy is reduced. The second is that because the pores remain air filled and hence unsaturated, they are able to absorb the stresses induced by frost. However, as the concrete is capable of absorbing these stresses, placing air-entrained concrete by pumping is generally not normally feasible.

The amount of air entrained is influenced by a number of factors. An increase in mixing time or concrete temperature reduces the amount of air entrained. The type of cement and the cement content both influence the volume of air entrained, with high cement contents or cements with high specific surfaces resulting in problems with air entrainment. The presence of organic matter can also affect the performance of the air entrainment.

There are two standardised methods of determining air content. BS EN 480:2005 and ASTM C457-80 are both used to determine the air content in hardened concrete. Similarly, there are two standard tests for determining the freeze–thaw resistance of a concrete. ASTM C666-84 describes two methods that both involve applying rapid freezing. Currently, there is a European standard being drafted (DD CEN/TS 12390-9:2006, *Testing hardened concrete. Freeze–thaw resistance. Scaling*). Basically, the standard methods subject a sample to a number of freeze–thaw cycles and produce comparisons of tested and untested samples.

3.5 Exfoliation

Exfoliation is caused by a similar mechanism to frost damage, in that the pore structure becomes filled with a material that subsequently expands. In this case it is a solution containing soluble salts. As the moisture content decreases, the solution becomes more and more concentrated until, ultimately, the soluble salts recrystallise. The recrystallised salts may be greater in volume than the original solution, and so tensile stresses are imparted into the concrete. Figure 3.3 shows an example of exfoliation.

It should be noted that the surface of concrete that is exposed to the solution may not be the one that suffers from exfoliation. If one face of a concrete surface is adjacent to a highly concentrated solution, and the other is exposed to a drying environment, the salt solution can be drawn a significant distance to the drying face, where the recrystallisation takes place. This can be a problem in immersed tunnels. Typically, the solution on the exterior of the tunnel is chloride contaminated, and

Fig. 3.3 Exfoliation of a seawall

41

Table 3.1 Solubilities and densities of various salt solutions

Chemical	Solubility: g/l	Density: g/l
Sodium chloride	359	2160
Calcium chloride (anhydrous)	745	2150
Calcium chloride (dehydrate)	745	835
Ammonium nitrate	1900	1700

the capillary absorption that takes place will cause corrosion as well as exfoliation.

The first thing to note is that concrete already contains a supersaturated solution of calcium hydroxide. However, calcium salts typically have a relatively low solubility, and so as the moisture dries out, the salts that are precipitated do not have a large volume expansion associated with them. For exfoliation to occur, a more soluble salt is usually present. In many cases in natural environments this is sodium chloride based, but any highly soluble salt can produce a similar effect. Table 3.1 provides an illustration of the range of solubilities and densities found. It is fairly simple to understand that when a saturated solution of ammonium nitrate is dried out, there will be an increase in volume: 1900 g will dissolve in 1 litre of water, but the volume of 1900 g is greater than a litre.

This effect has been reported for concrete contaminated with sodium chloride subsequently overcoated, due to the nature of the pore structure of cement that contains other soluble products that will also precipitate.

Coatings or waterproofing for concrete are commonly vapour permeable to prevent osmotic blistering. As a result, this can cause a drying out of the coated concrete. If the concrete is contaminated with a soluble salt, the concentrations in the pores under the coating will increase, until the salts precipitate. This can cause the surface to exfoliate, which will result in coating failure.

As a general rule of thumb, all nitrates and sodium, potassium and ammonium salts are highly soluble. Due to its high solubility, significant problems can be encountered when trying to contain ammonium nitrate solutions in concrete tanks.

It should also be noted that solubilities typically increase with temperature. As a result, exfoliation may also occur due to a temperatures change. Also note that the density, and consequently the volumetric change that occurs on drying out, of compounds such as calcium chloride depends on the amount of chemically combined water. This can also affect the solubility of a salt, and can further complicate the temperature dependence of solubility. For example, sodium sulfate decahydrate loses its chemically combined water at 32°C, to form a more soluble anhydrous salt.

In order to control the risk of exfoliation or frost attack, the basic approach is to either prevent significant salt ingress or prevent drying of the surface. Typically, a coating system would provide protection against salt ingress, but note that a vapour-permeable coating can increase drying out of the concrete, and may therefore cause exfoliation if the substrate is heavily contaminated. As with any attempt to repair concrete, a full understanding of the problems is required in order to carry out a successful repair.

3.6   Fire

The first thing to note about concrete is that it forms a basic method of fire protection for steel structures. After the great fire in Chicago, all steel structures were coated with a basic form of concrete to provide fire protection. When cement hydrates, it chemically combines with a significant amount of water. In the case of fire, as the temperature increases, the hydrated compounds break down and release water, thus dissipating heat and energy. This means that whilst the surface of the concrete may have been subject to high temperatures, the depth of damage may be limited.

After a fire, the appearance of any structure can give the impression of a scene of total destruction. Due to the smoke, everything will be blackened. There will be areas where concrete has cracked and may have spalled. The levels of damage may appear significant, but due to ability of concrete to dissipate heat, the actual damage may well be recoverable.

If the temperature of the concrete has not exceeded 300°C, the residual strength, on cooling, is commonly considered to not be significantly affected(see CS TR 33[1]). The reduction in strength is often offset by factors that resulted in the concrete having higher strengths than required by the original structural capacity. A word of caution should be raised, since it is often assumed that the concrete strength increases with time beyond that required by the design. For older cements this was the case, as in order to achieve the required strength at 28 days the ultimate strength needed to be higher. Today, cements are more finely ground, and cement-producing companies have improved quality control systems, which mean the factors of safety included to ensure

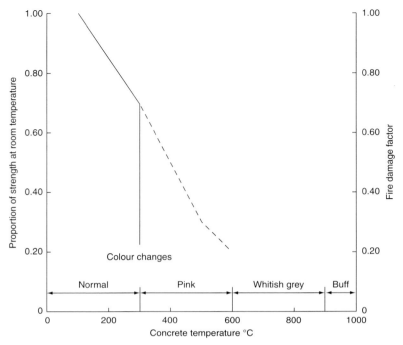

Fig. 3.4 Typical effect of heat on the compressive strength of dense aggregate concrete after cooling. Colour changes shown for concrete containing most types of siliceous aggregate may not apply to concrete with calcareous or crushed flint aggregates. (Reproduced with permission from 'Technical Report 68: Assessment, design and repair of fire damaged concrete structures', © The Concrete Society)

that concrete meets the required 28 day strength are now lower. The ultimate strength achieved may not be significantly greater than that obtained at 28 days.

The temperature at which significant structural deterioration occurs is often considered to coincide with a colour change in many concretes, where the material develops a pink hue (Fig. 3.4). This discoloration is due to the presence of ferrous salts in the aggregates, and may not occur if these are not present. It is important to note this does not apply in every case, and the levels of heat damage and degradation of the cement should be confirmed using petrographic analysis on samples. Where temperatures exceed 500°C, the strength is usually severely compromised.

The actual degradation curve depends on a range of factors, including the aggregate type. Limestone aggregate starts to decompose at a relatively low temperature compared with a siliceous aggregate, but decomposes more gradually and over a greater temperature range, and this will obviously have an effect on the performance of concrete in fires. CS TR 33 provides a range of examples produced using a variety of software packages that illustrate the thermal properties of concrete. In one example, a 380 mm square column is subject to a standard furnace test for 1 hour. At a depth of 25 mm the temperature has reached 300°C, at 50 mm deep the temperature has reached 150°C and at 75 mm deep the temperature is only 100°C. In this case, only 25 mm of the concrete is likely to have suffered significant degradation, requiring replacement. BS EN 1992-1-2 provides tables for a range of member types of minimum dimensions which satisfy specified periods of fire resistance.

For concrete subject to fire, the free water will be heated and will expand. If the pore structure is relatively open, the pressures may be able to dissipate. In a dense impermeable concrete there is no means of dissipating the pressure, and the concrete may rapidly spall due to a phenomenon often termed explosive spalling. This degradation results in more rapid loss of concrete, and therefore an increased level of damage.

This mechanism has been postulated to have occurred in recent tunnel fires, such as that in the Channel Tunnel in October 2008, where it was suggested the high-strength concrete used resulted in explosive spalling. It is worth noting that this particular fire is reported as reaching temperatures in excess of 1000°C, and burned for 16 hours. It is unlikely that any concrete exposed to this temperature for this duration would survive. Explosive spalling is generally considered to occur early in the life of a fire, as the free water in the near-surface concrete expands rapidly. If the concrete cannot dissipate the pressures involved, it will spall. There is much debate with regard to the use of direct fire protection such as sprinkler systems in tunnels. The fires themselves are statistically rare events, and so it can be difficult to justify the cost of installation and ongoing maintenance requirements for active fire control measures. In order to reconcile the cost of this, it should be noted that the major cost associated with the aftermath of a fire is not that of repair. For any infrastructure, downtime is difficult to cost. Even for structures such as the Channel Tunnel that for which the reduction in service can be directly measured and a cost associated with it, there are additional lost revenues due to the reduction in public confidence that will inevitably follow a major fire.

It is important to consider the effects of heat on the steel reinforcement as well. Typically, heating steel up to 450°C and allowing it to return to room temperature naturally is unlikely to affect most common structural steelworks. Above 550°C, steel typically begins to deform plastically, and if heated still further, internal phase changes may occur, which may get locked in if the steel is subsequently quenched during fire fighting. For prestressed and post-tensioned steels this may not be the case. Typically, these steels will have undergone specific treatment to achieve the higher strengths which will effectively be removed by the fire. The increased temperatures will result in accelerated creep and stress relaxation in the tendons, and ultimately may result in their capacity reducing to that of conventional steel. This is important to understand when assessing the levels of fire damage in a structure.

In addition to this, recent research[2] suggests that post-tensioned slabs designed with a 2 hour fire resistance collapsed in 66 minutes.

If it is considered that the structure has suffered to the point where its structural integrity is brought into question, it will be necessary to carry out coring to obtain the strength of concrete. Whilst surface techniques such as the sclerometer or internal fracture test can provide useful strength information for concrete structures, the fire will have obviously modified the surface concrete significantly, and so coring will be required to obtain strengths. For steelwork or embedded reinforcement, it is possible to carry out surface hardness tests, but these only provide information regarding the surface strength of the steel. This may not be representative of the entire section due to the fire and subsequent fire fighting that will have taken place.

In general, repair of fire-damaged structures will simply involve the removal and replacement of the deteriorated material, typically using cementitious repair products. Prior to undertaking any repair project it will be necessary to confirm the integrity of the structure, with regard to the extent of the concrete requiring replacement. The repair process will involve the removal of concrete beyond the damaged areas, which may affect the stability of the structure. It may therefore be appropriate to consider the size of any individual repair area to be limited. The repair material should obviously be of a similar structural capacity to that required, and can be applied using a range of techniques (e.g. sprayed concrete, flowable or self-compacting concrete or hand-applied repair products), depending on the extent or position of the repair. It should be noted that some high-build repair mortars have a lower strength than similarly named products. In December 2008, the European standard EN 1504 was introduced, which covers the products, processes and approaches to be used when carrying out a concrete repair.

3.7 Summary

There are a limited number of causes of physical degradation of concrete. Typically, the simplest way of avoiding degradation is to have a proper understanding of the working performance required of the concrete and use a properly selected material. The selection of the right mix is only the start of the process, and installation and curing are of equal if not greater importance since all the physical degradation mechanisms start at the surface of the concrete, and it is the surface that is most vulnerable to placing and curing effects.

As always, in order to carry out an effective repair it is necessary to understand the processes involved, and standards are available that provide a route to achieve a durable repair in the form of EN 1504. In the case of coatings to address chloride contamination of concrete, the repair itself can sometimes result in further physical degradation of concrete. When inappropriate repair materials are selected, this can also result in the repair causing physical damage to the original concrete.

References

1. Concrete Society (1990) *Assessment and repair of fire-damaged concrete structures*. Technical Report 33. Concrete Society, Camberley.
2. Kelly, F. and Purkiss, J. (2008) *New Civil Engineer*, 23 Oct.

4

Deterioration mechanisms – chemical

Concrete can suffer chemical damage from contaminants added during manufacturing or by the action of external aggressive agents, or by a combination of both. Deterioration in concrete due to chemical reactions manifest itself in different ways, such as an increase in porosity and permeability, a decrease in strength, cracking, dissolving and spalling. In this chapter, the chemical deterioration of concrete by sulfate attack, acid attack, the alkali–aggregate reaction and the corrosion of reinforcing steel is briefly reviewed.

4.1   Sulfate attack

Mustafa Sahmaran, Gaziantep University, Gaziantep, Turkey

Chemical reactions that generally involve the formation of expansive products in hardened concrete can lead to severe deterioration. Sulfate attack, which is associated with those expansive chemical reactions, is one of the most important chemical reactions that take place in concrete. The term 'sulfate attack' is used to define the attack suffered by concrete structures from salts and solutions of sulfate-bearing compounds, most commonly sodium sulfate (Na_2SO_4) and magnesium sulfate ($MgSO_4$). Chemical sulfate attack is caused by the ingress of sulfates into concrete, and involves chemical reactions with the hydration products of cement. The products of these reactions occupy a greater volume than the host reactants, and thus cause an internal expansion of the cement paste, which then produces localised tensile stresses in the hardened concrete. Sulfate attack ultimately manifests itself in the form of cracking, spalling and mass loss; all of which lead to loss in cross-section area or complete deterioration of the structural concrete elements. Although complete deterioration due to sulfate attack can be reached in a few months in accelerated laboratory tests, it typically takes several years in the field.

4.1.1　Mechanism of sulfate attack

Most experts believe that sulfate attack is generally attributed to the formation of expansive ettringite ($3CaO \cdot Al_2O_3 \cdot 3CaSO_4 \cdot 32H_2O$) and gypsum (calcium sulfate dihydrate, or $CaSO_4 \cdot 2H_2O$) which may be accompanied by the expansion or softening of concrete. The other effects of sulfate attack include the formation of thaumasite ($CaCO_3 \cdot CaSO_4 \cdot CaSiO_3 \cdot 15H_2O$), the decalcification of the calcium silicate hydrate gel ($3CaO \cdot 2SiO_2 \cdot 3H_2O$, or C–S–H in cement notation), and eventually the loss of cementitious structure. The details of the mechanism and the associated reactions are presented in the following sections.

4.1.1.1　Attack by sodium and magnesium sulfates

Hydration of the principal compounds of cement, dicalcium silicate ($2CaO \cdot SiO_2$, or C_2S in cement notation) and tricalcium silicate ($3CaO \cdot SiO_2$, or C_3S in cement notation), with water (H_2O, or H in cement notation) produces calcium silicate hydrate gel and calcium hydroxide ($Ca(OH)_2$, or CH in cement notation):

$$2(2CaO \cdot SiO_2) + 4H_2O \rightarrow 3CaO \cdot 2SiO_2 \cdot 3H_2O + Ca(OH)_2 \quad (4.1)$$

$$2(3CaO \cdot SiO_2) + 6H_2O \rightarrow 3CaO \cdot 2SiO_2 \cdot 3H_2O + 3Ca(OH)_2 \quad (4.2)$$

Among the products that are formed, calcium silicate hydrate gel, also known as $C_3S_2H_3$, is the primary binding component of hydrated cement, and calcium hydroxide is the water-soluble by-product that has no cementitious value.

Sulfate solutions attack various hydration products of cement. As a result of the sulfate attack, each individual hydration product can produce a new compound. Sodium sulfate and magnesium sulfate can react with the calcium hydroxide formed in the hydration of calcium silicates, to form gypsum, sodium hydroxide ($NaOH$) and magnesium hydroxide ($Mg(OH)_2$, or brucite) according to the following approximate reactions:

$$Ca(OH)_2 + Na_2SO_4 + 2H_2O \rightarrow CaSO_4 \cdot 2H_2O + 2NaOH \quad (4.3)$$

$$Ca(OH)_2 + MgSO_4 + 2H_2O \rightarrow CaSO_4 \cdot 2H_2O + Mg(OH)_2 \quad (4.4)$$

The deterioration of hardened concrete by the formation of gypsum ($CaSO_4 \cdot 2H_2O$) goes through a process leading to a reduction in stiffness and strength; this is followed by expansion.[1] Gypsum formation is also known to be the first step of ettringite formation, which can be considered as the principal cause of deterioration of concrete due to sulfate attack.

The gypsum that is formed can further react with hydrated calcium aluminates ($4CaO \cdot Al_2O_3 \cdot 13H_2O$), hydrated calcium sulfoaluminates ($3CaO \cdot Al_2O_3 \cdot CaSO_4 \cdot 12H_2O$), or unhydrated tricalcium aluminates ($3CaO \cdot Al_2O_3$, or C_3A in cement notation) to produce ettringite according to the following equations:

$$4CaO \cdot Al_2O_3 \cdot 13H_2O + 3(CaSO_4 \cdot 2H_2O) + 14H_2O$$
$$\rightarrow 3CaO \cdot Al_2O_3 \cdot 3CaSO_4 \cdot 32H_2O + Ca(OH)_2 \tag{4.5}$$

$$3CaO \cdot Al_2O_3 \cdot CaSO_4 \cdot 12H_2O + 2(CaSO_4 \cdot 2H_2O) + 16H_2O$$
$$\rightarrow 3CaO \cdot Al_2O_3 \cdot 3CaSO_4 \cdot 32H_2O \tag{4.6}$$

$$3CaO \cdot Al_2O_3 + 3(CaSO_4 \cdot 2H_2O) + 26H_2O$$
$$\rightarrow 3CaO \cdot Al_2O_3 \cdot 3CaSO_4 \cdot 32H_2O \tag{4.7}$$

This phenomenon is called ettringite corrosion.[2] The formation of ettringite according to the reactions in equations (4.5)–(4.7) primarily leads to an increase in solid volume, although there is still an ongoing debate in the literature over the mechanism of expansion. There are many theories on the mechanisms of expansion associated with ettringite formation during sulfate attack. Most of these theories can be classified under one of two major schools of thought: (1) crystal growth pressure during the formation of ettringite and (2) swelling of ettringite by water absorption.[3,4]

The ettringite crystals are generally formed in air voids that are filled with air and water at the beginning. When the void spaces in the matrix cannot accommodate any further expansion, internal pressure results in tensile strains that may cause cracking in the hardened concrete, thus causing loss of section and deterioration that is typically the result of sulfate attack in concrete. A typical micrograph depicting well-formed and randomly oriented, rod-like ettringite crystals in the pre-existing pores with diameters of about $150 \, \mu m$ is shown in Fig. 4.1.

In addition to the formation of expansive ettringite and gypsum, sulfate ions can also react with the calcium silicate hydrate gel formed in the hydration of dicalcium silicate and tricalcium silicate. The decalcification of the calcium silicate hydrate gel is only due to the action of magnesium sulfates.[1] The reaction proceeds as shown:

$$3CaO \cdot 2SiO_2 \cdot 3H_2O + 3MgSO_4 + 8H_2O$$
$$\rightarrow 3(CaSO_4 \cdot 2H_2O) + 3Mg(OH)_2 + 2(SiO_2 \cdot H_2O) \tag{4.8}$$

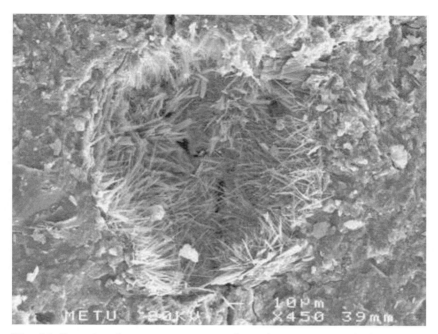

Fig. 4.1 Scanning electron microscope analysis shows an air void almost filled with ettringite crystals. (Courtesy of Mustafa Şahmaran)

Thus, this reaction produces more gypsum ($CaSO_4 \cdot 2H_2O$) that could go through another cycle of the reactions (4.5)–(4.7) described earlier, to produce more ettringite. Moreover, the $Mg(OH)_2$ (brucite) and the silica hydrate ($SiO_2 \cdot H_2O$) formed according to equation (4.8) can further react to produce non-cementitious magnesium silicate hydrate (M–S–H).[5] This process results in a gradual decline of the binding capability, and thus, in a strength loss of the hardened concrete. The decomposition of calcium silicate hydrate gel into a non-cementitious magnesium silicate hydrate can only be achieved by magnesium sulfate, and therefore, the magnesium sulfate is more damaging than sodium sulfate.

4.1.1.2 Thaumasite form of sulfate attack

The occurrence of thaumasite ($CaCO_3 \cdot CaSO_4 \cdot CaSiO_3 \cdot 15H_2O$) formation in sulfate attack has also been reported in the literature.[6–8] Thaumasite is reported to form in concrete during sulfate attack, as a result of the reaction of sulfate with calcium silicate hydrate gel in the presence of CO_3^{2-} ions or dissolved atmospheric carbon dioxide (CO_2), and water at low temperatures (below 15°C) in concrete.[8]

Thaumasite can also form as a result of the reaction of ettringite with calcium silicate hydrate gel in the presence of carbon dioxide. In such cases, thaumasite often seems to exist in a solid solution with ettringite.[9] The reactions leading to thaumasite formation may be represented in a very simplified way, as follows:

$$3CaO \cdot 2SiO_2 \cdot 3H_2O + 2(CaSO_4 \cdot 2H_2O) + 2CaCO_3 + 24H_2O$$
$$\rightarrow 2(CaCO_3 \cdot CaSO_4 \cdot CaSiO_3 \cdot 15H_2O) + Ca(OH)_2 \quad (4.9)$$

$$3CaO \cdot Al_2O_3 \cdot 3CaSO_4 \cdot 32H_2O + 3CaO \cdot 2SiO_2 \cdot 3H_2O$$
$$+ 2CaCO_3 + 4H_2O \rightarrow 2(CaCO_3 \cdot CaSO_4 \cdot CaSiO_3 \cdot 15H_2O)$$
$$+ CaSO_4 \cdot 2H_2O + Al_2O_3 \cdot 3H_2O + 4Ca(OH)_2 \quad (4.10)$$

By comparison with ettringite, thaumasite contains silicate and carbonate instead of aluminate. As calcium silicate hydrate gels are directly involved in thaumasite formation, the reaction is especially deleterious. Thaumasite has not been studied extensively, as most of the reported sulfate attack cases have been for semi-arid environments where the prevalent warm conditions are not considered conducive for the formation of thaumasite. However, with the increasing use of limestone dust as filler, and limestone aggregates in concrete production, a ready source of CO_3^{2-} ions is available within the concrete for the formation of thaumasite whenever low-temperature conditions exist. Thaumasite forms slowly, and results, eventually, in a soft, white and pulpy mass. This causes total disintegration of the concrete, and exposes the reinforcement to consequent rust formation. Magnesium sulfate attack further reinforces the thamausite form of sulfate attack, and makes the overall deterioration worse. Since the thaumasite formation does not depend on the level of calcium aluminate hydrates, sulfate-resistant Portland cement is also vulnerable to this form of aggressive attack.

The thaumasite form of sulfate attack has commonly been found in buried concrete.[10] Figure 4.2 presents a subsurface concrete pier affected by the thaumasite form of sulfate attack in the UK. Thaumasite-related deterioration has been identified in many historical buildings in cold climates as well, where hydraulic cement-based mixtures with a high water-to-cement ratio were used for restoration purposes.[11]

4.1.2   Types of sulfate attack

Sulfate attack can be divided into two groups, depending on the source of the sulfate: internal sulfate attack and external sulfate attack.

Fig. 4.2 Severe thaumasite form of sulfate attack at the base of a 29-year-old bridge column. (© 2003 Elsevier. Courtesy of Ian Longworth)

Internal sulfate attack refers to the cases where the source of sulfate is the concrete itself. The source of sulfate can be the cement, supplementary cementing materials such as fly ash or ground granulated blast furnace slag, the aggregate, the chemical admixtures and/or the water. External sulfate attack occurs when sulfate penetrates the concrete from an external source such as groundwater, seawater or sewage water. In this section, an engineering view of the various mechanisms of sulfate attack in accordance with the sulfate source is discussed.

4.1.2.1 Internal sulfate attack

Skalny *et al.*[7] prefer to talk about the two internal sulfate attack mechanism: excess sulfate-generated expansion at ambient temperature and heat-treatment-generated expansion, and assign them different names as composition-induced internal sulfate attack and heat-induced internal sulfate attack, respectively. The latter mechanism is also called delayed ettringite formation.

Composition-induced internal sulfate attack
Controlled amounts of calcium sulfate, in the form of gypsum, are added to Portland cement during its production to control its setting and early-age behaviour by moderating the rate of reaction of tricalcium aluminate (C_3A) with water. Another important source of composition-induced internal sulfate attack is due to the contamination of aggregates by sulfates – this issue has been on the rise with the increase use of recycled building materials as aggregate. In composition-induced internal sulfate attack at ambient temperatures, the expansion occurs fairly rapidly, within about 6 months. For this reason, the amounts of sulfate in the cement and aggregates are limited by most standards to prevent the occurrence of composition-induced internal sulfate attack. In the ASTM C150 standard,[12] the highest acceptable overall sulfate (SO_3) content for Portland cements varies between 2.3% and 4.5%, depending on the cement type.

Heat-induced internal sulfate attack (delayed ettringite formation)
Another form of internal sulfate attack is delayed ettringite formation. It may be defined as the formation of ettringite in a cementitious material by a process that begins after hardening is substantially complete. Delayed ettringite formation usually occurs in concrete that has experienced elevated temperatures during curing (above ~70°C), either through the external application of heat as in steam curing, or from internal temperature rise due to the heat generated during hydration.[13] Therefore, the danger of damage due to delayed ettringite formation can be eliminated by adjusting the curing temperature, controlling the heat of hydration and limiting the clinker sulfate levels.

Stages in the deterioration of concrete due to delayed ettringite formation are listed by Fu and Beaudoin[14] as follows:

(a) the presence of an internal sulfate source
(b) the high adsorption of sulfate by calcium silicate hydrate gel cured at high temperatures (the critical temperature for sulfate adsorption by tricalcium silicate hydrates appears to be above ~70°C) and the slow desorption of sulfate at later ages
(c) the diffusion of reactants into pre-existing cracks through concrete pore solutions

Fig. 4.3 Typical delayed ettringite formation associated damage in a precast concrete beam. (Reproduced by permission of the American Concrete Institute from Livingstone et al.[15] Courtesy of Richard Livingstone)

(d) water curing, which accelerates delayed ettringite formation by providing a medium for diffusion of reactants, i.e. satisfying the requirements of a thorough process
(e) the preferred crystallisation of ettringite in cracks (preferred crystallisation of ettringite in cracks results in the extension of cracks and the damage of concrete).

The resulting damage to concrete from delayed ettringite formation is very similar to the damage caused by the alkali–aggregate reaction because both produce internal expansion reaction within brittle concrete. Figure 4.3 shows an example of cracking damage in a concrete structure due to delayed ettringite formation.

4.1.2.2 External sulfate attack

The sulfate responsible for external sulfate attack can originate from a variety of sources. Sulfur and sulfide sulfur are found in minerals from the earth's crust in the form of sulfides such as iron or copper. Organic sulfur originates from animal and vegetable manner. Oxidation of minerals and bacterial action on organic material transform sulfides into sulfates.[16] Sulfates are likely to be found in high concentrations in shallow lakes, mining pits, marshes and reservoirs. Moreover, man-made sources of sulfates include run-off containing fertilisers, rain near urban areas, which may contain high levels of sulfur dioxide,[16] and run-off from industrial wastes.

Seawater is another source of sulfate: however, due to the chloride-content of seawater, it poses slightly different problems than solutions of just sulfate. The main ions presented in seawater are chloride, sulfate, sodium and magnesium. The presence of a large quantity of sulfates in seawater could lead to the expectation of sulfate attack, and the reaction of hardened concrete with the sulfate ion in seawater is similar to that with sulfate ion in groundwater, but the effects are different.[17] The presence of chloride ions alters the extent and nature of the chemical reaction so that less expansion is produced by a cement of a given calculated tricalcium aluminate content than would be expected of the same cement in a groundwater exposure where the water has the same sulfate ion concentration. This is attributable to the formation of ettringite in a chloride environment, rather than in an alkaline environment, which is essential for the swelling of ettringite by water adsorption.[18] The magnesium ions in seawater also participate in the reactions as in equation (4.8), and the resulting $Mg(OH)_2$, often called brucite, precipitates in the pores at the surface of the concrete, thus forming a protective surface layer which impedes further reaction.[19]

In external sulfate attack, sulfates from an external source penetrate into the hardened concrete, causing its degradation. Concrete elements that are particularly susceptible to external sulfate attack by ground-water are footings, foundation walls, retaining walls, piers, piles, culverts, pipes and slabs-on-grade. The types of structures that are at high risk are those that have faces open to evaporation such as piles, canals, pipelines, transmission tower footings and highway pavements. Two types of external sulfate attack can occur: The first type is favoured by a high tricalcium aluminate content in the cement and by a low sulfate concentration, and is characterised by the formation of ettringite. The other, favoured by a low tricalcium aluminate content and by a high sulfate concentration, is characterised by gypsum formation. The transition between the two types of attack is gradual, and both can occur simultaneously. Similar conclusions are drawn that the formation of ettringite leads to expansion, and that the formation of gypsum leads to loss of cohesion and softening of the cement matrix, but to no

Fig. 4.4 Map pattern cracks due to external sulfate attacked in over-site concrete slab. (Courtesy of Ian Longworth)

more than a marginal degree of expansion. The loss of cohesion and softening may also be attributed more directly to the incorporation of sulfate ions into the calcium silicate hydrate gel.

Typical cracks induced by external sulfate attack in concrete slabs, placed directly on moist soils of high sulfate content, are shown in Fig. 4.4. Crack patterns and expansion due to external sulfate attack are typically characterised by map cracks, which are similar to patterns of cracks due to the alkali–aggregate reaction and delayed ettringite formation.

4.1.3 Control of sulfate attack

The extent to which concrete is affected by sulfates depends on several factors, including the chemistry of the cement, the quality of the concrete (i.e. permeability, type of materials used, water-to-cement ratio and quality of placement and curing), the concentration of external sources of sulfate ions and the severity of the concrete exposure to the sulfate environment.[20] Assuming similar environmental conditions, two factors will tend to control the resistance of a given concrete to sulfate attack: the chemistry of the cement and the quality of the concrete.

4.1.3.1 Chemistry of Portland cement

The property of concrete with the greatest impact on sulfate resistance is generally considered to be the chemistry of the Portland cement. The primary compounds that constitute Portland cement are tricalcium silicate, dicalcium silicate, tricalcium aluminate (C_3A in cement notation), tetracalcium aluminoferrite ($4CaO \cdot Al_2O_3 \cdot Fe_2O_3$, or C_4AF in cement notation) and a sulfate hydrate compound called gypsum that is added to Portland cement clinker to control the setting time. Among these, tricalcium aluminate and gypsum compounds have the largest impact on sulfate attack, since they are directly involved in the formation of ettringite. Today, most of the investigations identify tricalcium aluminate as the primary measurable component of Portland cement that influences sulfate resistance. Specifically, damage due to sulfate attack has been found to decrease as the tricalcium aluminate content is lowered. Harboe[21] presented the results of Smith[22] to demonstrate the effect of tricalcium aluminate content on expansion due to sulfate attack. These results are reproduced in Fig. 4.5, and show how the expansion increases as the tricalcium aluminate content is increased. The exposure condition for the test data shown in Fig. 4.5 was a laboratory accelerated test in which concrete specimens were alternately soaked for 16 hours in a 2.1% sodium sulfate solution at 23°C and dried for 8 hours at 55°C.

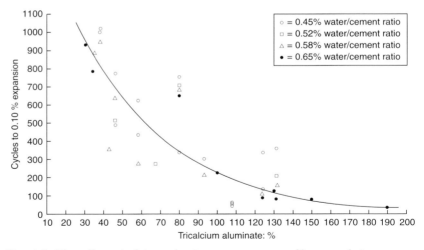

Fig. 4.5 *The effect tricalcium aluminate content on sulfate attack in concrete.* (© 1982 *the American Concrete Institute. Reproduced with permission from Harboe*[21])

To control the cement chemistry, American standards (ASTM C150)[12] suggest a limit on the tricalcium aluminate content for each cement type. Therefore, this limit addresses the fact that the tricalcium aluminate content is the critical compound of cement that affects the sulfate resistance of concrete. The suggested limits on tricalcium aluminate content in ASTM C150 are 8% for moderate sulfate resistance (Type-II cement), and 5% for high-sulfate-resistance cements (Type V cement). This standard also limits the tetracalcium aluminoferrite content plus twice the tricalcium aluminate content ($2C_3A + C_4AF$) of the cement to 25%, and the SO_3 content of the cement to 2.3% for sulfate-resistant Portland cements. The British standard for sulfate resistance cement specifies a maximum tricalcium aluminate content of 3.2% and a maximum SO_3 content of 2.5% in BS 4027.[23]

On the other hand, cements with a low content of tricalcium aluminate and tetracalcium aluminoferrite compounds generally tend to have a higher tricalcium silicate to dicalcium silicate ratio (C_3S/C_2S), and an increase in the tricalcium silicate content of cement generates a significantly higher amount of calcium hydroxide, as the hydration of tricalcium silicate produces nearly 2.2 times more calcium hydroxide than the hydration of dicalcium silicate. Calcium hydroxide is known to be responsible for the formation of gypsum, and gypsum is known to be the first step in the formation of ettringite, which can be considered as the principal cause of deterioration. Moreover, gypsum formation leads to the reduction of stiffness and strength; this is followed by expansion and cracking, and eventual transformation of the hardened concrete into a mushy or non-cohesive mass.[24,25] Irassar et al.[26] found that mortar bars containing a low tricalcium aluminate and a low tricalcium silicate cement showed about 10 times less expansion than those with a low tricalcium aluminate and a high tricalcium silicate content.

4.1.3.2 *Quality of the concrete*

While a reduced tricalcium aluminate content can help to create a sulfate-resistant mixture, many researchers have reported that for maximum resistance the concrete must also be of good quality and low permeability. To enhance the quality and permeability of concrete, it is recommended to lower the water-to-cement ratio, increase the cement content, consolidate the concrete adequately during placement, provide proper and effective curing, and replace a percentage of the cement by cementitious materials with known properties.[19,27] The goal of using low-permeability concrete is to minimise the penetration of sulfate ions into the concrete. Keeping the sulfate ion concentration in hardened concrete low prevents the deleterious formation of ettringite. According to Mehta,[3] for the prevention of sulfate attack in concrete, control of the permeability of concrete is more important than control of the chemistry of the cement. It is well known that a low water-to-cementitious materials ratio results in a low-permeability concrete if compaction and curing have been carried out adequately.

While the role of sulfate-resistant Portland cement in preventing sulfate attack has long been recognised, it has been found recently that the use of supplementary cementitious materials such as ground granulated blast furnace slag, fly ash, natural pozzolan, calcined shale and silica fume can provide equivalent or better resistance to sulfate attack than sulfate-resistant Portland cement alone. Supplementary cementitious materials may reduce not only the permeability but also the tricalcium aluminate content when they are a partial replacement of cement. Moreover, the use of supplementary cementitious materials or the use of blended cements (Portland blast furnace slag cements with more than 50% slag, and Portland pozzolan cements with at least 25% pozzolan) in general, reduces the quantity of calcium hydroxide due to pozzolanic reactions which would otherwise react with sulfates to form gypsum. However, special care should be taken to select a suitable source of fly ash and/or slag. Information on the influence of these two types of supplementary cementitious materials indicates that the influence of these materials on the performance of concrete tends to vary significantly from one source to another.

4.1.4   Current guidelines and specifications for sulfate-resistant concrete

In this section, current guidelines and specifications for engineers to use when designing a concrete for use in a sulfate environment are outlined. Several agencies have developed their own classification systems for determining the risk of sulfate attack based on the concentrations of sulfate, including the US Bureau of Reclamation, the American Concrete Institute (ACI), the Canadian Standards Association (CSA) and the British Standards Institution (BSI).

The US Bureau of Reclamation classified the severity of sulfate attack on concrete by the concentrations of sulfates in solution, and in soils as follows: negligible at less than 0.015%, moderate at 0.015–0.100%, severe at 0.100–0.200% and very severe at over 0.200%.[28]

The ACI has also developed recommendations for the cement type, maximum water-to-cementitious (w/cm) materials ratios and minimum compressive strength values to be used in different exposure conditions, which are given in *Building Code Requirements for Structural Concrete and Commentary* by ACI Committee 318-05.[29] Table 4.1 provides current ACI building code requirements for concrete exposed to sulfate-containing solutions.

The negligible sulfate environment according to ACI Committee 318-05 guidelines has sulfate so slight as in the US Bureau of Reclamation guideline that sulfate attack is not a concern, and there is no restriction on the material selection and the water-to-cementitious materials ratio. For moderate sulfate environments, ASTM Type II Portland cement (moderate sulfate-resistant cement) or Portland pozzolan or Portland slag cement is recommended for this level of sulfate resistance. The mix proportions for a moderate sulfate environment are limited such that a water-to-cementitious materials ratio of 0.50 or less for normal-weight concrete is required to ensure the use of good-quality concrete. The ACI building code also designates a minimum compressive strength of about 28 MPa for normal-weight and light-weight aggregate concrete.

For severe sulfate environments, sulfate-resistant Portland cement (ASTM Type V) with a water-to-cementitious materials ratio of less than a 0.45 is recommended. For a very severe environment, ACI Committee 318-05 recommends a Type V cement augmented with a pozzolan or slag and a maximum water-to-cementitious materials ratio of 0.45. For severe and very severe sulfate environments, the guideline also requires a minimum compressive strength of 31 MPa for normal-weight and light-weight aggregate concrete.

CSA Standard A23.1-00, *Concrete Materials and Methods of Concrete Construction*,[30] also has three classifications of sulfate attack designated S-1, S-2 or S-3 based on the standards initially developed by the US Bureau of Reclamation, and has guidelines for selecting the cement type, the water-to-cementitious materials ratio and minimum compressive strength values for each of these exposure classes.

The new European standard for concrete (BS EN 206-1)[31] contains three classifications of sulfate attack (XA1, XA2 and XA3), based on the concentration of sulfate in water or in soil. The BS EN 206-1 requirements are shown in Table 4.2. As seen from Table 4.2, the standard gives guidance on the minimum cement content, maximum water-to-cement (w/c) ratio, the minimum compressive strength class and the cement type for the different exposure classes.

Table 4.1 ACI recommendations for sulfate resistance[29]

Severity of potential exposure	Water-soluble sulfate %: w/w	Sulfate (SO_4) in water: ppm	Cement type	Maximum w/cm ratio, by mass[a]	Minimum 28 day concrete compressive strength: MPa
Not applicable	$SO_4 < 0.10$	$SO_4 < 150$	No special requirements for sulfate resistance		
Moderate	$0.10 \leq SO_4 \leq 0.20$	$150 \leq SO_4 \leq 1500$	C150 Type II or equivalent	0.50	28
Severe	$0.20 \leq SO_4 \leq 2.00$	$1500 \leq SO_4 \leq 10\,000$	C150 Type V	0.45	31
Very severe	$SO_4 > 2.00$	$SO_4 > 10\,000$	Type V plus pozzolan or slag[b]	0.45	31

[a] The amount of the pozzolan or slag to be used shall not be less than the amount that has been determined by the service record to improve sulfate resistance when used in concrete containing Type V cement.
[b] These values are applicable to normal-weight concrete. They are also applicable to light-weight aggregate concrete except that the maximum w/cm ratios 0.50 and 0.45 should be replaced by specified 28 day compressive strengths of 28 MPa and 31 MPa, respectively.

Table 4.2 BS EN 206-1 requirements to protect against damage to concrete by sulfate attack[31]

Severity of potential exposure (exposure class)	Sulfate (SO_4) in groundwater samples: mg/l	Sulfate (SO_4) in soil samples: mg/kg	Minimum cement content: kg/m^3	Maximum w/c ratio, by mass	Minimum compressive strength: MPa[a]
Slightly aggressive (XA1)	$200 \leq SO_4 \leq 600$	$2000 \leq SO_4 \leq 3000$	300	0.55	30
Moderately aggressive (XA2)	$600 \leq SO_4 \leq 3000$	$3000 \leq SO_4 \leq 12\,000$	320[b]	0.50	30
Highly aggressive (XA3)	$3000 \leq SO_4 \leq 6000$	$12\,000 \leq SO_4 \leq 24\,000$	360[b]	0.45	35

[a] When SO_4 leads to exposure classes XA2 and XA3, moderate or high sulfate-resisting cement should be used in exposure class XA2, and high sulfate-resisting cement should be used in exposure class XA3.
[b] The minimum compressive strength at 28 days of 150 mm diameter by 300 mm cylinders.

Different types of surface treatments can also be applied to a concrete structure to protect the concrete from sulfate attacks. There are four main classes of treatment: coatings and sealers, pore-liners, pore blockers and renderings. A report by ACI Committee 515, *A Guide to the Use of Waterproofing, Damp-proofing, Protective, and Decorative Barrier Systems for Concrete,*[32] contains a wealth of information on different protective systems against over-aggressive agents.

4.2 Acid attack

Mustafa Sahmaran, *Gaziantep University, Gaziantep, Turkey*

Concrete is chemically basic, having pH values ranging between approximately 12.5 and 13.5, depending upon the mixture proportion. Since it has high alkalinity, it is readily attacked by various acids, which have pH values of less than 7. Dry concrete is generally immune to attack by dry chemicals. Also, it is highly resistant to many chemicals in solution. However, a considerable number of chemicals in solution may attack concrete. Concrete can be subjected to attack by various inorganic and organic acids, including sulfuric, sulfurous, carbonic, humic, hydrochloric, nitric, phosphoric, acetic and lactic, among others.[33]

Acid attack on concrete can occur in a variety of ways. Buried concrete structures can be attacked by acids present in groundwater or soil, either naturally occurring or due to the dumping of chemical wastes from industrial processes. Leakage and random spillage of acids can also occur in industrial environments.[34] The disposal of colliery waste, ash from coal combustion in power plants or from refuse incineration, or gasworks waste, in which sulfuric acid is an important component, can also give rise to acidic groundwater. Acid rain can affect the long-term durability of concrete structures exposed to the atmosphere.[35] Another form of acid attack on concrete, which is especially severe in hot climates, occurs in sewage systems. Seawater is also one of the sources of salts which affect concrete structures. Chemical degradation of concrete structures exposed to aggressive agricultural chemicals may also lead to serious technical and economical problems.[36] Therefore, in practice, acid attack is most commonly a problem in industrial processes, in sewers and in circumstances where the concrete is exposed to rapid flows and considerable volumes of acid.

4.2.1 Mechanism of acid attack

The spectrum of acidic media is broad, and the mechanism of acid attack varies according to the type of acids. Among them, sulfuric, carbonic and humic acids are the acids most commonly encountered by concrete, since they are found in natural ground waters.[37] Sulfuric acid is a highly ionised mineral acid, and may result in a pH value lower than 2. Carbonic and humic acids are only moderately aggressive, and will not produce a pH value of below ~3.5. The effect is generally the dissolution of cement hydrates and calcium hydroxide as calcium salts. This weakens the affected concrete, but is distinct from sulfate attack, as the degradation does not involve significant expansion. Similar to cement hydration products, carbonaceous aggregate such as limestone and dolomite used in concrete production is also susceptible to acid attack.

Unlike other acids, sulfuric acid (H_2SO_4) can cause both dissolution and swelling of concrete. Several researchers reported a loss in the weight of concrete when exposed to sulfuric acid. When sulfuric acid is present with concrete, it reacts with the calcium hydroxide and the calcium silicate hydrate gel in concrete to form gypsum.[38] This phenomenon is called gypsum corrosion, and causes the dissolution and swelling of concrete (the term 'corrosion' used in Sections 4.1 and 4.2 is used to indicate that the deterioration occurs in the concrete matrix rather than involving a steel corrosion mechanism). When gypsum corrosion occurs, the alkalinity of the concrete pore solution is reduced, which is favourable to microbiologically induced corrosion (see the next paragraph). The product of corrosion is a weak compound with no cementitious properties, and its active formation eventually leads to early deterioration, loss of strength and, in extreme cases, the destruction and collapse of the concrete structure. In the second stage of the corrosion of concrete by sulfuric acid, gypsum could go through another cycle of the reactions described in the sulfate attack section, to produce ettringite and/or thaumasite. However, ettringite and thaumasite are not stable in an acidic environment,[39,40] and therefore the reaction product from sulfuric acid attack will be primarily gypsum.

Domestic sewage does not usually have detrimental effects on good concrete, as evidenced by the satisfactory performance of the thousands of miles of concrete sewage pipes and conduits in widespread use. However, under a combination of special conditions of a high concentration of sewage, a low velocity of flow or a high temperature, hydrogen sulfide gas (H_2S) may be evolved as a result of the oxidising action of anaerobic bacteria (*Desulfovibrio desulfuricans*) on organic or inorganic sulfur

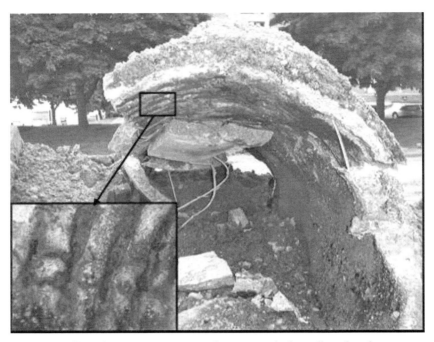

Fig. 4.6 Collapsed concrete sewer pipe due to microbiologically induced corrosion. (Reproduced by permission of the American Society of Civil Engineers from Vaidya et al.[43] Courtesy of Erez N. Allouche)

compounds present in the sewage.[41] Hydrogen sulfide gas forms relatively weak acid solutions with only a mild corrosive action on concrete when dissolved in water. However, under certain conditions, principally in sewers carrying organic wastes, hydrogen sulfide can be oxidised by aerobic bacterial (Thiobacillus) action to sulfuric acid, which can attack concrete.[42] Sulfuric acid attack deterioration following these reactions is extremely destructive. This phenomenon is also called microbiologically induced corrosion. The primary product produced during concrete decomposition by microbiologically induced corrosion is calcium sulfate ($CaSO_4$), which when hydrated becomes gypsum. This material has low structural strength, especially when it is wet. It is usually present in corroded sewers as a pasty white mass at the crown region above the water line.[43] An example of severe microbiologically induced corrosion in a 27-year-old prefabricated 900 mm concrete sewer pipe collapsed by microbiologically induced corrosion is shown in Fig. 4.6.

Acid attack of concrete can also be caused by dissolved carbon dioxide in soft water, which behaves as a solution of carbonic acid (H_2CO_3).

When cement-based materials are exposed to carbonic acid, a reaction producing carbonates (calcium carbonate ($CaCO_3$) and calcium bicarbonate ($Ca(HCO_3)_2$) takes place which is accompanied by carbonation shrinkage. This process is usually referred to as carbonation. The reaction of concrete with atmospheric carbon dioxide is generally a slow process and highly dependent on the external exposure conditions (humidity and temperature), the quality of the concrete and the concentration of aggressive carbon dioxide.[44] The results of the carbonation can either be beneficial or harmful depending on the extent to which it occurs. Limited carbonation of the surface layer of concrete is known to seal the pores by forming calcium carbonate, which may have the beneficial effect of reducing the permeability and increasing the strength of the surface layer. Carbonate attack is only significant with high flows of soft water saturated with carbon dioxide, and is generally a surface effect since precipitation of silica gel as a result of cement hydrate dissolution inhibits further attack. For example, water draining from mountains is often nearly free of dissolved salts, but may nevertheless be acidic because of its carbonic acid content derived from carbon dioxide coming from the atmosphere. This nearly pure water can be corrosive to concrete, particularly lean and therefore permeable concrete. Moreover, continued carbonation may also cause a reduction in the alkalinity of the cement paste from about 12.5 to 8.5. Because this pH value is no longer inimical to micro-organisms, bacteria may start to grow, and by acid production further reduce the pH, leading to further acid attack on concrete. However, even if no additional acid attack ensues, at a pH value of 8.5, the steel reinforcement in concrete becomes susceptible to corrosion because the passivation by a high pH value no longer exists (see Section 4.4.2).

Humic acid, which is the product of microbial decay and the decomposition of dead organisms and vegetables, has also been considered to have corrosive properties when it comes into contact with hardened concrete. When humic acid reacts with hardened concrete, it forms a mixture of calcium salts whose solubilities vary with the molecular weight of the individual acids. All natural waters, marine sediments and terrestrial soils contain varying concentrations of various organic compounds, of which the humic compounds constitute a major part. The effect of humic acid on concrete has been investigated by Robertson and Rashid.[45] They immersed hardened concrete blocks in fresh and salt water to which known concentrations of humic acid had been added. It was found that the surface of the concrete became discoloured (usually brown), and a protective coating (calcium humate) on the surface of the concrete formed by reaction with the concrete. The protective coating formed is nearly insoluble, and humic acid attack on concrete is superficial. This coating prevented any further serious attacks.

4.2.2   Control of acid attack

When concrete is attacked by an acid solution, the rate and extent of deterioration depends on the nature and concentration of the acids in the solution in contact with the concrete, the temperature and pressure of the solution, and, most importantly, on the quality of the concrete as measured primarily by its permeability. To obtain high resistance to acid attack, the same rules must be followed as in sulfate attack. The concrete must be hard and non-porous, of suitable 'richness', must be adequately compacted and must have a low water-to-cement ratio.[44]

Concrete pipes, or other conduits, to carry sewage should be of low permeability to minimise penetration by liquid. In practice, this means that the concrete should have a high cement content and a low water-to-cement ratio, and that fabrication should be by a process that results in a dense and homogeneous structure. If acid attack is likely to occur, the service life of the concrete may be increased by the use of good-quality calcareous aggregate (limestone or dolomite), which is attacked by the acid, thereby preventing a more concentrated acid attack on the hardened cement binder. There is also a more uniform loss of surface compared with the rough exposed aggregate surface left with insoluble aggregates. For these reasons, the use of calcareous aggregate has been accepted in South Africa as a means of prolonging the service life of sewage pipes that are likely to suffer attack by sulfuric acid.[46] This is not considered a solution to the problem but a means of retarding the rate of attack. Resolution of the problem can occur only through methods that prevent the formation of sulfuric acid in deleterious concentrations. These include designing the sewer system to provide a sufficient flow rate, ventilation to remove hydrogen sulfide, use of chemicals to prevent sulfur compounds from being converted to hydrogen sulfide, or using toxic materials to decrease or eliminate the activity of aerobic bacteria.[41]

Although the degree of resistance varies depending on the solution chosen, the use of supplementary cementitious materials such as silica fume and fly ash in concrete has also been found to improve the resistance of concrete to acid attack because of the reduced quantity of calcium hydroxide due to the pozzolanic reactions, which is the hydration product most vulnerable to acid attack.[47,48]

As mentioned before, three exposure classes (XA1, XA2 and XA3, corresponding to slightly, moderately and highly aggressive, respectively) are described in BS EN 206-1,[31] depending on the aggressiveness of the chemical attack on concrete, which takes into account, in addition to sulfate concentration (see Table 4.2), also pH and the presence of carbon dioxide and ammonium and magnesium ions in the case of ground water. An informative annex in BS EN 206-1 gives indicative limits for the minimum cement content, the maximum water-to-cement ratio and the minimum strength class for each of these exposure classes (see Table 4.2).

Concrete can also be protected from attack by acid solutions by a number of surface treatments.[32] These may block the pores and render the concrete less permeable. They may react with the free lime or they may consist of a surface coating which physically prevents contact between the concrete and the aggressive solution. Some of these surface treatments also harden the surface and prevent dusting. Surface treatment with a water-repelling agent represents only a provisional remedy that is able to delay concrete attack. Plasticised polyvinyl chloride (PVC) linings can also be used to control deterioration and erosion of concrete in acid environments. This technique has been used commercially for many years. The process uses a PVC lining in a concrete pipe, which offers the advantages of size versatility and lower cost when compared with plastic pipe.

4.3 The alkali–aggregate reaction (AAR) damage to concrete

Alan Poole, A. B. Poole & Associates, UK

Concrete as used in civil engineering, if properly specified, produced, placed and cured, can be expected to have a long design life. Nevertheless, examples of the premature deterioration of concrete structures have been reported for many years. There are numerous causes of these failures: some are due to inadequate design or substandard or inappropriate materials, and others to poor quality production, placing and curing.

Several mechanisms involving chemical damage within the fabric of the concrete have been discussed in earlier sections of this chapter. However, one particular materials problem arises from a direct reaction between the cement and aggregate, and can cause cracking, spalling and disruptive expansion of the hardened concrete. This particular type of damage was first observed in California in the 1920s and 1930s, and was first described by Stanton.[49,50] Since then, examples of the AAR have been reported from around the world, and have been the subject of a large volume of published research. Perhaps the most important studies are those reported in the series of proceedings of international conferences. The first of these was held in 1974, and the most recent, the 13th International Conference on Alkali–Aggregate Reaction (ICAAR) was held in Trondheim, Norway in June 2008. The research studies presented at these conferences have been important for the publication of a number of national and international standards, guidance documents and codes of practice that are concerned with appropriate test procedures and with the avoidance or minimisation of the effects of AAR. There are also many case study reports in the proceedings giving examples of repair and remediation of affected concrete structures. Important among the published reviews and guidance documents are the (BRE) Digest 330,[51] *The Diagnosis of Alkali–silica Reaction*,[52] Concrete Society Technical Report 30[53] and Helmuth *et al.*[54] in the Strategic Highway Research Programme in the USA.

4.3.1   Types of AAR in concrete

There are two main forms of the AAR, by far the most common is the alkali–silica reaction (ASR), the second, much rarer, form is the alkali–carbonate reaction (ACR). A possible third type, alkali–silicate reaction has been reported, but is now generally considered to be a more complex form of alkali–silica reactivity.

The ASR is a chemical reaction within a hardened concrete between specific siliceous constituents in the aggregate particles and the alkali hydroxides released into the pore space in the hydrated Portland cement matrix. The reaction product formed is an alkali–silica gel of variable composition and properties, but it has the ability to absorb more water and exert internal stresses, typically of the order of 3 MPa or 4 MPa, but may range up to 10 MPa.[54] In many cases, such stresses are sufficient to crack and disrupt the concrete.

The three essential requirements for this reaction to initiate are, firstly, the presence in the concrete of critical amounts of a reactive form of silica, present as a mineral constituent in the aggregate particles. Secondly, a critical concentration of alkali hydroxide (sodium or potassium), usually derived from the Portland cement used, is necessary. The alkali is calculated in terms of 'sodium equivalent', that is $(Na_2O + 0.658K_2O)$ wt%. The third essential is the presence of adequate water, usually defined as a relative humidity of at least 85% in the concrete. Water is important as a carrier for the alkali metal and hydroxide ions that feed the reaction and also because it is absorbed by the hygroscopic gel reaction product, producing the expansive pressure.

The reactions can be summarised as

$$4SiO_2 + 2NaOH \rightarrow Na_2Si_4O_9 + H_2O \qquad (4.11)$$

$$silica + alkali \rightarrow alkali-silica\ gel + water \qquad (4.12)$$

$$SiO_2 + 2NaOH \rightarrow Na_2Si_3 + H_2O \qquad (4.13)$$

However, in reality the alkali–silica gel reaction product is both variable and indefinite in composition. Also, the swelling pressures developed involve absorption of additional water, so these equations only represent an idealised situation.

Reactive forms of silica can include cryptocrystalline or microcrystalline forms of quartz with large surface areas, siliceous volcanic glass, tridymite, opal, chalcedony, some forms of chert and other amorphous or poorly crystallised quartz. Sometimes these are present as part of the cementing agent between mineral grains in the aggregate particles, for example in some greywackes and sandstones. Certain granites, gneisses, hornfels and greywacke aggregates have been found to be alkali reactive in concrete, but the reactive constituent has not been precisely identified. However, the crystalline quartz they contain shows optical strain shadows when viewed under the polarising microscope, and the severity of this strain has been used to provide a rough indicator of the observed degree of alkali reactivity, though there are many exceptions to this correlation. The initial conclusion, that lattice defects, as indicated by the strain shadows in the quartz grains, allowed the direct reaction of the silica with alkalis, is now discounted. Provided no annealing has occurred, the observed strain is generally considered to be a measure of the severity of the geological stresses of tectonism and metamorphism that the rock has undergone. Such processes are usually accompanied by mineral recrystallisation which, for quartz, begins with incipient recrystallisation at the grain margins. It is now thought that this crypto-crystalline silica formed at the grain margins is the reactive component in these cases.[56]

The alkalis from the cement accumulate as a solution in the minute pore spaces within the cement hydrates that form when the concrete sets and hardens.[57] Though the alkali is usually derived from the Portland cement itself, there is also some evidence that alkalis may be derived in part from external sources, or from the partial breakdown

71

of altered alkali-rich minerals such as feldspar and mica present in certain aggregates. Typically, water is always present in hardened concrete, trapped within the pore structure. Even in desert conditions, where the outer layer may dry out to a depth of several centimetres, the centre of a concrete element will retain a relative humidity of 80–90%. It has also been suggested that calcium hydroxide, a major product of cement hydration, is also necessary for the expansive reaction product to form.[58]

Natural mineral aggregates usually contain several mineral species, and many aggregate rock types contain silica as one component. A few of these may contain reactive forms of silica, perhaps only in small amounts as cement between mineral grains, or as a minor component in the rock aggregate. Nevertheless, these may be present in sufficient amounts to react with alkali solutions within the concrete, and may cause disruptive damage. Consequently, because the reactive component is only one constituent that may or may not be present in a particular rock type, a very wide variety of geological rock types world-wide have been reported as 'alkali reactive'. The list is extensive, even including types such as granite, sandstone and basalt, although, of course, the vast majority of these rock types make entirely satisfactory and stable aggregates.

Unfortunately, if a reactive component is contained within a particular aggregate type, only a small percentage needs to be present for the reaction to occur. In many case study examples in the UK and elsewhere, this is typically lower than 10%, but is sufficient to produce a damaging reaction. This relatively small percentage has been called 'the pessimum proportion'. The idea arises from the results of a variety of test investigations and research studies.[59] These show that there is a particular proportion of reactive component which when reacted with alkali produces a maximum expansive effect in the concrete. Higher and lower percentages than this 'pessimum proportion' have smaller, sometimes even negligible, expansive effects. Figure 4.7 shows three hypothetical expansion curves: curve A has a pronounced pessimum at about 6%, typical of some UK examples; curve B has a typical broad pessimum, and peaks at about 30%; while curve C has no pessimum, and is unusual, but may be the situation pertaining to certain slowly reacting aggregates. It has been suggested that the pessimum mechanism may be a consequence of a critical level of reactive silica within the aggregate which will react with an excess of alkali so as to form the maximum proportion of gel product with a composition that on absorbing water produces the most expansive forces. Above this

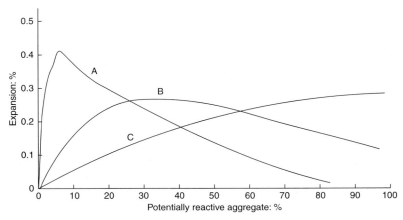

Fig. 4.7 Curves illustrating the concept of the pessimum proportion of reactive component against expansion

concentration, there is an excess of reactive silica in the system, which leads to a reduction in the expansion, either by a more widespread, but reduced, reaction and expansive effect, or by changes in the gel properties.

The ACR differs from the ASR in that it is a damaging expansive reaction between certain types of limestones and impure dolomitic limestones with alkalis in the concrete but does not develop an expansive gel. It is very rare by comparison with the ASR, but case study examples have been reported from Kingston, Ontario,[60] the Middle East,[61] and China.[62,63]

In detail, at least three variants have been recognised:[64] all produce clear observable reaction rims around the affected limestone aggregate particles and, in some examples, rims can develop in as little as 30 days after casting. Rims up to 0.5 mm wide may be observed at the margins of some calcitic limestones. They are more porous than the interior, but are not expansive, and may be the result of the conversion of calcite in the aggregate to calcium hydroxide. The other two variants involve dolomitic limestones; one involves a distinct dedolomitised rim, but again does not appear to be expansive. The third type produces marked expansion effects in the concrete that may develop within a matter of months. In this type, the fine-grained dolomitic limestone aggregate also contains clay: in the Kingston example, this is a mix of illite and chlorite.[65]

Although a de-dolomitisation reaction is involved, and the dolomite crystal size appears to affect the expansion, no distinct reaction product

73

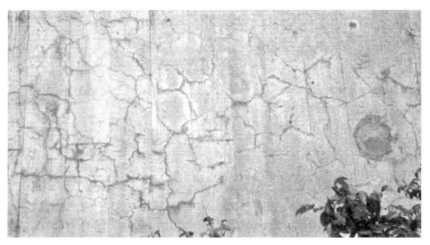

Fig. 4.8 Typical map cracking in a bridge abutment wing wall. Note that there is discoloration associated with some of the cracks

has been identified, and the actual cause of expansion has not been definitively established. Some concretes supposedly affected by the ACR have been found to have evidence of the ASR within the carbonate aggregate, suggesting that the reactions within the carbonate aggregate may be complex.[66,67]

4.3.2 The effects of the AAR on concrete structures

Typically, the first observable macroscopic effect of the AAR is the development of cracking on the surface of the concrete. With the rare ACR, cracking and expansion becomes noticeable after 1–3 years.[65] Usually, cracking due to the ASR takes between 5 and 12 years to become apparent, though there are reported examples where damage appears earlier, and a number of other case studies of slowly reacting concrete where damage took 20–30 years to develop. The cracking associated with the ASR has characteristic patterns on concrete surfaces, and typically tends to appear first, and more severely, on the weather faces of structures. There are many reported cases of ASR-affected concrete structures that are continuing to expand after 50 years.

On unconstrained elements, for example a wall panel, or floor slab, cracking develops as a random network, reminiscent of the political boundaries on a map: this is so called 'map cracking' (Fig. 4.8). In many examples, the cracks themselves have characteristic bleached or

Fig. 4.9 A central bridge support column: note vertical cracks, the DEMEC monitoring points and the remedial measure of steel restraining hoops at 0.5 m centres

pinkish margins, probably the result of strong alkali solutions leaking from them and destroying the lichens and other microbial material on the surface adjacent to the crack. Where the concrete element is under a stress, for example a column carrying a bridge deck, cracking tends to be parallel to the direction of maximum stress (Fig. 4.9).

Since reaction sites are associated with aggregate particles distributed through the concrete, the expansions caused by the gel reaction product will form at reaction sites throughout the affected concrete. Consequently, the observed surface cracking is only an expression of the cracking which may extend through the whole concrete mass. As an example, a $3 \, m^3$ multiple-pile concrete cap that formed a viaduct column base was found to have cracking that had developed and propagated right through the cap after a period of about 15–20 years as a result of the ASR. One of the cracks was found to have a width of 3 mm.

75

Fig. 4.10 A 50 mm displacement upstream (arrowed) of the bay nearest the camera: Val de la Mare Dam, Jersey. (Reproduced with permission from Poole.[68] *© Alan Poole)*

The expansive gels formed as a result of the ASR cause the concrete to expand unless the constraints due to the steel reinforcement prevent this. The progress of the reaction depends on a number of variable factors, including the reactivity of the silica, the locations of reactive aggregate particles distributed in the concrete, the local availability of alkalis and water, the porosity and permeability of the cement matrix, temperature and moisture movements. Consequently, the expansive effects will vary widely from element to element in the same structure. The result of this is to displace one part of the structure with respect to the next, as shown in Fig. 4.10. Such expansive displacements have in some examples caused serious difficulties with moving metal structures, such as sluice gates set between concrete supports.

It is rare for the gel reaction product to be observed exuding to the surface from cracks, and although transparent and resinous when fresh, it readily carbonates and turns white. Consequently, it appears similar to, and may be mistaken for, calcium carbonate resulting from the normal and common carbonation of calcium hydroxide which has

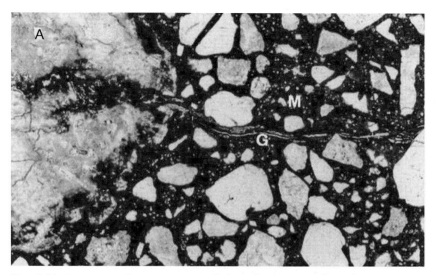

Fig. 4.11 A concrete thin section viewed through a petrographic polarising micro-scope. Alkali–silica gel (G) formed in the coarse aggregate particle (A) on the left is exuding along a crack in the cement matrix (M)

leached to the concrete surface. However, if ASR gel is present, it is best identified by careful examination of freshly broken surfaces of a concrete sample, and since it is hygroscopic, damp patches can sometimes be observed close to a reacting particle. A more detailed analysis using petrographic thin sections of the concrete is usually able to confirm the presence of gel in cracks and to identify the reacting component in the concrete aggregate (Fig. 4.11).

A further consequence of the development of an expansive gel on a reactive aggregate particle close to a concrete surface is that the resultant cracking will spall off a portion of the surface concrete, as is shown in cross-section in Fig. 4.12. Typically, such spalls are circular or oval in shape, but they are comparatively rare, whereas map cracking is the most common first indication of the ASR in a concrete. This polygonal map cracking develops as a 'star' of three or more cracks propagating from each reacting aggregate particles close to the concrete surface, which then join up to form the familiar polygonal pattern. Since some particles react more quickly than others and some are further from the surface, the critical stress needed to initiate cracking at the surface differs and develops over time. Consequently, the initial pattern of map cracking is gradually augmented by later cracks, thus intensifying the initial pattern.

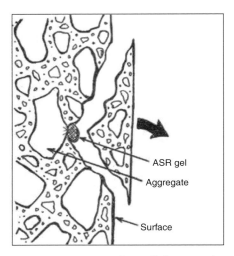

Fig. 4.12 A diagrammatic cross-section of a spalled portion from a concrete surface

4.3.3   The specialised testing for the AAR in concrete

The problems associated with the AAR in concrete are now recognised worldwide, and measures to evaluate aggregates, cements and concretes have been devised in many countries. These methods include evaluating the potential reactivity of any aggregates to be used, testing the proposed actual concrete mix designs and also evaluating the potential for expansion of concretes sampled from existing structures.

Methods concerned with aggregates rely firstly on a petrographic examination to determine whether components in the aggregate are of types that may react. The second stage is an evaluation of the effects of treating the aggregate with alkali, or by preparing a mortar or concrete prisms containing the aggregate under test. The list of minerals within aggregates that may exhibit alkali reactivity is now well established for most countries around the world. Consequently, a simple petrographic analysis (e.g. BS 812-104[69] or ASTM C295[70]) will identify suspect aggregate components that can then be checked by a number of other tests.

One of the simplest (BS 7943[71]) involves exposing particles of the aggregate in the surface of a cement disc that is stored in alkali solution. After a period of days, alkali–silica gel exudations appear on the surface of any alkali–silica reactive particles present. However, to check whether this reactivity will case damaging expansion in a concrete, a mortar bar (ASTM C227[72]), or a concrete prism (RILEM AAR-4,[73] BS 812-123[74] or CSA A-23.2-94-14A[75]) is prepared containing the aggregate and stored in humid conditions in the laboratory. Length changes are monitored over a period of months, and any expansions plotted and compared against national scales that interpret the potential risk of damaging expansion in actual structures. Another approach (ASTM C289[76]) is to measure the loss in silica from finely crushed aggregate when treated with strong alkali solution, but anomalies have been reported for some of the results that appear in either of the 'fields' designated as non-reactive, or reactive, and others lie ambiguously close to the boundary between the fields, so results obtained by this chemical test should be treated with caution.

The damaging ACR is rare, but methods of evaluating the potential risk using impure fine-grained dolomitic limestones as aggregate have been developed in Canada (CSA A23.2-26A[77]) and China, and are similar in some respects to tests for the ASR. These also begin with a petrographic assessment, which is then followed by a chemical analysis, and by noting expansions in small cylinders of the suspect aggregate stored in alkali solution (ASTM C586[78]), or concrete prisms stored in humid conditions.[79]

Assessing the risk of expansion with concrete mixes relies mainly on variants of monitoring expansions of laboratory-prepared concrete prisms. The principle differences between them relate to details of dimensions and the storage environment, which may range from humid conditions, to storage under alkali solutions and to storage temperatures which range from 20°C to 40°C and even higher. The advantage of increasing the storage temperature is that any expansion will occur more rapidly, allowing a more rapid test. However, increasing the temperature makes the test less reliable as a true indicator of the likely expansions in a real concrete structure.

4.3.4 Codes of practice and specifications for the avoidance of the AAR

The three essential requirements for AAR in a concrete are a reactive component in the aggregate, a high alkali concentration in the pore solution and moisture. Removing any one of these from the system will prevent the reaction from occurring. Consequently, since moisture is always present, most codes of practice for preventing the reaction in concrete either suggest tests to check the aggregates used, recommending that reactive aggregates are not used, or are only used with a low-alkali cement, and/or set limits on the maximum concentration of alkali permissible in the concrete or the cement that is to be used.

Research has found clear evidence that cement replacements such as fly ash or ground granulated blast furnace slags can reduce or remove the potential for AAR expansion in concretes using these admixtures. Although in detail the effects of using these admixtures is complex, to a first approximation they might be considered as diluents, thus reducing the alkali contribution to the concrete from the cement used. The cement is usually the main source of alkali, but, as noted previously, contributions to the total alkalis can be derived from diffusion into the concrete from external sources, from the admixture itself, or from the breakdown of alkali-rich minerals present in the aggregate, particularly if they have been degraded or altered by metamorphism or by weathering processes.

There are numerous national codes and guidelines concerned with the avoidance of the AAR in concrete. Recently, the International RILEM Committee has published an extensive series of studies and proposed specifications for avoiding AAR.[80] They suggest three categories of risk relating to the type of concrete structure concerned (S) and three concerning the severity of environmental conditions (E). This gives a 3×3 matrix for the level of precaution necessary (P), as summarised in Table 4.3. RILEM also proposes three levels of potential alkali reactivity (low, medium and high) and recommend alkali limits for minimising the effect of these in terms of kg/m^3 as Na_2O equivalent in the concrete. Respectively, these are: P1, none required; P2, typically $3.0–3.5\,kg/m^3$; and P3 (high risk), a lower limit is recommended (e.g. $2.5\,kg/m^3$ or less). Diluting the cement with a cement replacement is another approach, where RILEM recommends minimum proportions of fly ash ranging from 25% to 50% in order to reduce the risk of expansion due to alkali reactivity.

Other informative guidelines concerned with avoiding the AAR in concrete include Concrete Society Technical Report 30,[53] recommendations of the Danish Code of Practice DS 411,[81] the Canadian standard practice CSA A23.2-27A,[82] CSA A23.2-88M[83] and BRE Digest 330.[51] The recommendations made in these national approaches to the avoidance of damage to concrete by AAR all focus on one or more of the following possibilities for limiting risk:

- limiting the alkali content of the concrete
- avoiding potentially reactive aggregate materials
- assessing the potential severity of the reaction for a particular aggregate
- assessing the risk for the type of structure

Table 4.3 A summary of the proposals of RILEM TC 191-ARP[80]

Structures	Environment		
	E1 Internal concrete within buildings Protected external concrete, e.g. by cladding	E2 Internal concrete where humidity is high, e.g. laundries, tanks, swimming pools Concrete exposed to atmosphere or non-aggressive ground	E3 Internal/external concrete exposed to de-icing salts Concrete exposed to seawater/salt spray Concrete exposed to freeze–thaw when wet Concrete exposed to prolonged elevated temperatures when wet
S1 Non-load-bearing element inside buildings Temporary or short service life structures Easily replaceable elements	P1	P1	P1
S2 Most building and civil engineering structures	P2	P2	P2
S3 Long service life or highly critical structures where AAR damage is judged unacceptable such as: • nuclear installations • dams • tunnels • important bridges and viaducts • structures retaining hazardous materials	P2	P3	P3

Recommended precautions: P1, none; P2, normal precautions; P3, maximum precautions.

- evaluating the risk from environmental factors local to the structure
- using a cement replacement or admixture to reduce the potential for reactivity.

The approach recently recommended by RILEM is the result of an extensive international review and programme of testing evaluation. It takes account of most of the factors listed above, and is perhaps the most complete set of recommendations currently available.

4.3.5 The diagnosis and prognosis of alkali reactions in existing structures

As has already been suggested, the first observations indicating that a structure might be affected by the AAR is the appearance of cracking on the concrete surface, and/or evidence of the differential expansion of elements within the structure. The next stage depends on the nature of the concrete structure and the location and severity of the damage. In cases of limited incipient damage, setting up a programme of crack or dimension monitoring may be all that is required.

In the investigation of more serious cases, it may be necessary to take samples from the concrete structure. Usually this is done by cutting cores, preferably 100 mm in diameter or larger. They are immediately wrapped in cling film to avoid moisture loss, and prepared for petrogaphic analysis in the laboratory. The petrographer will initially look for evidence of the ASR, such as gel development, when cores are unwrapped. The cores are then cut and prepared as thin sections so that they can be examined using a petrographic microscope for evidence of gel either associated with aggregate particles, or in microcracks. Parts of the remainder of the core may be broken, to provide fresh surfaces, again for examination for the presence of damp patches indicative of gel.

Another series of tests may involve storing sample cores in humid conditions in closed containers for a period of months in order to monitor them for expansion with time. The interpretation of all these tests taken together usually gives a clear indication as to whether or not the concrete is being damaged as a result of the AAR. However, there may be other contributory causes for the observed damage to the structure, and these may in some cases be of greater significance than the AAR itself. A clear review of the appropriate methodology of field examination, sampling and testing is given in the BCA[52] report on the diagnosis of the AAR.

Although evidence of the AAR can usually be identified unambiguously, the prognosis for the continuance and severity of the reaction and the estimation of the consequent future damage it a matter of experienced interpretation.

In recent years, computer modelling of the progress of a reaction in real structures has had some success in specific case studies.[84–88] However, the models are of necessity simplifications of the real situation, and the assumptions made cannot take account of all the varying factors that influence the progress of the reaction over time and the consequent damage. Typical assumptions made include the uniformity of the concrete, the supply of moisture over time, a stable temperature regime, constancy of the concrete diffusion coefficients and alkali concentration over time, and levels of constraint provided by steel reinforcement within the structure. Assumptions such as these are either explicit or implicit in all models, and some of the factors are very sensitive in controlling the model equations. Although, in reality most of them are variable with time, they are usually taken as constants in the equations. Consequently, the value of modelling the predicted future course of the AAR reaction in a real concrete structure is limited, and at best it can only provide a crude assessment of the severity and whether the damage will continue. These techniques are probably little better than the alternative of monitoring expansions in a structure and extrapolating the results forward in time. Perhaps a better dual approach would be to use a modelling technique coupled with a programme of monitoring. The results from the monitoring could then be used to continually correct and update the model predictions.

4.3.6 Possible remedial measures for affected structures

The development of damage to a concrete structure as the result of the AAR is relatively rare, but if the engineer has responsibility for a particular structure that is affected, then dealing adequately with the problem becomes a pressing requirement.

The ASR, by its nature, is distributed though the mass of the concrete, so there is no major long-term advantage to be gained by treating the surface of the structure only. In some cases, a water-excluding membrane on the surface has been tried. This may reduce moisture movement resulting from wetting and drying by rain, and/or prevent the ingress of additional salts. The effect may be to slow the reaction in the short term, but the reaction will not be arrested, and the coating will soon fail and become permeable.

It is almost impossible to completely remove water from the matrix of the concrete. Also, once it has been placed, it is not possible to remove a reactive aggregate, or to add a cement replacement material. This implies that only two options remain: the first is to reduce the alkali content of the concrete and the second is to repair the damaged elements or to remove and replace the structure itself.

Electrolytic treatments (similar to the cathodic protection of steel reinforcement) applied to AAR-affected structures have been explored as a means of removing alkalis, and limited success has been claimed. These methods pass electric current from anodic to cathodic plates each side of the concrete element, with the concrete acting as the electrolyte. This technique induces the positive sodium and potassium ions to migrate to the cathode, thus reducing the alkali levels in the pore solution of the concrete. One difficulty with this de-alkalisation process is that the passivity of steel reinforcement to corrosion is reduced as a result of the reduction in pH. In some studies, electrolysis has also been used to re-alkalise the concrete in order to arrest corrosion of the steel reinforcement. However, this procedure should also be considered with care, because it may risk initiating a dormant AAR in the concrete.

A promising system, still in part experimental, uses electrolysis to replace sodium and potassium ions with lithium ions, because lithium gels formed as a result of the AAR are non-expansive.[89,90] In some studies, the steel reinforcement is used as the cathode. This rapid lithium migration technique (RLMT) is claimed to significantly reduce expansion, and may provide an effective means of repair. This treatment has been applied to a number of structures in the USA, but no long-term results are available at present.

The replacement of a number of damaged structures has been undertaken in many parts of the world. The requirement for replacement depends on a risk analysis relating to the nature of the structure and which parts of the structure are critically affected. As an example, a decorative non-structural wall may look unsightly if cracked by the AAR, but the only difficulties it creates are cosmetic; by contrast, abutment supports for the load of a bridge deck that become damaged by the AAR could initiate a catastrophic failure and risk to life.

The rational procedure that has been generally adopted worldwide proceeds as a series of steps. The first stage is to confirm the cause of the observed damage by a site and laboratory investigation. Then, an appropriate monitoring programme is usually installed. Finally, a critical risk analysis of the affected concrete components may be undertaken. This analysis is done by assembling all the information available from the laboratory tests, the monitoring data and also the possible effects resulting from environmental conditions local to the structure. The analysis will allow a prediction of the likely continued development of the deterioration and its severity to be made. Depending on the outcome of this assessment, decisions can be implemented: these may range from merely continuing the monitoring programme, undertaking limited repair measures, planning a strategy for a future scheme of repair, undertaking remedial action immediately or, in extreme cases, replacement of the whole or part of the structure as a matter of urgency.

4.4   Steel reinforcement corrosion
John Broomfield, Broomfield Consultants, UK

4.4.1   How concrete protects reinforcing steel from corrosion
This section is concerned with the corrosion of steel in concrete. This is one of the most important deterioration mechanisms of reinforced concrete, and has major financial implications around the world. While concrete is a very durable material, reinforcement corrosion is a major cause of deterioration of the built infrastructure, costing most major economies billions of pounds for damage to highway bridges, marine structures, buildings and industrial plant.

4.4.1.1   The chemical and physical structure of concrete and implications for its durability
When considering why steel corrodes in concrete, the most important question is the converse: why does steel not corrode in concrete in most situations? Concrete is a porous material, typically of 10–20% porosity. This is made up of fine gel pores, coarser capillary pores and the porosity of the coarse aggregate. Water and oxygen easily diffuse into concrete, and can reach reinforcing steel regardless of the concrete cover over the steel. Depending on the pore sizes and the amount of water in the environment, pores may be relatively empty of water, lined with water or filled with water. However, despite this porosity, most concrete exposed to the outdoor environment does not corrode. This is because of the chemical rather than the physical composition or barrier properties of concrete.

The reason why steel does not corrode in concrete is the alkalinity of the cement paste. The pore water in hardened Portland cement usually has a pH in the range 12.5–13.5. This is due to large reserves of calcium hydroxide, $Ca(OH)_2$, which is pH 12.4, supplemented by far smaller amounts of sodium and potassium oxides or hydroxides, with sodium hydroxide having a pH of 14. In this pH range, steel does not corrode due to passivation, as discussed in the following subsection.

4.4.1.2 Passivation of steel in concrete

There are many oxides of iron and steel formed in different conditions, from the friable rust we see on unpainted ironwork to dense protective oxides formed at high temperatures in boilers and other industrial plant. Iron can corrode at a rapid linear rate, a controlled parabolic rate or a very slow logarithmic rate, depending on the properties and quality of the oxide formed and the environment in which it forms.

In atmospheric conditions, a number of factors influence the nature of the corrosion products, and hence the rate of corrosion. One of the most important is the pH of the steel surface. If the pH is higher than about 10.5, a passive oxide layer will form on the steel surface. This passive layer is generally agreed to be a dense film of γFeOOH of the order of 1 nanometer (nm) thick. However, in concrete it is complicated by the formation of a dense mineral layer many micrometers (μm) thick on the steel surface, which may supplement, or substitute for, the action of a 'true' passive layer. This passive layer is dynamic – it breaks down and reforms – and will last indefinitely as long as the correct conditions persist.

Concrete is if course vulnerable to attack. For instance, as stated in Section 4.2.1, acids with a pH of less than 7 will attack concrete, and soft peaty water can also attack it. This form of attack is obvious, and the embedded reinforcing steel will be vulnerable to corrosion once it is exposed and no longer protected by a pH of greater than 10.5. However, there are two chemical species that can break down the passive layer without destroying the concrete first. They are the chloride ion and the carbon dioxide molecule, as will be discussed in the next two subsections.

4.4.2 Carbonation attack

The atmosphere typically contains between 0.03 and 1% carbon dioxide (CO_2). Like most gasses at room temperature, CO_2 is an acid gas. It forms carbonic acid when dissolved in water. This weak acid will react with the calcium, potassium and sodium hydroxides in the pore water, reducing the pH to below the critical 10.5 threshold for steel corrosion.

4.4.2.1 The carbonation mechanism

Chemically, the reaction is a simple one familiar in high-school chemistry. In practice, gaseous CO_2 enters the pores in the concrete and dissolves in the pore water to form carbonic acid, which then reacts with dissolved calcium hydroxide to form insoluble calcium carbonate, which precipitates out of solution and lines the pores. There are also some minor reactions with other constituents and some shrinkage of the cement matrix: however, there is no significant deleterious effect of the carbonation process on the concrete itself. The damage is the reduction of pH, allowing the steel to depassivate and start to actively corrode.

4.4.2.2 *Carbonation rates*

All concrete will carbonate at some rate or other. The carbonation rate will be determined by the amount of moisture present, and particularly the amount of wetting and drying, which will allow CO_2 into the concrete when dry, and then dissolve it to form carbonic acid when wet. The carbonation rate is also governed by the porosity of the concrete and the amount of alkaline reserves in the concrete.

The rate of carbonation follows a roughly parabolic relationship:

$$x = kt^{1/2} \tag{4.14}$$

where

$x =$ carbonation depth in mm
$k =$ carbonation coefficient in mm/year$^{1/2}$
$t =$ time in years

This equation is an approximation, and the power law exponent can vary from about 0.4 to 0.6, depending upon exposure conditions and the concrete.[91]

Carbonation depths can be measured by a range of techniques. The most popular is the phenolphthalein indicator test, as described in Section 4.8.3. In an investigation of a number of field surveys, Trend 2000 et al.[92] found the following range of k values:

- for 11 buildings 8–24 years old, the constant k ranged from 1.2 to 6.7 mm/year$^{1/2}$, with an average of 3 mm/year$^{1/2}$
- for seven car parks 14–41 years old, the constant k ranged from 2.2 to 7.6 mm/year$^{1/2}$, with an average 4.27 mm/year$^{1/2}$
- for a 10-year-old jetty and a 90-year-old bridge, k was 1.8 mm/year$^{1/2}$ for the jetty and 1.6 mm/year$^{1/2}$ for the bridge.

It is quite common to find office buildings constructed in the 1960s and 1970s with severe carbonation-induced corrosion while bridges of the same era will show negligible carbonation depths and no carbonation-induced corrosion. This is due to bridges being built with a higher cement content, higher and better-controlled cover to the reinforcement and better compaction of the concrete, ensuring dense, low-porosity concrete around the steel.

The carbonation advances as a 'front' into the concrete. This can be somewhat uneven and irregular, but it is useful as it is possible to define a maximum, minimum and average carbonation depth and compare it with the cover depth. Using the simple $x = kt^{1/2}$ equation, it is possible to determine a range for the carbonation coefficient k and to estimate the time to corrosion.

4.4.3   Chloride attack

As stated in Section 4.4.1.2, the other chemical agent that can initiate corrosion without attacking the concrete first is the chloride ion. While almost all concrete exposed to the environment will eventually carbonate, only those exposed to chlorides will suffer from chloride attack. However, our ability to protect our structures from chloride ingress has been very poor in the past, and therefore the present cost of repairing and maintaining bridges and marine structures is very high due to chloride ingress and subsequent reinforcement corrosion.

4.4.3.1   Sources of chloride

The chloride ion is of course 50% of the constituents of common salt, NaCl (60% by weight). It is used in de-icing salt and is present at a concentration of about 3.5% in seawater. Therefore, any structure exposed to seawater, de-icing salts or any other source of chloride is vulnerable to chloride-induced corrosion.

Chlorides will migrate into concrete by a number of mechanisms discussed in Section 4.4.3.2 unless adequate protection is put in place during construction. However, there are also problems of chlorides being cast into concrete due to contamination of the mix constituents. Up until the 1960s it was thought by many engineers that chemical binding of chloride ions occurred with tricalcium aluminate (C_3A), to form chloroaluminates. While this undoubtedly happens, it is not sufficient to prevent chlorides from attacking the passive layer when the concrete is made with seawater or where calcium chloride was used as a set accelerator for cold weather concreting until it was banned in the 1970s.

4.4.3.2   *Transport of chloride through concrete*

Like carbonation, the rate of chloride ingress is often approximated to Fick's law of diffusion. There are further complications here. The initial mechanism appears to be suction, especially when the surface is dry, i.e. capillary action. Salt water is rapidly absorbed by dry concrete. There is then some capillary movement of the salt-laden water through the pores followed by 'true' diffusion. There are other opposing mechanisms that slow the chlorides down. These include a chemical reaction to form chloroaluminates, and absorption onto the pore surfaces. The detailed transport mechanisms of chloride ions into concrete are discussed by Kropp and Hilsdorf.[93]

Classical diffusion modelling is widely used to predict chloride transport into concrete. Fick's second law of diffusion can be applied, and there is a range of models used, as discussed by Broomfield.[94]

4.4.3.3   *Chloride thresholds*

In principle, there is a chloride threshold for corrosion of steel in concrete. Hausmann[95] showed both theoretically and in practice in calcium hydroxide solutions that above a chloride/hydroxide ion ratio of 0.6 the probability of corrosion rises rapidly. However, in real concrete, this threshold is harder to pin down for the following reasons:

- Concrete pH varies with the type of cement and other mix constituents. Since pH is a logarithmic scale, a pH change from 12 to 13 is a factor of 10 change in the hydroxide ion concentration.
- Chlorides can be bound chemically by aluminates and physically by absorption to pore surfaces. This can temporarily or permanently remove chlorides from the pore water.
- Carbonation will free chemically bound chlorides so that there is a 'synergistic' effect, leading to the onset of corrosion in chloride-contaminated concrete due to the progression of chlorides into the concrete, initiating depassivation.
- There must be sufficient moisture present for corrosion to progress, so even depassivated steel may not corrode.
- Corrosion proceeds by the formation of anodes and cathodes, as described in Section 4.4.4. Therefore, some areas of steel will not corrode as they are cathodic sites which must sustain a non-corroding reaction for the corrosion reaction to occur at the anodic sites.

- Corrosion initiates at voids on the steel surface where the alkaline cement paste/pore water is not in contact with the steel, so chlorides can more easily initiate corrosion.[96] More voids mean a lower threshold; fewer voids mean higher threshold.

It is generally accepted that corrosion can occur at chloride levels from 0.1–1% chloride by mass of cement. A commonly used threshold is 0.4%, although the UK Highways Agency prefers 0.3%, and others consider 0.2% to be the best value.

4.4.4 The corrosion process after depassivation

Regardless of the source of the depassivation of the steel, once it has occurred, the mechanism is the same. Iron dissolves at the anodic site of depassivation, initially producing soluble ferrous ions (Fe^{2+}) and electrons (e^-). The ferrous ions reacts with hydroxide ions, then water and oxygen, to ultimately form rust. At an adjacent cathodic site, the electron reacts with water and oxygen, to hydroxide ions. This is illustrated in Fig. 4.13. As can be seen from the figure, there is no initial requirement for oxygen at the anode, only at the cathode. Water is required to dissolve the Fe^{2+} ion but not for the anodic reaction. However, water and oxygen are required to form rust. This corrosion product has a volume of about 6.5 times that of the steel consumed.[27]

4.4.4.1 Expansive oxide formation

The formation of the expansive oxide leads to the cracking and spalling of concrete. This will initiate in corners, in areas of low cover, and may

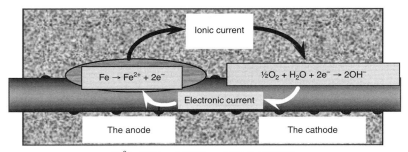

$$Fe^{2+} + 2OH^- \rightarrow Fe(OH)_2 \quad \text{Ferrous hydroxide}$$
$$4Fe(OH)_2 + O_2 + 2H_2O \rightarrow 4Fe(OH)_3 \quad \text{Ferric hydroxide}$$
$$2Fe(OH)_3 \rightarrow Fe_2O_3 \cdot H_2O + 2H_2O \quad \text{Hydrated ferric oxide (rust)}$$

Fig. 4.13 The anodic, cathodic, oxidation and hydration reactions of corroding steel in concrete. (Reproduced with permission from Broomfield[94] © 2006 Taylor and Francis)

Corner spall due to doubled
access for water, oxygen,
chlorides, CO_2

Delamination leading to spall of cover and
horizontal corrosion-induced cracks due to
'plane of weakness' at rebar level

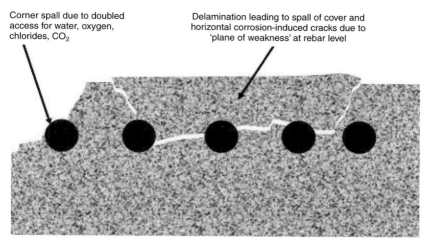

Fig. 4.14 Schematic illustration of corner spall and delamination of concrete due to corrosion

occur in sheets or delaminations as cracks propagate between adjacent reinforcing bars, as shown schematically in Fig. 4.14.

In laboratory tests, Rodrígues *et al.*[97] found that 15–40 µm section loss generated enough corrosion product to crack the concrete on bars with a cover/bar diameter ratio between 2 and 4.

4.4.4.2 Non-expansive oxide formation

From the anodic reaction and subsequent oxide-forming reactions in Fig. 4.13, we can see that there must be sufficient moisture to initially dissolve the ferrous ion (Fe^{2+}) in the presence of sufficient oxygen and water to allow the subsequent reactions to occur. If the concrete is saturated or close to saturation, but with an oxygenated cathode close by and a low electrical resistance path through both the steel and the concrete connecting them, the iron could dissolve and stay in solution. This form of corrosion, often called 'black rust' or 'anaerobic corrosion', can be very rapid. Broomfield[94] gives examples of section losses of greater than 0.5 mm/year for a steel reinforcing bar in a bridge deck in the UK under a damaged waterproofing membrane which trapped salt-laden water. A more extreme example was a precast post-tensioned marine bridge pile in Florida, where tendon failures showed corrosion rates of 3 mm/year.

4.4.4.3   Typical corrosion rates

Methods of measuring corrosion rates for steel in concrete will be discussed more fully in Section 4.4.5.4. It is important to understand the ranges of corrosion rates found in concrete. Rates of less than 1 µm section loss per year are generally considered to reflect passive, non-corroding conditions. As stated at the end of Section 4.4.4.1, 15–40 µm steel section loss was found to lead to cracking and spalling of the cover concrete when the cover/bar diameter ratio was between 2 and 4. Generally, greater than 10 µm/year section loss is considered to be active corrosion, leading ultimately to cracking and spalling.

4.4.5   Testing for corrosion

There is a wide range of tests available for assessing the condition of deteriorated concrete. This section will concentrate on those more technically sophisticated tests directly related to corrosion. In any systematic survey of a deteriorating reinforced-concrete structure, a detailed visual survey will be conducted, and if there is corrosion damage, the cracks, spalled and delaminated concrete, and exposed rebars will be noted, along with a 'make safe' survey to remove any loose concrete that could cause injury or damage. Further guidance on testing and surveying is provided by Broomfield,[94] Concrete Society Technical Report 60[98] and the Concrete Bridge Development Group.[99]

4.4.5.1   Measuring the chloride content in hardened concrete

If there is a source of external chlorides or any chance that chlorides were cast into the concrete, then the chloride content of the concrete should be measured in all three dimensions, i.e. horizontally, vertically and with depth. The method is destructive, requiring the collection of drilled powder from the structure. Cores may also be taken, but will need to be crushed to powder for analysis. Samples may be collected at progressive increments into the concrete. Three increments is the minimum to be useful, but sampling at five depths or more will give sufficient points on a graph to allow curve fitting for modelling chloride ingress rates and back-calculating the effective diffusion coefficient.

In Europe, the standard BS EN 14629[100] is used by UKAS/NAMAS (the UK Accreditation Service/National Materials Service) or other comparable accredited test houses for the determination of acid-soluble chlorides from powder samples. Approximately 25 g of sample is required for accurate analysis. This can be obtained from drilling a pair of 25 mm diameter holes 15–25 mm deep in the concrete. Depth measurements of 15 mm increments can be collected with care, to minimise cross-contamination of successive samples.

In North America, two different approaches are used for analysis. There are several ASTM standards for the analysis of chlorides, and fall into two separate groups. One group is for acid-soluble chlorides, like BS EN 14629. The other is for water-soluble chlorides. This latter test is less reproducible as it requires a solid sample such as a core to be broken up into chunks and then boiled, to dissolve out the chlorides. However, the results can vary with the size grading of the chunks and powder. This test is important for certain concrete-containing aggregates that contain well-bound chlorides that show up on an acid-soluble analysis. These aggregates are found in certain parts of Canada and, to a lesser extent, in the USA.

There are field kits for measuring chloride contents using Quantab strips or using a selective chloride ion electrode on site. These require considerable experience and expertise as well as an understanding of their limitations. In the UK, the speed of turn around and the modest cost of laboratory chloride analysis mean that field analysis is rarely used.

4.4.5.2 Measuring carbonation depths

Carbonation depths are measured using chemical indicators that change colour in the range pH 7–9. The most popular and effective method is the use of phenolphthalein: the indicator turns pink above pH 9.5, which is below pH 10.5, where steel starts to corrode. The whole carbonation front will be some 10–15 mm wide, with the pH falling from 12.5 to 8, so there is some inaccuracy in the measurement, but given the variability in the front itself, the measurement is useful. BS EN 14630[101] describes the method of preparing a phenolphthalein solution, spraying it onto a freshly broken surface and measuring the carbonation depth.

4.4.5.3 The half cell or reference electrode potential survey

The half cell or reference electrode is a pure metal in a solution of its own ions such as copper in copper sulfate or silver in silver chloride. When coupled to another metal surrounded by its own ions such as a corroding reinforcing bar in damp concrete, a small electrochemical cell is formed, and a voltage difference exists between the reinforcing bar and the reference electrode. Therefore, if we move our reference

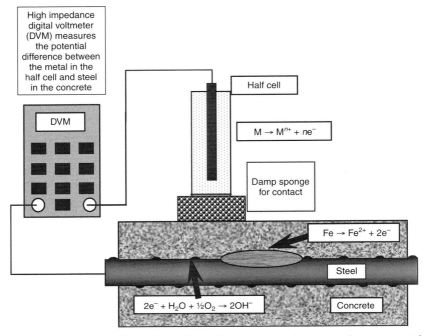

Fig. 4.15 Schematic illustration of a reference electrode set up for a potential survey

electrode across the surface of the concrete, touching it down on a regular grid using a damp sponge for contact, the bigger the voltage we measure, the more active is the corrosion cell. We can therefore map the anodes and cathodes (see Fig. 4.13) by plotting lines of constant potential on the concrete surface. The basic arrangement is shown schematically in Fig. 4.15.

The methodology of conducting and interpreting a reference electrode potential survey is described in ASTM C876,[102] and in Concrete Society Technical Report 60.[98]

4.4.5.4 Corrosion rate measurement

Corrosion rates are measured using the polarisation resistance or linear polarisation (LPR) technique. A simple explanation of the LPR corrosion rate-measuring technique for steel in concrete is that the equipment consists of a reference electrode or half cell and an auxiliary electrode which are combined in a single head and placed in contact

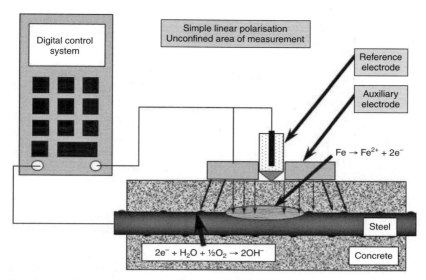

Fig. 4.16 Schematic illustration of a typical LPR probe arrangement

with the concrete surface via a conductive sponge. A schematic illustration of the arrangement is shown in Fig. 4.16.

After storing the value of the electrical potential (voltage) between the steel and reference electrode, a small current is applied from the LPR meter. This changes the potential reading in the meter. The higher the corrosion rate, the more current is needed to shift the potential by a fixed amount (up to 20 mV). The corrosion rate is proportional to the applied current divided by the voltage change.

The system will record the corrosion current in microamperes (μA). However, that corrosion current is coming from a certain area of reinforcing steel. The area is determined from the bar diameter, the size of the head and any system for confining the measurement to specific areas. The current can therefore be turned into a current density in μA/cm^2. Knowing that the corrosion reaction is $Fe \rightarrow Fe^{2+} + 2e^-$, we can use Faraday's law of electrochemical equivalents to convert the current density into a section loss. The conversion tells us that $1\,\mu$A/cm$^2 = 11.5\,\mu$m/year section loss. The corrosion rate values discussed in Section 4.4.4.3 can therefore be used to interpret the results. Corrosion rate measurements and interpretation are discussed more fully by Broomfield[94] and Concrete Society Technical Report 60.[98]

4.4.5.5   Concrete resistivity

Since corrosion is an electrochemical phenomenon, the electrical resistivity of the concrete will have a bearing on the corrosion rate of the concrete, as an ionic current (electric current in the form of a flow of charged ions in the pore water) must pass from the anodes to the cathodes for corrosion to occur (see Fig. 4.13).

The four-probe resistivity meter or Wenner probe was developed for measuring soil resistivity.[103] Specialised modifications of the Wenner probe are frequently used for measurement of concrete resistivity on site. The measurement can be used to indicate the possible corrosion activity if steel is depassivated. A proprietary version of the system is shown in Fig. 4.17. The system uses four probes. The outer probes pass a current through the concrete while the inner probes detect the voltage difference. This approach eliminates any effects due to surface contact resistances.

Interpretation of resistivity results is empirical. The following interpretation of resistivity measurements from the Wenner four-probe system has been cited when referring to depassivated steel:[104]

>20 kΩ cm	Low corrosion rate
10–20 kΩ cm	Low to moderate corrosion rate
5–10 kΩ cm	High corrosion rate
<5 kΩ cm	Very high corrosion rate

Further information on resistivity measurement is provided by Broomfield[94] and Concrete Society Technical Report 60.[98]

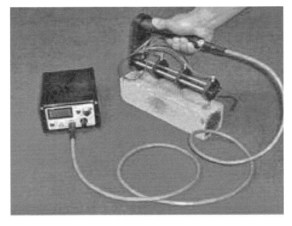

Fig. 4.17 A four-probe resistivity meter with push-on contacts. (Courtesy of CNS Farnell Ltd)

4.4.5.6   Cover depth testing

While the cover depth is not strictly a corrosion-related measurement, the cover is critical in assessing the likelihood and severity of corrosion. Correlating the cover depth with the carbonation depth and the chloride content at typical cover depths will inform the engineer of the extent of existing corrosion and allow prediction of the time to corrosion. It will also help to inform the engineer on the most cost-effective repair options, by determining the time to corrosion using modelling techniques as discussed in Sections 4.4.2 and 4.4.3 and then looking at the available repair options as discussed by Broomfield,[94] e.g. whether it is worth trying to keep chlorides out with a coating or other barrier or whether the chloride level is so high at the reinforcement depth that corrosion has started and therefore active corrosion control is necessary.

The electromagnetic cover meter is widely used in the construction industry. These are now digital devices which can log results and measure bar diameters as well as cover depths. Scanning devices that plot the bar locations and depths are now available. These are discussed by Broomfield[94] and the Concrete Bridge Development Group.[99] The only standard for applying a cover meter is BS 1881.[105]

4.4.5.7   Other test techniques

There is a huge range of other test methods, especially for locating flaws, cracks and delaminations in concrete. These include ground-penetrating radar, infrared thermography, impact echo, radiography and ultrasonic pulse velocity. None are as widely used or easily applied as the techniques discussed above, although they may yield wider information about the properties of the concrete and what is embedded in it.

There is also a far wider range of tests that will inform the engineer about concrete properties which will impact its ability to resist ingress or corrosion. These include permeability and surface absorption. Finally, petrographic analysis is an important part of any investigation of a concrete structure.

4.4.6   Corrosion monitoring

The electrical equipment for measuring the corrosion rate, potential and concrete resistivity described in the previous sections is portable and battery operated, with probes or heads that are placed on the concrete surface to take a measurement. It is also possible to install these probes permanently in the concrete, either during construction or after problems are identified. They can then be linked either to a monitoring location where the meters can be connected to take repeat readings, or to a logging system that takes measurements automatically at fixed intervals.

4.4.6.1 Corrosion-monitoring probes

The following permanent monitoring probes can be embedded in concrete:

- corrosion rate (LPR) probes
- concrete resistivity probes
- reference electrodes
- temperature probes
- moisture (relative humidity) probes.

These can be linked to other probes, such strain gauges, for long-term assessment of a structure.

Figure 4.18 shows a permanent monitoring probe consisting of a reference electrode, an LPR probe and a temperature sensor. It can also measure concrete resistance.

Figure 4.19 shows a 'ladder probe', which measures the development of the galvanic current between carbon steel 'rungs' and a stainless steel cathode to monitor the progress of the 'corrosion front' into the concrete.

Reference electrode (with cap on and notice to remove before pouring concrete)

Auxiliary electrode to pass current

Working electrode of known area

Connection to reinforcement for readings of unknown area

Conduit containing cables to a junction box on formwork exposed after striking shutters for connection to a monitoring system

Fig. 4.18 A permanent corrosion rate-monitoring probe fixed into a reinforcement cage prior to casting concrete

Fig. 4.19 Ladder probes for measuring the galvanic current prior to casting concrete. (Courtesy of Dr-Ing M. Raupach)

4.4.6.2 Corrosion-monitoring systems

Corrosion-monitoring systems have been installed during construction on a number of major bridge and tunnel projects in Europe, including the Dartford Tunnel Redecking project and the Great Belt Bridge/Tunnel project. Data from the Dartford tunnel project are shown and discussed by Broomfield *et al.*[106] It has been collected from over 60 probes since 2001.

References

1. Cohen, M.D. and Mather, B. (1991) Sulfate attack on concrete: research needs. *ACI Materials Journal*, 88, 62–69.
2. Cohen, M.D. and Bentur, A. (1988) Durability of Portland cement–silica fume pastes in magnesium sulfate and sodium sulfate solutions. *ACI Materials Journal*, 85, 148–157.
3. Mehta, P.K. (1983) Mechanism of sulfate attack on Portland cement concrete – another look. *Cement and Concrete Research*, 13, 401–406.

4. Brown, P.W. and Taylor, H.F.W. (1999) The role of ettringite in external sulfate attack. In: J. Skalny and J. Marchand (eds), *Material science of concrete–sulfate attack mechanisms*. American Ceramic Society, Westerville, OH, 73–98.

5. Bonen, D. (1992) Composition and appearance of magnesium silicate hydrate and its relation to deterioration of cement based materials. *Journal of the American Ceramic Society*, 75, 2904–2906.

6. Erlin, B. and Stark, D.C. (1965) Identification and occurrence of thaumasite in concrete. *Highway Research Record*, 113, 108–113.

7. Skalny, J., Marchand, J. and Odler, I. (2002) *Sulfate attack on concrete. Modern Concrete Technology*. Taylor and Francis, New York, vol. 10, 217.

8. DETR (1999) *The thaumasite form of sulfate attack: risks, diagnosis, remedial works and guidance on new construction*. Report of the Thaumasite Expert Group, Department of the Environment, Transport and the Regions, London.

9. Bensted, J. (1999) Thaumasite – background and nature in deterioration of cements, mortars, and concretes. *Cement and Concrete Composites*, 21, 117–121.

10. Longworth, T.I. (2003) Contribution of construction activity to aggressive ground conditions causing the thaumasite form of sulfate attack to concrete in pyritic ground. *Cement and Concrete Composites*, 25, 1005–1013.

11. Collepardi, M. (1999) Thaumasite formation and deterioration in historic buildings. *Cement and Concrete Composites*, 21, 147–154.

12. ASTM (2000) *Standard specification for Portland cement*, ASTM C150. American Society for Testing and Materials, Philadelphia.

13. Taylor, H.F.W., Famy, C. and Scrivener, K.L. (2001) Delayed ettringite formation. *Cement and Concrete Research*, 31, 683–693.

14. Fu, Y. and Beaudoin, J.J. (1996) Mechanisms of delayed ettringite formation in Portland cement systems. *ACI Materials Journal*, 93, 327–333.

15. Livingston, R.A., Ormsby, C., Amde, A.M., Ceary, M.S., McMorris, N. and Finnerty, P.G. (2006) Field survey of delayed ettringite formation-related damage in concrete bridges in the state of Maryland. In: V.M. Malhotra, (ed.) *7th CANMET/ACI Conference on Durability of Concrete, Montreal*, SP-234. American Concrete Institute, Farmington Hills, MI, 251–268.

16. Mehta, P.K. (1992) *Sulfate attack on concrete – a critical review. Materials science of concrete III*. American Ceramic Society, Westerville, OH.

17. Mather, B. (1966) Effects of seawater on concrete. *Highway Research Record (Transportation)*, 113, 33–42.

18. Mehta, P.K. (1986) *Concrete: structure, properties, and materials*. Prentice-Hall, Englewood Cliffs, NJ, 105–169.

19. Neville, A. (1996) *Properties of concrete.* John Wiley, New York, 4th edn.
20. Skalny, J. (1999) In: J. Skalny and J. Marchand, (eds), *Sulfate attack issues: an overview. Material science of concrete–sulfate attack mechanisms.* The American Ceramic Society, Westerville, OH, 49–63.
21. Harboe, E.M. (1982) Longtime studies and field experiences with sulfate attack. In: *Sulfate Resistance of Concrete (George Verbeck Symposium)*, ACI SP-77, 1–20.
22. Smith, F.L. (1958) *Effect of cement type on the resistance of concrete to sulfate attack.* USBR Report No. C-828. Bureau of Reclamation, Washington, DC.
23. BSI (1996) *Specification for sulfate-resisting Portland cement, BS 4027.* British Standards Institution, London.
24. Moon, H.Y. and Lee, S.T. (2003) Influence of silicate ratio and additives on the sulphate resistance of Portland cement. *Advances in Cement Research*, 15, 91–101.
25. Şahmaran, M., Kasap, Ö., Duru, K. and Yaman, İ.Ö. (2007) Effects of mix composition and water–cement ratio on the sulfate resistance of blended cements. *Cement and Concrete Composites*, 29, 159–167.
26. Irassar, E.F., Gonzalez, M. and Rahhal, V. (2000) Sulfate resistance of type V cements with limestone filler and natural pozzolana. *Cement and Concrete Composites*, 22, 361–368.
27. Mehta, P.K. and Monteiro, P.J.M. (2006) *Concrete–microstructure, properties, and materials.* McGraw Hill, New York, 3rd edn.
28. Kalousek, G.L., Porter, L.C. and Benton, E.J. (1972) Concrete for long-time service in sulfate environment. *Cement and Concrete Research*, 2, 79–89.
29. ACI (2005) *Building code requirements for structural concrete and commentary*, ACI 318-05. American Concrete Institution, Farmington Hills, MI.
30. CSA (2000) *Concrete materials and methods of concrete construction*, CSA A23.1-00. Canadian Standards Association, Toronto.
31. BSI (2000) *Concrete, part 1: specification, performance, production and conformity*, BS EN 206-1. British Standards Institution, London.
32. ACI (1990) *A guide to the use of waterproofing, damp-proofing, protective, and decorative barrier systems for concrete*, ACI 515. American Concrete Institution, Farmington Hills, MI.
33. Allahverdi, A. and Škvára, F. (2000) Acidic corrosion of hydrated cement based materials. *Ceramics*, 44(1), 114–120.
34. Harrison, W.H. (1987) Durability of concrete in acidic soils and waters. *Concrete (London)*, 12, 18–24.
35. Sersale, R., Frigione, G. and Bonavita, L. (1998) Acid depositions and concrete attack: main influences. *Cement and Concrete Research*, 28, 19–24.
36. De Belie, N., Verselder, H.J., De Blaere, B., Van Nieuwenburg, D. and Verschoore, R. (1996) Influence of the cement type on the resistance of concrete to feed acids. *Cement and Concrete Research*, 26, 1717–1725.

37. BRE (2005) *Concrete in aggressive ground*. Special Digest 1. Building Research Establishment, Watford, 3rd edn.
38. Attigobe, E.K. and Rizkalla, S.H. (1988) Response of concrete to sulfuric acid. *ACI Material Journal*, 85, 481–488.
39. Biczók, I. (1972) *Concrete corrosion, concrete protection*. Akadémiai Kiadó, Budapest, 8th edn.
40. Gaze, M.E. and Crammond, N.J. (2000) The formation of thaumasite in a cement:lime:sand mortar exposed to cold magnesium and potassium sulfate solutions. *Cement and Concrete Composites*, 22, 209–222.
41. Thornton, H.T. (1978) Acid attack of concrete caused by sulfur bacteria action. *ACI Journal Proceedings*, 75, 577–584.
42. Padival, N.A., Navnit, A., Weiss, J.S. and Arnold, R.G. (1995) Control of *Thiobacillus* by means of microbial competition: implications for corrosion of concrete sewers. *Water Environmental Research*, 67, 201–205.
43. Vaidya, S., Montes, C. and Allouche, E.N. (2007) Use of nanomaterials for concrete pipe protection. In: L. Osborn and M. Najafi (eds), *Proceedings of the Pipelines 2007 International Conference*. Boston, MA, 1–11.
44. ACI (1997) *Guide to durable concrete*, ACI 201.2R-97. American Concrete Institution, Farmington Hills, MI.
45. Robertson, K.R. and Rashid, M.A. (1976) Effect of solutions of humic compounds on concrete. *ACI Journal Proceedings*, 73, 577–580.
46. Barnard, J.L. (1967) *Corrosion of sewers*. Research Report 250. Council for Scientific and Industrial Research, Cape Town.
47. Durning, T.A. and Hicks, M.C. (1991) Using microsilica to increase concrete's resistance to aggressive chemicals. *Concrete International*, 13, 42–48.
48. Kazuyuk, T. and Mitsunor K. (1994) Effects of fly ash and silica fume on the resistance of mortar to sulphuric acid and sulphate attack. *Cement and Concrete Research*, 24, 361–370.
49. Stanton, T.E. (1940) Influence of cement and aggregate on concrete expansion. *Engineering News Record*, 124, (5), 59–61.
50. Stanton, T.E. (1940) Expansion of concrete through reaction between cement and aggregate. *Proceedings of the American Society of Civil Engineers – Papers*, 66, 1781–1811.
51. BRE (1997) *Alkali aggregate reactions in concrete*. BRE Digest 330. Building Research Establishment, Watford, 2nd edn.
52. BCA (1992) *The diagnosis of alkali–silica reaction*. British Cement Association, Wexham Springs, Slough, UK, 2nd edn.
53. Concrete Society (1999) *Alkali–silica reaction, minimising the risk of damage to concrete. Guidance notes and model specification clauses.* Technical Report 30. Concrete Society, Crowthorne, 3rd edn.
54. Helmuth, R., Stark, D., Diamond, S. and Monranville-Regourd, M. (1993) *Alkali–silica reactivity: an overview of research.* SHRP-C-342. National Research Council, Strategic Highways Research Program, Washington, DC (reprinted 1994).

55. Diamond, S. (1989) ASR – another look at mechanisms. In: K. Okada (ed.), *8th International Conference on Alkali–aggregate Reaction (ICAAR)*, *Kyoto*. Society of Materials Science, Kyoto, 83–94.

56. Grattan-Bellew, P.E. (ed.) (1987) Is high undulatory extinction in quartz indicative of alkali-expansivity of granitic aggregates? *Proceedings of the 7th International Conference Concrete Alkali–aggregate Reactions, Ottawa, Canada.* P.E. Noyes Publications, Bracknell, 434–439.

57. Diamond, S. and Barneyback, Jr. R.S. (1976) A prospective measure for the extent of alkali–silica reaction. In: A.B. Poole (ed.), *The Effect of Alkalis on the Properties of Concrete. Proceedings of the Symposium, London.* Cement and Concrete Association, Wexham Springs, Slough, 149–162.

58. Chatterji, S. (1979) The role of Ca(OH)$_2$ in the breakdown of Portland cement concrete due to alkali–silica reaction. *Cement and Concrete Research*, 9(2), 185–188.

59. Ozal, M.A. (1976) The pessimum proportion as a reference point in modulating alkali–silica reaction. In: A.B. Poole (ed.), *The Effect of Alkalis on the Properties of Concrete, Proceedings of the Symposium, London.* Cement and Concrete Association, Wexham Springs, Slough, 113–123.

60. Gillott, J.E. (1975) Alkali–aggregate reactions in concrete. *Engineering Geology*, 9, 303–326.

61. French, W.J. and Poole, A.B. (1974) Deleterious reactions between dolomites from Bahrain and cement paste. *Cement and Concrete Research*, 4(6), 925–937.

62. Deng, M., Lan, X.H. and Xu, Z.Z. (2004) Petrographic characteristics and distributions of reactive aggregates in China. In: M. Tang and M. Deng (eds), *Proceedings of the 12th International Conference on Alkali–aggregate Reaction in Concrete.* International Academic Publishers, Beijing World Publishing, Beijing, 87–98.

63. Tang, M., Liu, Z. and Han, S. (1986) Mechanism of alkali-carbonate reaction. In: P.E. Grattan-Bellew (ed.) *Proceedings of the 7th International Conference Concrete Alkali–Aggregate Reactions, Ottawa, Canada.* P.E. Noyes Publications, Bracknell, 275–279.

64. Poole, A.B. (1981) Alkali–carbonate reactions in concrete. In: *Proceedings of the 5th International Conference on Alkali–Aggregate Reaction in Concrete.* CSIR S252/34. Cape Town, 1–9.

65. Gillott, J.E. (1963) Petrology of the dolomitic limestones, Kingston, Ontario, Canada. *Geological Society of America, Bulletin*, 74, 759–758.

66. Katayama, T. (2004) How to identify carbonate reactions in concrete. *Materials Characterization*, 53, 85–104.

67. Katayama, T. (2006) Modern petrography of carbonate aggregates in concrete–diagnosis of so-called alkali–carbonate reaction and alkali-silica reaction. Supplement to the Conference Proceedings. *Marc-André*

Bérubé Symposium on Alkali–Aggregate Reaction in Concrete, 8th CANMET/ACI International Conference on Recent Advances in Concrete Technology. Montreal.

68. Poole, A.B. (2008) Under the microscope. *Materials World*, 1 May, 32–33.

69. BSI (1994) *Testing aggregates, methods for qualitative and quantitative petrographic examination of aggregates*, BS 812-104. British Standards Institution, London.

70. ASTM (2003) *Standard practice for petrogaphic examination of aggregates for concrete*, ASTM C295. American Society for Testing and Materials, West Conshohocken, PA.

71. BSI (1999) *Guide to the interpretation of petrographical examinations for alkali–silica reactivity*, BS 7943. British Standards Institution, London.

72. ASTM (1990) *Standard test method for potential alkali reactivity of cement–aggregate combinations (mortar bar method)*, ASTM C227. American Society for Testing and Materials, PA.

73. RILEM (2000) *Detection of potential alkali-reactivity of aggregates: accelerated (60°C) concrete prism test, AAR-4. Test method TC-106-4.* Committee draft. RILEM, Bagneux.

74. BSI (1999) *Testing aggregates, method for the determination of alkali–silica reactivity – concrete prism method*, BS 812-123. British Standards Institution, London.

75. CSA (1994) *Methods of test for concrete. 14. Potential expansivity of cement–aggregate combinations (concrete prism expansion method)*, CSA A-23.2-94-14A. Canadian Standards Association, Toronto.

76. ASTM (1994) *Standard test method for potential alkali–silica reactivity of aggregates (chemical method)*, ASTM C289. American Society for Testing and Materials, PA.

77. CSA (1994) *Methods of test for concrete. Determination of potential alkali–carbonate reactivity of quarried carbonate rocks by chemical composition*, CSA A23.2-26A. Canadian Standards Association, Toronto.

78. ASTM (1992) *Standard test method for potential alkali reactivity of carbonate rocks for concrete aggregates (rock cylinder method)*, ASTM C586. American Society for Testing and Materials, PA.

79. Rogers, C.A. (1990) *Inter-laboratory study of the concrete prism expansion test for alkali–carbonate reaction. Canadian developments in testing concrete aggregates for alkali–aggregate reactivity.* Report EM-92. Ministry of Transportation, Ontario, Canada, 136–149.

80. Nixon, P. *et al.* (2003) Developing an international specification to combat AAR, proposals of RILEM TC 191-ARP. In: M. Tang and M. Deng (eds), *Proceedings of the 12th International Conference on Alkali–aggregate Reaction in Concrete.* International Academic Publishers, Beijing World Publishing, Beijing, 8–16.

81. DS (1984) *Danish code of practice for the structural use of concrete*, DS 411. Danish Standards, Copenhagen, 3rd edn.

82. CSA (2000) *Standard practice to identify degree of alkali-reactivity of aggregates and to identify measures to avoid deleterious expansion in concrete*, CSA A23.2-27A. Final draft versions of CSA A23.1-00/A23.2-0. Canadian Standards Association, Toronto.

83. CSA (1986) *Concrete materials and methods of concrete construction, methods of test for concrete*, CSA A23.2-88M. Canadian Standard Association, Toronto.

84. Seignol, J.-F., Barbier, F., Multon, S. and Toutlemonde, F. (2004) Numerical simulation of ASR affected beams comparison to experimental data. In: M. Tang and M. Deng (eds), *Proceedings of the 12th International Conference on Alkali–Aggregate Reaction in Concrete*. International Academic Publishers, Beijing World Publishing, Beijing, 198–206.

85. Poyet, S., Sellier, A., Capra, B., Foray, G., Torrenti, J.-M., Cognon, H. and Bourdarot, E. (2004) Modelling of alkali–silica reaction in concrete, part 1: influence of aggregate size range on chemical modelling of ASR. In: M. Tang and M. Deng (eds), *Proceedings of the 12th International Conference on Alkali–Aggregate Reaction in Concrete*. International Academic Publishers, Beijing World Publishing, China, 173–184.

86. Poyet, S., Sellier, A., Capra, B., Foray, G., Torrenti, J.-M., Cognon, H. and Bourdarot, E. (2004) Modelling of alkali–silica reaction in concrete, part 2: influence of water on ASR. In: M. Tang and M. Deng (eds), *Proceedings of the 12th International Conference on Alkali–Aggregate Reaction in Concrete*. International Academic Publishers, Beijing World Publishing, Beijing, 185–190.

87. Poyet, S., Sellier, A., Capra, B., Foray, G., Torrenti, J.-M., Cognon, H. and Bourdarot, E. (2004) Modelling of alkali–silica reaction in concrete, part 3: structural effects induced by ASR. In: M. Tang and M. Deng (eds), *Proceedings of the 12th International Conference on Alkali–Aggregate Reaction in Concrete*. International Academic Publishers, Beijing World Publishing, Beijing, 191–197.

88. Poole, A.B., Eden, M.A. and Lawrence, D.F. (2004) The effects of ionic mobilities on alkali–silica reaction progression in concretes. In: M. Tang and M. Deng (eds), *Proceedings of the 12th International Conference on Alkali–Aggregate Reaction in Concrete*. International Academic Publishers, Beijing World Publishing, Beijing, 163–172.

89. Lee, C., Chien Liu, C. and Su, M.H. (2004) Feasibility of using rapid lithium migration technique to repair concrete damaged by Alkali–Aggregate reaction. In: M. Tang and M. Deng (eds), *Proceedings of the 12th International Conference on Alkali–Aggregate Reaction in Concrete*. International Academic Publishers, Beijing World Publishing, Beijing, 1262–1270.

90. Whitmore, D. and Abbott, S. (2000) Use of an applied electric field to drive lithium ions into alkali–silica reactive structures. In: M.A. Bérubé, B. Fournier and B. Durand (eds), *11th International Conference on*

Alkali–aggregate Reaction, Quebec, Canada. ICON/CANMET, Ottawa, 1089–1108.

91. Parrott, L.J. (1987) *A review of carbonation in reinforced concrete.* Carried out by the Cement and Concrete Association under a BRE contract. BRE, Garston, Watford.

92. Trend 2000 (2001) *Evaluation of life performance and modelling. Corrosion of steel in concrete.* Report D29; DTI DMI 5.1(Report 2), BRE. J. Broomfield Consultants and Risk Review Ltd, London.

93. Kropp, J. and Hilsdorf, H.J. (1995) *Performance criteria for concrete durability.* RILEM Report 12, RILEM Technical Committee 116-PCD. E&FN Spon, London.

94. Broomfield, J.P. (2006) *Corrosion of steel in concrete, understanding, investigation and repair.* Taylor and Francis, London.

95. Hausmann, D.A. (1967) Steel corrosion in concrete: how does it occur? *Materials Protection,* 6(19) 19–23.

96. Glass, G.K., Roberts, A.C. and Davison, N. (2004) Achieving high chloride threshold levels on steel in concrete. In: *Proceedings of Corrosion 2004.* NACE International, Houston, TX.

97. Rodríguez, J., Ortega, L.M. and García, A.M. (1994) Assessment of structural elements with corroded reinforcement. In: R.N. Swamy (ed.), *Corrosion and corrosion protection of steel in concrete.* Sheffield Academic Press, Sheffield, 171–185.

98. Concrete Society (2004) *Electrochemical tests for reinforcement corrosion.* Concrete Society, Camberley, Surrey, UK, Technical Report 60.

99. Concrete Bridge Development Group (2002) *Technical guide 2: guide to testing and monitoring the durability of concrete structures.* Concrete Society, Camberley.

100. BSI (2007) *Products and systems for the protection and repair of concrete structures. Test methods, determination of chloride content in hardened concrete,* BS EN 14629. British Standards Institution, London.

101. BSI (2006) *Products and systems for the protection and repair of concrete structures. Test methods, determination of carbonation depth in hardened concrete by the phenolphthalein method,* BS EN 14630. British Standards Institution, London.

102. ASTM (1991) *Standard test method for half-cell potentials of uncoated reinforcing steel in concrete,* ASTM C876. American Society of Testing and Materials, West Conshohocken, PA.

103. ASTM (2006) *Standard test method for field measurement of soil resistivity using the Wenner four-electrode method,* ASTM G57-06. American Society of Testing and Materials, West Conshohocken, PA.

104. Langford, P. and Broomfield, J. (1987) Monitoring the corrosion of reinforcing steel. *Construction Repair,* 1(2).

105. BSI (1988) *Recommendations for the use of electromagnetic cover meters. Testing Concrete,* BS 1881:204. British Standards Institution, London.

106. Broomfield, J., Davies, K. and Hladky, K. (2003) Monitoring of reinforcement corrosion in concrete structures in the field. In: M. Grantham *et al.* (eds), *Concrete Solutions. Proceedings of the 1st International Conference on Concrete Repair*, St.-Malo.

5

Durability performance tests

Alan Richardson, *Northumbria University, UK*

5.1 Factors affecting durability

5.1.1 Introduction

The mechanisms by which concrete suffers durability problems and subsequent damage are freeze–thaw; ion flow, absorption and moisture movement within concrete; all of which relate to aspects of durability. Durability is defined within Eurocode 2 (2006), which states: 'A durable structure shall meet the requirements of serviceability, strength and stability throughout its intended working life, without significant loss of utility or excessive maintenance.' The satisfactory durability of CEM 1 (Portland cement) concrete, with or without additives, is a major reason why it is the world's most widely used construction material.

Areas of concern that will adversely affect the durability of concrete are material limitations; design and construction practices; and severe exposure conditions that can cause concrete to deteriorate, which may result in aesthetic, functional or structural problems.

Concrete can deteriorate for a variety of reasons, and concrete damage is often the result of a combination of factors, starting with the initiation of cracks which lead to processes that involve deleterious chemical reactions and hydrostatic breakdown. The rate of crack propagation is controlled by ionic/molecular transport systems, producing microstructural changes degrading the physical properties of the concrete.

Concrete is subject to deterioration caused by the absorption of moisture and thermal expansion and contraction. Extreme temperature ranges of both hot and cold can cause spalling. Moisture absorbed by the concrete expands and contracts with temperature changes, and the resulting mechanical action can cause fractures and spalling. Airborne pollutants, such as acid rain and carbon dioxide, can cause adverse chemical reactions which can result in surface deterioration.

Concrete durability
978-0-7277-3517-1

Environmental factors such as seasonal temperature variations, cyclical freezing and thawing, rainfall and relative humidity changes, and the concentration of deleterious chemicals in the atmosphere/ water in contact with the concrete, are the main causes of degradation. Geographical location is an important consideration with regard to durability, as are multiple, severe freeze–thaw cycles, which are worse for the destructive stresses applied within the concrete than an extremely low constant temperature.

All processes affecting durability are reliant upon the ingress of water or ions into the concrete.

Therefore, three main areas for durability testing are:

- freeze–thaw damage
- ion transfer
- factors affecting diffusion and absorption.

Without proper concrete design and production control, deteriorating concrete can compromise structural integrity, pose serious liability issues, and create significant problems throughout a structure.

5.2 Mechanisms, factors and design that affect concrete durability and types of concrete degradation

The mechanisms of deterioration of concrete and their rate of deterioration are controlled by the environment, the microstructure and the fracture strength of the concrete.

Nixon[1] outlined design factors that are required to achieve performance for a planned service, and the main interactions are shown in Fig. 5.1. The interactions outline the key areas where problems can arise as a result of failure at any one of a number of stages throughout the construction process. This view is supported by Jones et al.,[2] who asks for 'the development of a holistic approach to ensure the durability of new concrete construction'.

Figure 5.1 shows factors affecting concrete structures, to achieve performance for a planned service life. The figure shows durability, such as material performance, loading, strength, etc.

Well-designed concrete in protected environments can last a substantial period of time. Concrete is made from various combinations of cement, water, and fine and coarse aggregates with additives to affect the properties of the concrete. Absorption is a key area in need of testing with regard to durability performance.

According to Karr:[4]

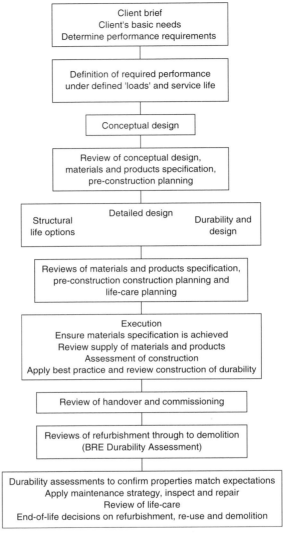

Fig. 5.1 *Design of concrete structures to achieve performance for a planned service life*[3]

Concrete deterioration is related to its permeability. Most researchers believe that a well designed and manufactured concrete is originally watertight, containing discontinuous pores and micro cracks. When subjected to extreme loading or weathering, concrete deteriorates through a variety of physical and chemical processes, which result in cracking. Cracks in concrete generally interconnect flow paths and increase concrete permeability. The increase in concrete permeability due to crack progression allows more water or aggressive chemical ions to penetrate into the concrete, facilitating further deterioration. Such a chain reaction of deterioration – cracking – more permeable concrete – further deterioration, may eventually result in destructive deterioration of the concrete structure.

Permeable concrete permits water ingress, which can leave the concrete subject to damage from freeze–thaw cycles. Surface scaling is perhaps the most evident as well as the most common form of freeze–thaw damage. Tensile stresses occur at the surface due to the action of ice plug formation and the subsequent expansion of water. This action causes the loss of small particles, and thus the aggregate is left exposed to the environment.

Pop out (type 1) deterioration is a term used to describe a saturated aggregate split in two due to the internal pressure exerted by the action of concrete freezing. Most aggregates have a much greater tensile strength then Portland cement paste: however, it is possible to encounter aggregates that are weaker than the surrounding cement paste, and these lead to severe pop out breakdown of the concrete when subjected to freeze–thaw conditions.

Pop out (type 2) is identified by the mortar cover being broken up by the hydraulic pressure formed due to the freezing of concrete and the expansion of the entrapped water.

'D-line' cracking is defined by the *ACI Manual of Concrete Practice* (2000) as a series of cracks in concrete near and roughly parallel to joints, edges and structure cracks. Hobbs[5] states 'surface cracking associated with freeze–thaw expansion ... within such concretes, cracks are present parallel to the exposed face which decrease in intensity with depth changing to a random distribution of cracks often about 100 to 200 microns or more in width'. D-line cracking develops along the joint or the edge of a concrete surface because the concrete near the joints is weaker and more susceptible to freeze–thaw damage.[6] In addition, stress concentrations at the corners and edges of the concrete slab due to curling and warping contribute to D-line cracking. Stark and Klieger[7] found that decreasing the maximum size of the coarse aggregate reduces the rate at which D-line cracking develops: however, it does not necessarily completely stop D-line cracking. Internal cracking can be caused if the concrete is subject to freezing prior to the initial set taking place. In such cases, the internal damage is often severe. In the case of fully cured concrete, internal cracking generally occurs when concrete is frozen when saturated.

The influence of environmental factors on the various deterioration mechanisms involved causes the micro-cracks to propagate until they become continuous.

In essence, the permeability of concrete influences the primary method of transport of moisture and aggressive ions into the concrete matrix, and subsequent increases in the permeation properties are responsible for the increased rate of damage. Thereafter, crack growth accelerates the penetration of aggressive substances into the concrete, and the spiral of deterioration continues.

Using more rigorous performance-based methods, it is possible to design structures for a specific service life with greater confidence. Broomfield and Millard[8] investigate ways of measuring resistivity in situ, which will help develop a model to predict water ingress into concrete, thus allowing the designer to predict life spans of the structure under various environmental conditions. Applications for explicit durability design are outlined in the *Concrete Society Discussion Document*,[9] which outlines the criteria of durable concrete design with regard to the length of service (long or short) and the performance requirements for 'normal' structures.

'The specifier may prefer to evaluate concrete performance at the point of placement'. While this is logically sound, it is logistically diffi-cult for several reasons. First, while some guidance exists in various documents, no standard method exists for sampling concrete at the point of placement, and methods in use vary widely. Questions remain about how to secure a representative sample, how to do so in a safe manner, and whether to attempt to sample immediately prior to impact at the point of placement, after impact, or before or after consolidation, and whether to retrieve and then re-compact a sample that has been placed, consolidated and, perhaps, finished.

Requirements for point-of-placement sampling and testing should therefore be accompanied by a specific sampling protocol. Further, it is to be expected that concrete properties, especially air content and density, will be affected by placing, consolidation, finishing and curing operations.[10] Clearly, on-site testing of concrete in its plastic state is subject to many vagaries.

With regard to durability performance testing, 'There is no agreed definition for durability performance test methods . . . a test that directly assesses the resistance to a standardised deterioration process e.g. the freeze/thaw test, or a test that directly assesses the resistance to one of the phases of a deterioration process, e.g. carbonation tests, or a test that directly assesses a performance related parameter e.g. measurement of cover'.[9] The location or point of testing involves two considerations: where to test and when to test.

5.3 Tests available to test for durability

5.3.1 Introduction

The tests reviewed are not meant to be an examination of every available test that may have an effect upon durability. Tests have been selected with regard to practical application and direct usefulness in establishing the parameters surrounding aspects of durability. Tests for durability are varied because concrete is a complex material with many aspects worthy of testing. All aspects of durability rely upon a transport mechanism to allow liquids or gasses into the concrete matrix that may cause physical damage or chemical/electrical instability. The main factors of moisture movement within concrete are capillary suction, absorption, water and gas permeability, which ultimately affect durability. The long-term performance of concrete depends on the interactions with the service environment, in which the penetration of deleterious substances is highly significant. The latter is controlled by permeation properties of the near-surface concrete. Durability tests are all essentially intended as a measure of the pore structure of concrete.[11] In general, 'the single most important parameter that leads to premature deterioration of concrete, is the ingress of moisture into the concrete'.[12]

In this respect, the tests as discussed herein are dealt with in terms of importance – the first ones being more significant than the later for the average case relating to durability. This may not be the case for every aspect of durability, and care should be taken to ensure specific problems are investigated and addressed to establish the root cause of durability problems; carbonation may be an example where this applies.

The application of a numerical modelling system to establish a defined value to quantify durability is still a distant objective. Aspects of durability relating to service life conditions can be evaluated and successfully dealt with by suitably qualified professionals, given sufficiently accurate data upon which a durability assessment can be made. Key areas specified to ensure minimum limits of durability are maximum water absorption as a percentage of the dry weight and maximum surface absorption in millmetres.

The tests as described herein will help assist the designer and specifier in making the correct choice of test for a service life situation.

5.3.2 Absorption testing

Absorption testing to BS 1881-122: 983 (*Method for Determination of Water Absorption*), requires a minimum of three concrete cores to be cut from the concrete to be tested. This test gives an insight into the internal absorption qualities of the concrete. The test is quick and cheap: once the samples have been prepared, the test time is 30 minutes. The test is reliable, and outliers are rare.

Dill[13] states:

General experience has shown that typical values for water absorption of concrete, as established by BS 1881, 30 minute immersion procedure are:

- Low absorption <3%
- Average absorption 3–4%.

Clearly, a 1% difference in absorption by weight is a significant change in absorption, when comparing plain and monofilament concrete regarding enhanced durability.

When comparisons are drawn with regard to water absorption, the unit weight of water when compared with the unit weight of concrete is different by a magnitude of approximately 2.3, and this is a key factor when interpreting the results. Small changes in weight represent significant amounts of absorbed water.

5.3.3 Surface absorption

The surface properties of concrete differ from the core in that laitance is present, which is generally of a different composition due to fines being drawn to the surface during trowelling or compaction. Surface absorption is measured in a uniaxial direction using BS 1881-208:1996 (*Recommendations for the Determination of the Initial Surface Absorption of Concrete – ISAT*). A water-tight cap is placed on the surface of the test specimen, and the concrete is subject to a head of 200 mm of water pressure over an area of 5000 mm^2. 'The method measures the rate at which water is absorbed into the surface and must not therefore be thought of as a test to measure the bulk permeability of concrete'.[14]

5.3.4 Other transport tests available

BS EN 12390/DIN 1048 examines effective water permeability, and is a true permeability test: however, according to Pocock *et al.*,[11] 'it is extremely unreliable'.

ASTM C1585-04 (2004), *Standard Test Method for Measurement of rate of Absorption of Water by Hydraulic-cement Concretes*, provides a unidirectional test for absorption.

5.3.5   Freeze–thaw tests

From work carried out by the Concrete Society's Specialist Materials Group,[9] conclusions were drawn by Haynes as to the postulated risk of damage, which depends on:

- the moisture state and distribution in the pore structure
- the availability during the freezing cycle of non-frozen water to form additional ice
- the lowest freezing temperature
- the length of the freezing cycle if external water is available.

Hobbs[5] suggests that freeze–thaw attack is one of three main causes of concrete degradation.

5.3.5.1   ASTM C666

ASTM C666-97 allows comparative statements to be made about test results when the concrete is made from one or two batches, as long as the standard deviation relating to the durability factor is within the acceptable range as shown in Table 1. To obtain the relative dynamic modulus as required by the ASTM test, a PUNDIT test to BS 1881:203 should be carried out within the test.

The procedures available are A and B – A being air thawed and B being water thawed. As concrete needs to be almost fully saturated before freeze–thaw damage can take place, procedure B is recommended.

Expansion tests may be carried out as they are an optional procedure (ASTM C666:325): however, final strain measurements may be used to evaluate internal cracking and subsequent expansion.

The following information is taken prior to commencement of the freeze–thaw tests:

- all nominal external dimensions
- saturated weight
- PUNDIT test, reading for transverse frequency time and relative elastic modulus
- air content of the fresh concrete (BS EN 12350-7, clause 7)
- pore and void spacing content as TR 32.[15]

5.3.5.2 Freeze–thaw procedure B
In this procedure, the freezing occurs in the air and the thawing is carried out in water until the centre of the sample is transformed from $-19.4°C$ to $6°C$ and the transition period between freezing and thawing phases of the cycle does not exceed 10 minutes.

ASTM C666 recommends the number of cycles be within the range 100–300 cycles, whereby the 'indexes of precision are valid at least over these ranges'.

The final report examined the following data:

- consistency limits per batch (slump)
- characteristics of the test specimen if found to deviate from the norm and surface condition
- original weight, ultrasonic pulse velocity and nominal size
- final weight, ultrasonic pulse velocity and size
- final number of freeze–thaw cycles to failure per concrete type
- the durability factor for all samples
- a paired comparison test on the control batches (*t* tables)
- a relative comparison test showing the average different performance, modulus of elasticity and standard deviation per concrete type.

The durability factor provides a final percentage evaluation against the concrete samples prior to the start of the freeze–thaw cycles. The ASTM C666 test allows a great deal of useful performance related durability information to be obtained.

5.3.5.3 Other freeze–thaw tests
ASTM C672 (1997), *Standard Test Method for Scaling Resistance of Concrete Surfaces Exposed to Deicing Chemicals*, provides a visual scale to determine surface scaling of concrete over 50 cycles. This test is subject to operator judgement.

5.3.6 Designing out latent defects due to poor durability
Design for the correct exposure class in accordance with BS EN 206-1/BS 8500 is critical to prevent breakdown and spalling of concrete due to poor design.

The normal means of assessing durability is to comply with BS EN 206, which defines durability in terms of minimum cement content, water/cement ratio and minimum cover. This simplistic approach is deemed unsatisfactory for modern durable structures: however, it is a good starting point.

Concrete cover to reinforcement is a critical aspect of design and production, with regard to durability. Whilst corrosion of reinforcement is not exclusively associated with the durability of concrete, the association between reinforcement corrosion and the condition of the concrete structure often leads to localised areas of spalling due to the corrosion of the steel rebar. Poor positioning of steel during the construction phase is often carried out, to tolerances less than the designer's requirements. Therefore, 'the protective capacity of a given concrete is broadly related to the square of the cover'.[9] This single aspect is related to many durability problems. Where there is a requirement to establish the depth of cover over rebar, an electromagnetic covermeter should be used (BS 1881-204:1988).

5.3.7 Gas permeability, resistivity, chloride transport, mechanisms

5.3.7.1 Introduction

The ionic diffusivity of concrete is a function of its microstructure at many length scales, ranging from nanometres to millimetres. 'The microstructure is largely controlled by the initial concrete mixture proportions and the ultimate curing conditions'.[16] With regard to the examination of electrical conduction in concrete, from an electrical point of view, concrete can be considered as a composite material comprising non-conductive aggregate particles embedded in an ionically conducting hardened cement paste matrix.[17]

'As the level of saturation of the capillary pores change, there is a corresponding change in the bulk electrical properties of the concrete which can vary over several orders of magnitude'.[17]

Concrete where saturation equals 0% has virtually zero conductivity whereas plain ordinary Portland cement concrete in a fully saturated state has a significant conductivity value.

The Nernst–Planck equation (flux = diffusion + migration + convection) determines the mass transfer of electrolytes, and is termed 'migration'. Migration is the transport of ions in electrolytes, due the action of an electrical field, as the driving force. In an electrical field, positive ions will move preferentially to the negative electrode, and negative ions to the positive one.[18]

The general equation states that the movement of charged species in an electrolyte is the sum of the diffusion component from concentration gradients, the migration component arising from potential gradients and the flow of the electrolyte itself due to capillary suction, permeation, etc.

The transport of media into concrete is in most cases not due to one single mechanism but several mechanisms acting simultaneously.[18] These mechanisms are interrelated, e.g. saturated concrete can adversely affect the migration of gas atoms within the concrete. Even at the best of times, liquid transport through porous media is a complex issue. It has been suggested that to fully understand the behaviour of fluids in porous media remains one of the biggest challenges in classical physics. Mulheron[19] states: 'It is well established that a major factor driving the deterioration ... is a combination of water ingress and the associated movement of chloride ions through the cover concrete'.

5.3.7.2   Tests

5.3.7.2.1 Resistivity

Direct resistivity measurements can be taken from samples, as shown in Fig. 5.2, which provide an indication of durability with regard to ion flow and water content. AC power supply is recommended, to avoid the capacitance effect.

If a signal generator is used to input a sine wave current (or voltage) at anywhere around 10–40 Hz and the resulting voltage (or current) measured, then a measurement free of polarisation effects should be obtained.

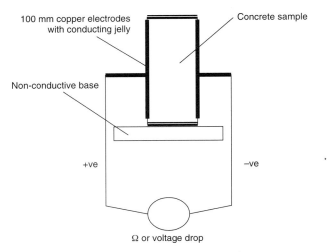

Fig. 5.2 Resistivity/conductance testing of concrete samples

Testing for resistivity

'The application of this applied voltage technique and interpretation of results is hampered by a lack of understanding of the mechanism of ion transport through concrete'.[14]

5.3.7.2.2 Chloride permeability

The onset of corrosion due to chlorides depends upon two factors: the rate at which the free chloride penetrates the concrete and the threshold value. The threshold level is the concentration of chloride ions tolerated before corrosion takes place. Oxygen and water are required before breakdown of rebar occurs due to oxidation.

Chloride levels required to initiate corrosion of steel range between 0.17 and 2.5 m/m of chloride atoms by mass of cement.

Test

Chloride permeability using ASTM C1202 expresses the results as coulombs of charge passed. There is no simple conversion to provide a chloride diffusion coefficient. To ensure that the results are useful, at least three test cubes should be tested from each sample. A high chloride permeability would be 4000 coulombs, whereas a low permeability would be 2000 coulombs.

Alternative to ASTM C1202

The specifier may state that the use of compressive strength results as an acceptance method are permitted to be used as an indicator of resistance to chloride ion penetration, based upon experience.

'In determining whether to permit the results of strength tests as an indirect indicator of resistance to chloride ion penetration, the specifier should consider the significance or criticality of the structure, the cost and time of performing these tests, and the reliability of the mixture-specific relationship between strength and the results of ASTM C1202 testing. It is recommended that compressive strength be only considered as an indicator of chloride ion penetration resistance for exposure classifications F1, S1, and P1 and not for the more severe levels of exposure in each category'.[10] 'Air content has a generally greater impact on compressive strength than on the permeability or resistance to chloride ion penetration of concrete, therefore it is important that air content be carefully controlled'.[10]

5.3.7.3 Gas permeability

5.3.7.3.1 Introduction

An aspect of concrete durability affected by gas permeability is car-bonation, and this can cause a breakdown of the cement due to the formation of carbonic acid. As a general rule, Darcy's law is used for both water and gas permeability measurements.

5.3.7.3.2 Tests

The tests used to determine gas permeability can be divided into in-situ and laboratory tests. The majority of tests use a similar apparatus to a drained tri-axial test, where a pressure is applied to one side of the sample and the gas is measured at the escape boundary. Other tests use migration in a sealed environment to evaluate flow.

In-situ tests by Figg, Montgomery and Adam, etc. are described in TR 31, *Permeability Testing of Site Concrete – A Review of Methods and Experience* (1987), as are laboratory tests.

5.3.7.3.3 Laboratory tests

Lawrence devised a permeability test as shown in Fig. 5.3. Gas permeability values of $5 \times 10^{-7}\,\mathrm{m}^2/\mathrm{s}$ are high, whereas values of $5 \times 10^{-8}\,\mathrm{m}^2/\mathrm{s}$ are low. The age and type of concrete affect the results due to the time taken to produce a closed-cell structure. Concrete with little curing will be more porous than fully matured concrete.

5.3.8   Structural performance

5.3.8.1   Compressive strength

Compressive strength is widely used as the key indicator of concrete quality, and therefore needs accurate determination. To evaluate the compressive strength of concrete test cubes, BS EN 12390-3:2002 can be used as a simple and readily available test that is widely used and accepted by industry. The load at failure is recorded and this is divided by the surface area of the cube in contact with the platens through which force is applied to the concrete cube. Units used are MPa or N/mm^2.

5.3.8.2   Abrasion resistance

Abrasion resistance may be specified for structural elements where surface wear is a concern for functional performance, such as for

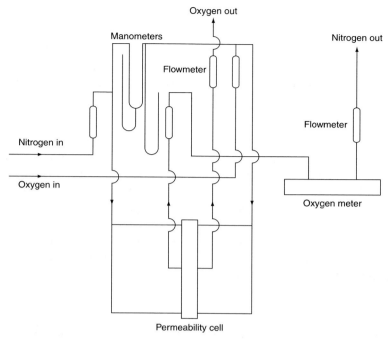

Fig. 5.3 Apparatus for measuring diffusion of oxygen in concrete. (Reproduced with permission from Lawrence)

industrial floor slabs. The specifier should identify elements of the structure to which abrasion resistance requirements apply. The specifier has the option to require a maximum abrasion loss and state the associated test method, and to state the age at which such values are to be measured. Multiple abrasion tests are available, but each varies in the type of wear pattern induced, and thus each may have varying relevance to the wear anticipated in service for the subject structure.

Tests available to test for abrasion are ASTM C418, ASTM C779, ASTM C944 and ASTM C1138.

5.3.8.3 Modulus of elasticity

The modulus of elasticity may be specified for structures or structural elements for which vertical or lateral deformation, overall stiffness or vibration control are critical to performance, such as electronics production facilities. ASTM C469 may be used to evaluate this property at 28 days. When deciding whether to specify and test for the modulus of elasticity, the specifier should recognise the approximate nature of the relationship between the elastic modulus and compressive strength.[10]

5.4   Conclusion

Certain tests such as flexural or compressive strength are definitive, whereas a test for durability has to be interpreted in the context of the performance required from the material or the whole structure. To reiterate the initial comments, transport mechanisms are the key to durable concrete, and absorption is a key element widely acknowledged as the main contributing factor where durability is concerned. Freeze–thaw durability is a strong contender for the second most significant test when examining the use of concrete in climates where freeze–thaw cycles are likely.

References

1. Nixon, P. (2002) Design of concrete structures to achieve performance for a planned service life. In: *Concrete Durability – Recent Advances, Society of Chemical Industry Conference*. London, 2–3.
2. Jones, A.E.K., Marsh, B.K., Clarke, L.A., Seymour, D., Basheer, P.A.M. and Long, A.E. (1997) *Development of an holistic approach to ensure the durability of new concrete construction*. British Cement Association, 1–81.
3. BRE (2000) *Constructing the future. Design build*, No. 7. British Standards Institution, London.
4. Wang, K., Jansen, D.C., Shah, S.P. and Karr, A.F. (1997) Permeability study of cracked concrete. *Cement and Concrete Research*, 27, 381–393.
5. Hobbs, D.W. (2002) The durability of concrete. In: *Yearbook 2002–2003*. Institute of Concrete Technology, Camberley, 55–69.
6. Cordon, W.A. (1966) *Freezing and thawing of concrete–mechanisms and control*. American Concrete Institute and Iowa State University Press, Ames IA, 2nd edn.
7. Stark, D. and Klieger, P. (1973) Effect of maximum size of coarse aggregate on D line cracking in concrete pavements. *Highway Research Records*, 441, 33–43.
8. Broomfield, J. and Millard, S. (2002) Measuring concrete resistivity to assess corrosion rates. *Concrete*, 128, 37–39.
9. Concrete Society (1996) *Developments in durability design and performance–based specification of concrete*. Discussion document prepared by a working party of the Materials Group. Concrete Society Special Publication CS109. Concrete Society, Camberley.

10. Bickley, J.A., Hooton, R.D. and Hover, K.C. (2008) *Guide to specifying concrete performance, phase II report of preparation of a performance-based specification for cast in place concrete.* RMC Research and Education Foundation and NRMCA P2P Steering Committee, Silver Spring, MD.

11. Pocock, D. and Corrans, J. (2007) Concrete durability testing in Middle East construction. Regional report – Middle East. *Concrete Engineering International*, 11(2), 52–53.

12. Long, A.E., Henderson, G.D. and Montgomery, F.R. (2001) Why assess the properties of near-surface concrete. *Construction and Building Materials*, 15, 65 –79.

13. Dill, M.J. (2000) *A review of testing for moisture in building element*, C538. CIRIA, London.

14. Concrete Society (1987) *TR 31, permeability testing of site concrete – a review of methods and experience.* Report by a working party of the Concrete Society, Camberley.

15. Concrete Society (1989) *TR 32, analysis of hardened concrete, a guide to test procedures and interpretation of results.* Report by a joint working party of the Concrete Society and the Society of the Chemical Industries, Camberley.

16. Bentz, D.P. and Garboczi, E.J. (1995) Water permeability and chloride ion diffusion in Portland cement mortars: relationship to sand content and critical pore diameter. *Cement and Concrete Research*, 25(4), 790–802.

17. McCarter, W.J., Chrisp, T.M., Butler, A. and Basheer, P.A.M. (2001) Near surface sensors for condition monitoring of cover zone concrete. *Construction and Building Materials*, 15(2), 115–132.

18. Kropp, J. and Hilsdorf, H.K. (1995) *Performance criteria for concrete durability.* RILEM Report 12. E&FN Spon, London.

19. Mulheron, M. (ed.) (2001) NMR applications to study concrete's durability – an engineer's viewpoint. EPSRC engineering network for the application of NMR techniques to improve concrete performance. *Network News*, 2, 1–4.

20. Hobbs, D.W. (1999) Aggregate influence on chloride ion diffusion into concrete. *Cement and Concrete Research*, 29, 1995–1998.

6

Modelling and predicting effects of deterioration mechanisms

Klaas van Breugel, Eric Schlangen, Oguzhan Copuroglu and Guang Ye,
TU Delft, The Netherlands

6.1   Role of models in durability and service life design

The majority of degradation processes in concrete generally proceed very slowly. The slow rate of deterioration processes and mechanisms is a complicating factor for modelling. The fact that many of these processes and mechanisms have multiple causes, often coupled chemical and physical processes, as well as the heterogeneous nature of concrete, further contribute to the complexity of modelling and predicting deterioration processes.

Depending on the particular purpose for which models are needed, one can distinguish science-oriented models and engineering models. Even though both models aim to mirror reality as accurately as possible, the different aims of these models may lead to completely different structures for each approach. The fact that models are developed for different purposes should be considered carefully when discussing and judging models. Ignorance of the specific aims models are developed for can easily lead to unfruitful debates, which demonstrate the short-comings of researchers rather than those of the individual models.

In spite of the aforementioned complicating factors, there is an increasing demand for reliable deterioration models and models for service life predictions. Contractors are now facing required service lives of 80, 100 or even 200 years, especially for infrastructure works. Currently used design codes generally do not give sufficient guidance for designing concrete structures for such long service lives. Even though modelling of deterioration processes is a complex task, it is only these models that have the power to simulate and predict the long-term performance of materials and structures. Of particular importance in this respect are the probability-based design concepts.[1] In the 1990s, such a probabilistic concept was developed in the European

Concrete durability
978-0-7277-3517-1

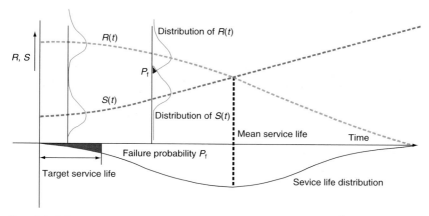

Fig. 6.1 Schematic representation of probabilistic durability design[2]

research project Duracrete.[2,3] The Duracrete code follows the structural design philosophy by stating that the service life is the period during which the structure's resistance $R(t)$ can withstand the environmental load $S(t)$, so it delivers a particular performance, with a predetermined maximum probability of failure. Schematically, the concept is shown in Fig. 6.1. Probabilities of failure for the serviceability limit state are usually in the range of a few per cent (reliability index $\beta = 1.5$–2).

6.2   Initial status of concrete structures

6.2.1 Modelling hydration and microstructure development

The initial status of the concrete, i.e. the status when hydration has almost ceased, determines the long-term performance of the structure to a large extent. Therefore, modelling of the long-term performance starts with the modelling of the hydration processes and development of the microstructure. In the first half of the 20th century, models were proposed for cement hydration and associated heat evolution. Most of these models were strongly phenomenological, and did not address the development of the microstructure explicitly. In the early 1950s, models for hardening concrete appeared with which the evolution of the degree of hydration was quantified. The degree of hydration could be used as basis for the determination of the volumes of the constituent components of the cement paste. These models lasted until the late 1970s, when models were developed with the aim to describe the evolution of the microstructure explicitly at the microscale. The increasing computation power of modern computers, as well as

125

deeper insight in the hydration processes, made it possible to describe the development of interparticle contacts and the associated evolution of the mechanical and transport properties.

Today, one of the most advanced models for simulating the evolution of the microstructure of cement paste is the NIST model.[4] It is a pixel-based model, each pixel representing a certain component of the cement paste to which the appropriate chemical and physical properties are assigned. The pixels, $1\,\mu m^3$ in size, are able to 'react' with other pixels and build up a virtual microstructure.

Navi's model[5] and the HYMOSTRUC model[6] start from a system of spheres randomly distributed in water, representing the cement or other powders in the paste. When reacting with water the spheres start to 'grow'. While growing, the spheres make contact with each other, and build up a virtual microstructure. The formation of interparticle contacts affects the rate of reaction of individual particles. Reactions which allow for the effect of physical interaction between particles on the rate of reaction have been termed integrated kinetics.[7] The DUCOM concept[8] is also based on spherical particle growth, but starting from a monosize-particle system.

After model-specific calibration, these models are able to simulate hydration processes, and hence the evolution of hydration heat, quite accurately. Examples of simulated microstructures and related pore structures are presented in Fig. 6.2.

For individual concrete mixtures, isothermal and/or adiabatic hydration curves can be produced, which can be used as the input for advanced software packages for the determination of the temperature distribution in hardening concrete structures. In these software packages, which are often incorporated in concrete curing control systems, the effect of the actual curing temperature on the rate of

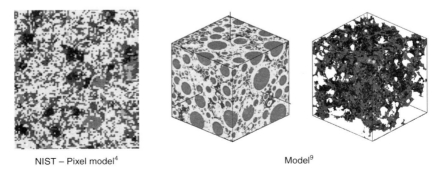

NIST – Pixel model[4]   Model[9]

Fig. 6.2 Examples of the virtual microstructure of cement paste

hydration is allowed for by using an Arrhenius function. In addition to the temperature distribution in hardening concrete structures, hydration fields can also be obtained.

Predicting the microstructural details of a cement paste is much more complicated than predicting temperature distributions or the degree of hydration. Detailed simulation of the stoichiometry of the reaction products and of the pore structure, however, is essential for modelling of the deterioration and transport processes.

6.2.2 Evolution of materials properties

The evolution of materials properties is based on the evolution of the microstructure. This holds for all materials properties, such as strength, stiffness, permeability, hydration-related volume changes (e.g. chemical shrinkage, autogenous shrinkage and thermal shrinkage) and time-dependent properties (e.g. creep and relaxation). During hydration, all these properties change with the progress of the hydration process. In order to ensure consistency in the quantitative analysis of young concrete, it is recommended that the evolution of materials properties and strains are associated with a single materials parameter. Basically, all materials properties depend on the properties of the microstructure. From the modelling point of view, the challenge is, therefore, to relate the materials properties to the properties of the virtual microstructure. Figure 6.3 shows a successful example of an attempt to relate the summarised contact area of a virtual cement paste to the compressive strength of cement paste. It was also found that the growth of the stiffness and changes in the ultrasonic pulse velocity in hardening cement paste correlate well with the intensity with which the particles of a virtual microstructure were connected.

Carol *et al.*[11] and Lokhorst[12] have proposed models for relating creep and relaxation in hardening concrete to the evolution of the microstructure explicitly. All these models illustrate that the direct coupling of materials properties to the properties of the microstructure is a challenging but, in the long-term, promising direction for modelling the performance of materials and structures. For the time being, however, it seems more realistic to associate the evolution of materials properties with the degree of hydration.

6.2.3 Stress development and risk of cracking in hardening concrete

Stress calculations in hardening concrete are carried out with the aim of predicting the probability of cracking in the early life time of a structure.

127

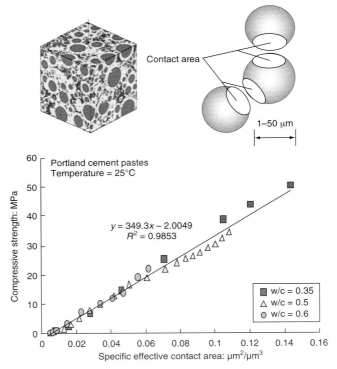

Fig. 6.3 Compressive strength as a function of the summarised contact area of a virtual cement paste for mixtures with a water/cement (w/c) ratio ranging from 0.35 to 0.6[10]

Calculations of the probability of cracking require information about the evolution of the materials properties, of the generated strains and, last but not least, the degree of restraint.

For the determination of the probability of cracking, both the mean values and their standard deviations of the relevant parameters, i.e. tensile stress, tensile strength or fracture energy, are required. Basically, the scheme presented in Fig. 6.1 applies, where $R(t)$ is the strength and $S(t)$ is the stress. In the particular case of hardening concrete, however, the strength $R(t)$ does not decrease but gradually increases with time. Parameter studies and sensitivity analysis have revealed that the degree of restraint and the time-dependent properties are very important parameters in view of the generated stresses and risk of cracking.[12,13]

The use of software packages to analyse the performance of hardening concrete has contributed significantly to our understanding of the state of stress and risk of cracking in young concrete. In this regard, it is important to realise that, in particular, the use of modern, low water/binder (w/b) ratio concrete mixtures has revealed the need for more advanced models. In the past, the performance and probability of cracking of traditional concrete mixtures could be determined satisfactorily by using experience-based temperature criteria. These criteria, however, fail when concrete mixtures with a low w/b ratio are used and which exhibit substantial autogenous deformations. In concrete mixtures with w/b ratios below 0.4, the contribution of autogenous strains to the risk of cracking can no longer be ignored.

6.3   Carbonation

6.3.1 *Carbonation mechanism*

The carbonation mechanism of concrete has been studied and described by many authors. The carbonation reaction occurs at the water–air boundary in the pore system of the cement paste. In a drying pore system, the reaction front penetrates deeper into the concrete unless the reaction product, i.e. calcium carbonate, blocks the pores and is not leached out. The calcium carbonate is formed according the reactions

$$CO_2 + H_2O \rightarrow H_2CO_3 \tag{6.1}$$

$$H_2CO_3 + Ca(OH)_2 \rightarrow CaCO_3 + 2H_2O \tag{6.2}$$

In fact, not only the calcium hydroxide reacts but also the calcium silicate hydrate (C–S–H) gel will be affected.[14] The major concern, however, is the consumption of the calcium hydroxide and the associated reduction of the pH, since this reduction poses a corrosion risk to the reinforcement. When the alkalinity decreases below pH 10, the steel is no longer passivated, and corrosion can occur over the total depassivated area, provided that all the preconditions for corrosion to occur are fulfilled.

The most often used formula for predicting the rate of carbonation is based on the solution of Fick's second law of diffusion:

$$X = A\sqrt{t} \tag{6.3}$$

where X the penetration depth of the carbonation front, A is a constant and t is the time. The constant A depends on the w/c ratio, the type of cement (for blended cements, A can vary quite substantially) and the degree of hydration. The degree of hydration will depend on the curing of the concrete. In particular, early drying out of the concrete will result in enhanced carbonation rates.

Because of the many parameters that affect the actual rate of carbonation, the validity of equation (6.3) is limited. The best results are obtained for a temperature of about 20°C and 65% relative humidity. At other values of these parameters, the carbonation is generally lower. In submerged conditions, concrete will not carbonate at all.

According to Audeneart et al.,[15] the use of blended cements affects the rate of carbonation, mainly due to associated changes of the capillary porosity and of the amount of carbonisable material. The author proposed, therefore, a model that allows for the amount of carbonisable material and the capillary porosity. The model is developed on the basis of work by Papadakis[16] and, for the description of the pore system, the work of Powers et al.[17] Thiery et al.[18] proposed a more comprehensive simulation model for carbonation based on coupled CO_2–H_2O ion transfers and chemical reactions. In their model, mass balances are explicitly addressed.

6.4   Chloride-induced corrosion

6.4.1 Mechanisms of chloride ion penetration

Today, corrosion mechanisms are well understood from the qualitative point of view. Yet, it is still a tough task to model rebar corrosion quantitatively. Up to now, most models have focused on the penetration of chloride ions and the numerical determination of the time needed to reach the chloride threshold value at the surface of the steel. Chloride ions penetrate into the concrete by diffusion, convection and migration. In all cases, gradients should be present in order to transport the ions into the concrete. Temperature gradients may further complicate the transport processes. In addition, binding of chloride ions by reaction products substantially affects the rate at which the chloride front penetrates into the concrete.

6.4.2 Engineering models for chloride penetration

Even though several transport mechanisms play a role in the ingress of chloride ions into concrete, the majority of models for chloride ion penetration are based on Fick's second law of diffusion. For a one-dimensional case, the following holds:

$$C(x,t) = C_s - (C_s - C_i)\mathrm{erf}\left(\frac{x}{\sqrt{4kD(t)t}}\right) \qquad (6.4)$$

where $C(x, t)$ is the chloride content at depth x and at time t, C_s is the surface chloride content, C_i is the initial chloride content, k is a correction factor and $D(t)$ is a time-dependent diffusion coefficient. Reasons for the time dependency of this coefficient are continuing hydration, binding of chloride ions, microcracking or, conversely, the self-healing of microcracks. To allow for this time dependency of the diffusion coefficient, it has been proposed (see also Maage et al.[19]) that

$$D(t) = D_0 \left(\frac{t_0}{t}\right)^n \tag{6.5}$$

where D_0 is the diffusion coefficient at reference time t_0 (e.g. 28 days) and n is the ageing coefficient. By combining equations (6.4) and (6.5), we get

$$C(x, t) = C_s - (C_s - C_i)\mathrm{erf}\left(\frac{x}{\sqrt{4kD_0(t_0/t)^n t}}\right) \tag{6.6}$$

The use of Fick's second law for the transport of chloride ions into concrete has often been criticised, since it presupposes diffusion to be the decisive, and in fact the only, transport mechanism, whereas in reality other mechanisms play a role as well. Depending on the exposure conditions, the dominating transport mechanism in the outer 10–15 mm is either capillary action or diffusion. Binding of chloride ions and other influencing factors mean that the diffusion coefficient must be considered as an apparent value. The coefficients k and n, which follow from curve-fitting analyses, partly reflect, or mask, the effect of a cocktail of influencing factors and different transport mechanisms. For offshore structures, transport mechanisms may apply, as indicated in Fig. 6.4. In the atmospheric zone, gas diffusion and water vapour diffusion will occur. In the splash zone, water vapour diffusion and ionic diffusion may dominate the transport processes. In the tidal zone, water absorption also occurs, whereas in the submerged zone, ionic diffusion and permeability have to be considered. In all these zones, different values for the model parameters apply. This is in addition to the effect of, for example, the type of cement on the transport properties.

An example of an engineering model that has adopted Fick's second law of diffusion as the core[2] of the analytical description of chloride ion profiles is the Duracrete model mentioned in Section 6.1. In that model the D_0 value is determined experimentally on 28-day-old concrete samples in a rapid chloride migration test. The surface chloride content is assumed to be independent of mix composition for reasons of

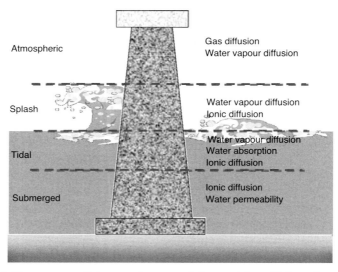

Fig. 6.4 Exposure conditions to be covered by model coefficients k and n in equation (6.6)

simplicity, i.e. 3.0% m/m of cement content for marine structures[20] and 1.5% m/m for land-based structures.[21] The initial chloride content was taken equal to 0.1% m/m.

The value of the ageing coefficient for a particular binder depends on the rate of hydration and on the extent of drying. Based on a large amount of data, n values were chosen for different binders in two groups of environmental classes: very wet (XD2/XS3) and moderately wet (XD1/XD3/XS1).[20,22] Table 6.1 shows the values of n used for

Table 6.1 Ageing coefficients n for different binders and different environmental classes[20,22]

Environmental class	Underground, splash zone	Above ground, marine atmospheric
NEN-EN 206	XD2, XS3	XD1, XD3, XS1
Type of binder		
CEM I	0.40	0.60
CEM I, 25–50% slag, II/B-S; or III/A, <50% slag	0.45	0.65
CEM III 50–80% slag	0.50	0.70
CEM I with 21–30% FA	0.60	0.70
CEM V/A (25% slag and 25% FA)	0.60	0.70

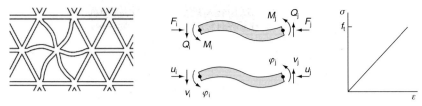

Fig. 6.8 Schematic representation of the lattice model[26]

solution for 14 days. The expansion of the specimen was registered at 0.8%, and showed significant cracking. Figure 6.9 shows microphotographs of a thin section taken from the prismatic bar.

In the model, the material is discretised as a lattice consisting of 42 000 small beam elements. Three spots are selected in the sample with a single aggregate in which swelling ASR gel is placed, i.e. inside the aggregate, inside the interface between the aggregate and the cement matrix, and inside the cement matrix. The three locations are shown as black dots in Fig. 6.10. It is then assumed that the ASR gel in the three locations swells over time at the same rate. Random values are assigned as input parameters for the beam elements in the specific locations for strength (and stiffness). The values are randomly chosen in the range given in Table 6.2. For the stiffness of each element the strength value is multiplied by 10.000.

Three different stages of cracking are shown in Fig. 6.10. The deformations are scaled with the same factor. The results of the simulations look realistic and compare very well with the experimental observations in Fig. 6.9.

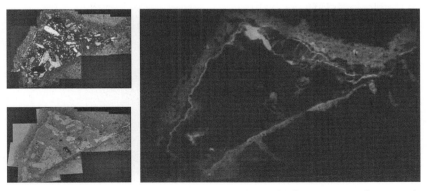

Fig. 6.9 Optical microphotographs of a damaged basaltoid aggregate and surrounding paste due to the ASR[28]

Table 6.2 Strength and stiffness of components used in lattice simulations

	Aggregate	Dissolved rim of aggregate	Interfacial transition zone	Cement matrix	ASR gel
Strength: MPa	7.0–13.0	1.0–2.0	1.5–2.5	3.0–5.0	1.0

Crucial input for the model, in order to make quantitative predictions of how much swelling and damage can be expected, is the properties of the gel. Experimental studies on the expansive properties of the gel and its potential to exert internal stresses have recently started.[27] However, even with incomplete input data, it has been possible to simulate the damage reasonably well.

6.6 Frost damage

6.6.1 Frost damage mechanisms

Freeze–thaw cycles, whether or not combined with the additional effect of salts that penetrate into the concrete, can lead to internal damage and surface scaling of concrete. The modelling of frost salt scaling (FSS) is a difficult phenomenon due to its complex physical and

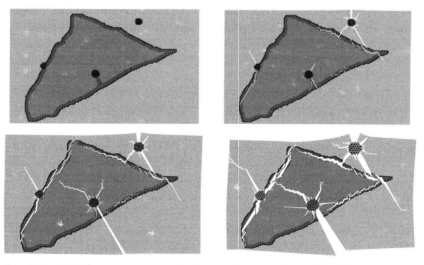

Fig. 6.10 ASR damage sequence simulated with a lattice model. Gel formation is initiated in various locations as observed in the microscopy study shown in Fig. 6.9

chemical mechanisms.[29] Plain frost action has attracted relatively more attention because of its less complicated mechanism. The work of Powers, Litvan, Pigeon, Marchand, Setzer and Schulson,[30-34] and many other researchers, has created a solid framework around this issue. Unfortunately, a similar argument cannot be made for frost salt attack, and there remain a number of questions which cannot be answered by a single theory.[35] An interesting theoretical explanation of the mechanism of frost salt attack has been put forward by Valenza and Scherer:[36,37] the researchers proposed a mechanism called glue–spall. According to this theory, the cracking of the ice/brine layer is the origin of FSS. The principle idea is that following the ice formation on top of the concrete surface, ice starts to shrink due to further cooling. The proposed mechanism also explains the greater damage of pessimum salt concentration under frost, which had remained unexplained until this theory. The shrinkage creates tensile stresses in the ice and causes three consequences, depending on the solute concentration of the liquid. These are:

(a) Weak salt concentration (i.e. 0.1%): due to ice formation and further cooling of the ice, the exerted tensile stress cannot exceed the tensile strength of the ice; hence no cracking occurs (in the ice or concrete).

(b) Pessimum salt concentration (1–3%): due to ice formation and further cooling of the ice, the exerted tensile stress exceeds the tensile strength of the ice and breaks the ice, which triggers surface scaling.

(c) Strong salt concentration (10–20%): in this case, the ice layer is too soft to exert enough stress on the underlying cementitious material, hence only the ice cracks and no scaling occurs.

Figure 6.11 shows a cement paste specimen with external loading exerted by a shrinking ice layer on top of the specimen. A pure tensile loading will cause a through-crack, whereas the shrinking ice layer on top of the specimen is theorised to cause scaling. Whether scaling really occurs and how severe the damage due to a glue–spall mechanism could be depends on the properties of the ice, the thickness of the ice layer and the bond between the ice and the substrate.

6.6.2 Numerical modelling of the glue–spall mechanism for frost salt scaling

For numerical modelling of the glue–spall phenomenon, a lattice model, as discussed in Section 6.5.2, can be used. A typical result of a

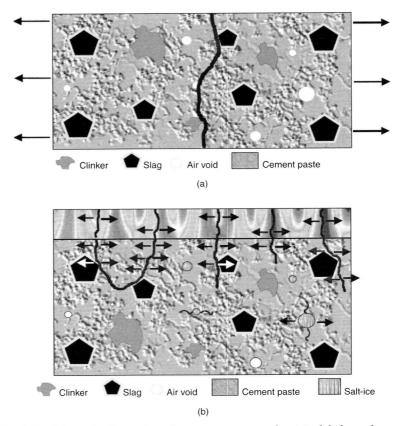

Fig. 6.11 Schematic illustration of a cement paste under (a) global tensile stress and (b) of the cracks induced by the cracking of an external ice layer[38]

simulation is shown in Fig. 6.12, where four subsequent stages of scaling of a mortar are presented. An arbitrary layer of 3 mm of ice on top of 12 mm of mortar is modelled. In the horizontal direction, periodic boundary conditions are assumed. At the bottom and at the top, the nodes are free. The materials properties used in the simulation are presented in Table 6.3.

In the analysis it is assumed that the temperature drops to −20°C and that the difference in thermal dilation between the concrete and the ice is 4×10^{-5}. No relaxation of the stresses is used in the analysis. In the simulations, the first crack in the ice has already started at 10% of the applied strain, and the ice cracks through at 13.5% of the applied strain. This correlates with a temperature change of only −2.7°C. This is somewhat less than the result of the linear elastic calculation by Valenza and

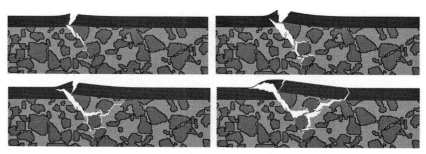

Fig. 6.12 Simulation of a frost salt scaling action by the Delft lattice model[38]

Scherer,[37] where a temperature decrease to $-4°C$ leads to cracking. However, if the ice layer is placed on a heterogeneous microstructure, the restraint of the ice is not uniform. Due to variations in stiffness of the supporting microstructure, stress concentrations in the uniform ice will appear, which results in earlier cracking than the linear elastic calculation gives.

In practice, a lot of the linear elastic stress will disappear due to relaxation of the ice and also of the cement based material under it. This is also described by Valenza and Scherer.[36,37] Due to this relaxation, the generated stresses for the complete temperature drop of $-20°C$ will not be enough to crack pure ice. It will be enough, however, to crack brine ice, which is weaker than pure ice. On cracking of the ice, the crack continues in the mortar, and kinks at a certain distance inside the material (Fig. 6.12). Afterwards, it continues until it meets the ice again, and a second crack in the ice is formed. Further continuation of the simulation would create another crack in the ice at another location, and the mechanism would repeat again. Of course, the cracking process is influenced by the choice of local materials properties and relaxation of stresses.

The simulations with the lattice model revealed that the effect of the thickness of the ice layer and the pessimum effect could be simulated

Table 6.3 Materials properties used in lattice simulations of concrete with ice layer

Material	E modulus: GPa	Tensile strength: MPa
Aggregate	70	10
Matrix	25	4
Interface	10	1
Ice	10	2.5

quite well. It was noticed, however, that the glue–spall mechanism is not the only mechanism to cause scaling. Other mechanisms, for example the ice lens theory as proposed by Powers[31] and modifications of that theory, may play a role as well. Moreover, in practice, thermal eigenstresses, particularly in the surface layer of a concrete element, may also influence the cracking process.

6.7 Conclusion and prospects

Deterioration of concrete is often a very slow process in which chemical, physical and mechanical mechanisms proceed simultaneously and may interact as well. This makes modelling a very complex task. Since deterioration processes, by definition, start small but may in the end have an overall effect on the structure, the need for a multiscale modelling approach is obvious. In the last two decades, much progress has been made in the field of modelling of hydration, microstructure development and prediction of materials properties. These models enable us to define the starting point for modelling deterioration processes. The increasing computation power of modern computers provides us with the required tools for modelling complex interactions in order to generate reliable predictions of deterioration processes and service life predictions.

References

1. Siemes, T., Vrouwenvelder, T. and van den Beukel, A. (1985) Durability of buildings: a reliability analysis. *HERON*, 30(3), 2–48.
2. Duracrete (2000) *Probabilistic performance based durability. Design of concrete structures.* Duracrete Final Technical Report R17, Document BE95-1347/R17, The European Union – Brite EuRam III, DuraCrete. CUR, Gouda.
3. Siemes, T., Schiessl, P. and Rostam, S. (2000) Future developments of service life design of concrete structures on the basis of DuraCrete. In: D. Naus (ed.), *Service life prediction and ageing management of concrete structures.* RILEM, Bagneux, 167–176.
4. Bentz, D.P. and Garboczi, E.J. (1993) Digital-image based computer modelling of cement based materials. In: J.D. Frost and J.R. Wright (eds), *Digital Image processing: techniques and application in civil engineering.* ASCE, New York, 63–74.
5. Navi, P. and Pignat, C. (1999) Three-dimensional characterisation of the pore structure of a simulated cement paste. *Cement and Concrete Research*, 29(4), 61–80.

6. van Breugel, K. (1991) *Simulation of hydration and formation of structure in hardening cement-based materials*. PhD thesis.

7. van Breugel, K. (1984) Models for prediction of microstructural development in cement based materials. In: H. Jennings, J. Krop and K. Scrivener (eds), *The modelling of microstructure and its potential for studying transport properties and durability*. NATO ASI Series E. Kluwer, Dordrecht, vol. 304, 91–106.

8. Ishida, T., Mabrouk, R.T.S. and Maekawa, K. (2001) An integrated computational framework for performance evaluation of cementitious materials and structures under various environmental actions. In: F.-J. Ulm, Z. Bazant and F. Wittmann (eds), *Proceedings of the 6th International Conference On Creep, Shrinkage and Durability Mechanics of Concrete and other Quasi-brittle Materials*. Elsevier, Oxford, 511–516.

9. Ye, G. (2003) *The microstructure and permeability of cementitious materials*. PhD thesis, Delft University.

10. Sun, Z.H, Ye, G. and Shah, S.P. (2005) Microstructure and early age properties of Portland cement pastes – effects of the connectivity of the solid phases. *ACI Materials Journal*, 102, 122–129.

11. Carol, I. and Bazant, Z.P. (1993) Solidification theory: a rational and effective framework for constitutive modelling of ageing viscoelasticity. *Proceedings of the 5th International Symposium on Creep and Shrinkage of Concrete*. RILEM Proceedings, 22, 177–189.

12. Lokhorst, S.J. and van Breugel, K. (1993) The effect of microstructural development on creep and relaxation of hardening concrete. *Proceedings of the 5th International Symposium on Creep and Shrinkage of Concrete*, RILEM Proceedings, 22, 145–150.

13. Nilsson, M. (2001) *Rotational boundary restraint factor*. European Research Project IPACS, Report BE96-3843/2001:69-9IPACS.

14. Bijen, J. (2003) *Durability of engineering structures – design, repair and maintenance*. CRC Press, Woodhead Publishing, Cambridge.

15. Audenaert, K., Boel, V. and De Schutter, G. Carbonation of filler type self-compacting concrete. https://archive.ugent.be/retrieve/6473/Carbonation+of+filler+type.pdf

16. Papadakis, V. (2000) Effect of supplementary cementing materials on concrete resistance against carbonation and chloride ingress. *Cement and Concrete Research*, 30, 291–299.

17. Powers, T. and Brownyard, T. (1946–1947) Studies of the physical properties of hardened cement paste. *Journal of the American Concrete Institute*, 43.

18. Thiery, M., Dangla, P., Villain, G. and Platret, G. (2005) A prediction model for concrete carbonation based on coupled CO_2–H_2O-ions transfers and chemical reactions. *10th DBMC International Conference on Durability of Building Materials and Components*. Lyon, Paper TTI-58.

19. Maage, M., Helland, S., Poulsen, E., Vennesland, O. and Carlsen, J.E. (1996) Service life prediction of existing concrete structures exposed to marine environment. *ACI Materials Journal*, 93(6), 602–608.

20. Rooij, M.R. de and Polder, R.B. (2005) *Durability of marine concrete structures*. CUR report 215. CUR, Gouda (in Dutch).

21. Gaal, G.C.M., Polder, R.B., Walraven, J.C. and van der Veen, C. (2003) Critical chloride content – state-of-the-art. In: M.C. Forde (ed.), *Proceedings of the 10th International Conference on Structural Faults and Repair*. Engineering Technics Press, Edinburgh.

22. Polder, R.B. and de Rooij, M.R. (2005) Durability of marine concrete structures – field investigations and modelling. *HERON*, 50(3), 133–154.

23. Kim, Y.-J. (2008) Production and construction technology of high-durable marine concrete in Busan-Geoje fixed link project. In: *Proceedings of the International Symposium on Advancement of Cement and Concrete Industries*. Seoul.

24. van Breugel, K., Polder, R.B. and van der Wegen, G.J.L. (2009) Service life design of structural concrete – the approach in The Netherlands. In: *Proceedings of the International Conference on Faults and Repair*. Edinburgh.

25. Li, K., Chen, Z. and Lian, H. (2008) Concepts and requirements of durability design for concrete structures: an extensive review of CCES01. *Materials and Structures*, 41, 717–731.

26. Schlangen, E. and Garboczi, E.J. (1997) Fracture simulations of concrete using lattice models: computational aspects. *Engineering Fracture Mechanics*, 57(2/3), 319–332.

27. Schlangen, E. and Çopuroğlu, O. (2007) Concrete damage due to ASR: a new method to determine the properties of the expansive gel. In: *Proceedings of the 6th International Conference on Fracture Mechanics of Concrete and Concrete Structures*. Catania.

28. Çopuroğlu, O. and Schlangen, E. (2007) Modelling of effect of ASR on concrete microstructure. *Key Engineering Materials*, 348/349, 809–812.

29. Çopuroğlu, O. (2006) *Frost salt scaling of cement-based materials with a high slag content*. Delft University of Technology, Delft.

30. Setzer, M.J. (2001) Micro-ice-lens formation in porous solid. *Journal of Colloid and Interface Science*, 243(1), 193–201.

31. Powers, T.C. (1965) The mechanisms of frost action in concrete. *Stanton Walker Lecture Series on the Material Science*.

32. Litvan, G.G. (1973) Frost action in cement paste. *Materials and Structures*, 34, Jul.–Aug.

33. Pigeon, M. and Pleau, R. (1995) *Durability of concrete in cold climates. Modern Concrete Technology*. E&FN Spon, London.

34. Schulson, E.M. (1998) *Ice damage to concrete*. Special Report 98-6. CRREL Technical Publications, Hanover, NH.

35. Lindmark, S. (1999) *Mechanisms of salt frost scaling of Portland cement bound materials: studies and hypothesis*. Lund University, Lund.

36. Valenza, J.J. and Scherer, G.W. (2004) Mechanism for salt scaling of a cementitious surface. In: *RILEM Spring Meeting 2004*. Chicago.
37. Valenza, J.J. and Scherer, G.W. (2006) Mechanism for salt scaling. *Journal of the American Ceramic Society*, 89(4), 1161–1179.
38. Çopuroğlu, O. and Schlangen, E. (2008) Modeling of frost salt scaling. *Cement and Concrete Research*, 38(1), 27–39.

7

Materials selection for improved durability performance

Construction materials that can be used to improve the durability performance of reinforced concrete structures include:

- Chemical admixtures
- Mineral admixtures:
    - Pulverised fly ash
    - Ground granulated blast furnace slag
    - Condensed silica fume (microsilica)
    - Metakaolin
- Alternative reinforcing materials to conventional steel reinforcement (black bars):
    - Epoxy-coated steel reinforcement
    - Galvanised steel reinforcement
    - Stainless steel reinforcement
    - Glass fibre reinforced polymers

The first section deals with chemical admixtures; their composition, the beneficial properties they impart on fresh and/or hardened concrete, and their mode of action in concrete.

The second section deals with supplementary cementitious materials, otherwise known as mineral admixtures or pozzolans, which can be used to improve the durability performance of reinforced concrete structures. Their method of manufacture, chemical composition, and their mode of action in concrete (fresh and early age properties and hardened properties especially with respect to improving the resistance of concrete to degradation mechanisms) is discussed.

The last section deals with four alternative reinforcing materials that can be used in concrete structures in place of conventional steel reinforcement (black bars). Their method of manufacture, how they

Concrete durability
978-0-7277-3517-1

provide improved durability, their properties that may affect design procedures that have been developed for conventional reinforcement, and, the reputation that they have based on past applications and performance in real structures is discussed.

7.1   Chemical admixtures

Robert Viles, Fosroc International Limited, UK

7.1.1   Introduction

The overall sustainability of a concrete structure depends upon both design and material parameters.[1] Admixtures have a potentially profound effect on the durability and sustainability of concrete structures. It is fairly well understood that 'plasticisers' may reduce the water/cement ratio, which in turn is an important determinant of concrete quality. This section discusses rheology/surfactant effects and also the resultant features that control the ingress of destructive elements, tolerance to the environment and the resistance of reinforcing steel to corrosion, whilst highlighting practical value.

7.1.2   Water-reducing admixtures

7.1.2.1   *Effect of water reduction by plasticiser types*

Water-reducing admixtures are commonly known as 'plasticisers', usually polymer compositions used to act as dispersants and workability aids for cementitious materials. The earliest forms to emerge were the sodium and calcium salts of lignosulfonate, which have been available since the early 1950s (Fig. 7.1).

Lignosulfonates are polymer by-products, occurring as a result of the sulfite bleaching of paper during its manufacture, and are examples of water-soluble anionic polyelectrolytes having surface active properties. These are regarded as a cost-effective base to improve and extend the workability of concrete, and may enable a similar slump to be achieved from the use of less mixing water. This, in turn, results in a reduction of porosity and permeability and

- either an increase in strength and reduction in permeability or
- reduction in cement content or
- both of the above.

Not all sources of lignosulfonate plasticiser perform the same, largely due to processing and refinement. The natural sugar or saccharide content determines, in particular, the early set and strength performance by retarding hydration. The sugar level can be controlled by processing.

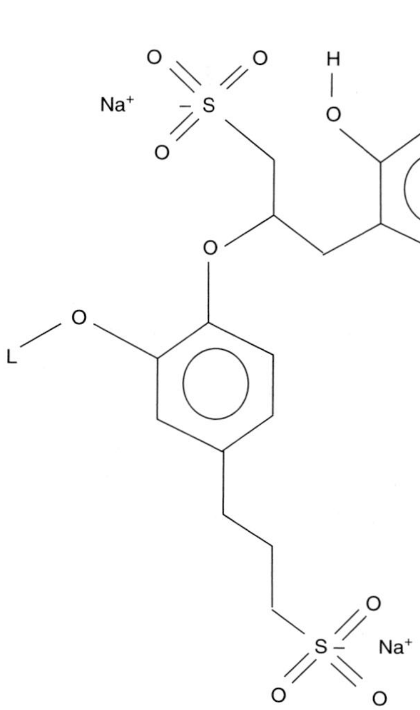

Fig. 7.1 Part of a lignosulfonate molecule

With formulation, other components can be included to improve performance according to needs, increasing or decreasing the set time.[2,3] The water reduction achieved by using lignosulfonates is generally about 10% relative to that of a control mix.

Improved synthetic water-reducing polyelectrolyte polymer admixtures were developed and introduced as cement dispersants from the 1960s and 1970s.

Two similar types are sulfonated naphthalene formaldehyde condensates (SNFs) and sulfonated melamine formaldehyde condensates (SMFs) (Figs 7.2 and 7.3). Both of these are referred to as superplasticisers, since they have a high water-reducing capability. Like

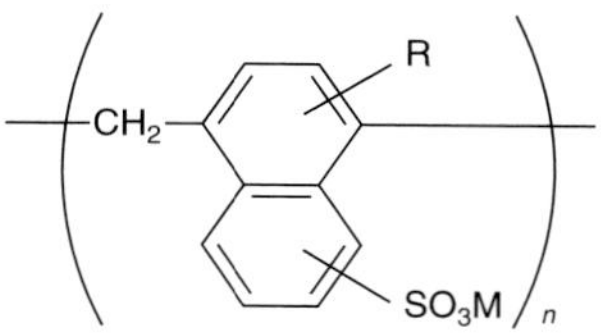

Fig. 7.2 Sulfonated naphthalene formaldehyde (SNF)

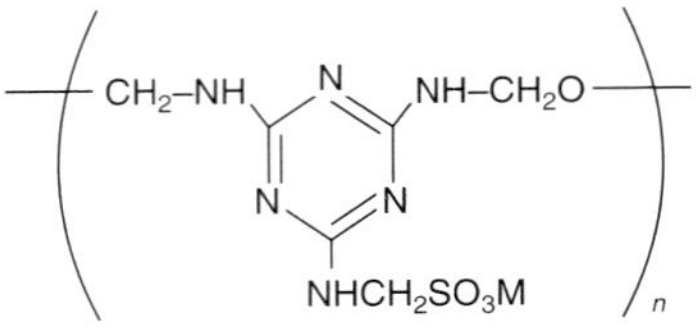

$$CH_2-NH \quad N \quad NH-CH_2O$$
$$N \quad N$$
$$NHCH_2SO_3M$$

Fig. 7.3 Sulfonated melamine formaldehyde (SMF)

most polymer admixtures genres, the chemistry was known well before the uptake of their use in concrete.[4]

The organic part of such solutions is negatively charged, the overall balance being kept by the cation in solution (usually sodium or calcium.) Dispersion is considered to be principally electrostatic repulsion, either caused by the anion adsorbed on the cement particle, resulting in the said particle to be negatively charged, or by adsorption of cations onto the cement, then associated with the anions. In both cases, the like charges repel, thus keeping the particles apart (Fig. 7.4).

Cement, of course, is not a simple model material, and dispersion effects are described in more detail elsewhere.[2]

Performance can be related to molecular weight, active matter contents, dosage rates and other constituents that may be added or present from the process. Crude SNFs in particular can contain significant sodium sulfate levels that not only reduce active matter but can also influence dispersion, in that it can affect adsorption of the plasticiser.[5]

SMFs and SNFs have similar potential water reduction capacity (10–25%, depending on the slump and cement content required) but vary in the intrinsic air content entrained in the concrete.

The next generation of design molecules that emerged are known as polycarboxylates (PCs) or the more commonly used derivatives, polycarboxylate ethers (PCEs) (Fig. 7.5).

Basic polycarboxylates were already known to be powerful dispersants, used in mineral pigment and ink dispersions, before their use in

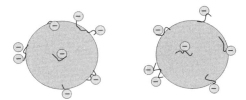

Fig. 7.4 Electrostatic dispersion of cement particles

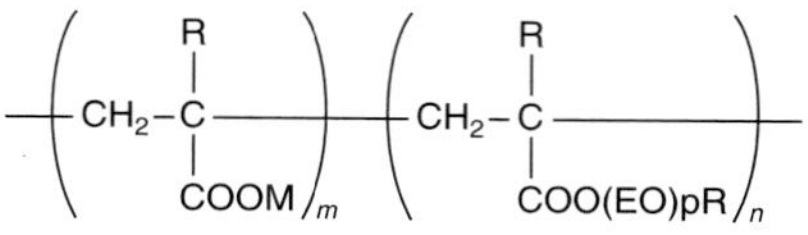

$$\left(\begin{array}{c} R \\ | \\ -CH_2-C- \\ | \\ COOM \end{array}\right)_m \left(\begin{array}{c} R \\ | \\ -CH_2-C- \\ | \\ COO(EO)pR \end{array}\right)_n$$

Fig. 7.5 Basic polycarboxylate molecule (EO, ethylene oxide)

cement was recognised. Some were even used with dispersants such as SMF.[6] After their very high dispersion power was realised, the chemistry was designed to work more effectively with cements and various concrete applications. Two effects on cement may be manipulated by design:

(*a*) workability
(*b*) workability retention.

Broadly, an increase in carboxylate functionality on the backbone of the polymer will increase adsorption onto the cement particle and increase workability retention. The attached polyether side-chain has a marked effect on dispersion power. In simple terms, the side-chains get in each other's way physically, and they tend to avoid entanglement (called steric hindrance). That, then, keeps the attached particles of cement apart, as shown schematically in Fig. 7.6. More powerful dispersants tend to have longer or even branched side-chains (e.g. $(CH_2O)_n$, where typically $n = 40-100$), but often either mixtures and/or hybrid molecules are used to balance workability with workability retention. Some molecules can also lead to undesirable rheological sensitivity or adverse filling/finishing effects such as stickiness. The detailed molecular design and formulation can therefore produce a wide range of properties. Soluble sulfate is an important determinant for how different

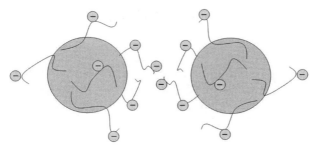

Fig. 7.6 Separation of cement particles is facilitated by steric hindrance

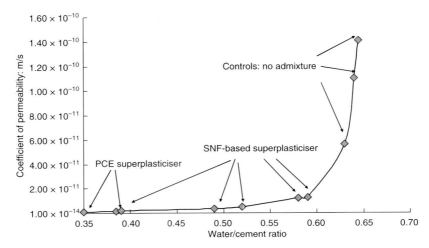

Fig. 7.7 The effect of plasticisers on water permeability

cements behave with such admixtures,[7] although the soluble sulfate often reflects cement chemistry, and admixtures can be formulated accordingly.

The benefit of strength improvement is seen in many instances,[8] and strength is considered to be the most important parameter that determines durability. However, it is also true to say that higher strength is broadly correlated with lower permeability. The reduction in permeability reduces the rate of ingress of aggressive elements such as chloride and carbon dioxide.

The full range of plasticisers and their effect on reducing permeability can be seen in Fig. 7.7.

Note that models of ingress that just look at simple diffusion can be misleading if concrete is cracked or exposed to non-steady state driving forces.

7.1.2.2 Effect on cement content

The reduction of cement has more than an effect on cost. As a first-stage contribution to sustainability, it has been argued that full use of admixtures could result in globally reducing consumption of cement by up to 20%, thus reducing carbon dioxide emissions by about 400 million tonnes per annum. The exact figure is arguable, but the order of magnitude is probably correct.[9] It is also worth noting that most aggregates have lower permeability than hardened cement paste.

In some cases, therefore, the simple specification of a minimum cement content can be contrary to durability and sustainability. Moreover, a high cement content will lead to high exotherm and greater chance of early thermal contraction cracking.

The benefits of cement reduction do have limits since the binder needs[10] to fill the space between aggregates. Good grading of aggregates to minimise cement content and maximise workability is possibly the most important aspect of concrete technology. There are also counter effects, such as a higher permeability of the interfacial zone between the cement paste and aggregate.[11]

7.1.2.3 Effect of air content on durability

As previously mentioned, SMFs tend to include less air than SNFs, and PCEs can result in even higher air contents. Most concrete technologists aim at a content of about 2% air to produce good workability and practical rheology in concrete. The use of antifoaming or air-detraining agents can control air to some extent. These are often included in formulated products, but care is needed on choice and amount since effective agents are often near to compatibility limits with the liquid admixture.

Air-entraining agents are sometimes added to confer beneficial freeze–thaw resistance (see the section on air entrainment-agents, below). Depending on its form, however, air can be deleterious to durability. It will reduce strength by about 5% for each 1% of air by volume. It will increase permeability. What is not often appreciated is that air can affect parameters such as the chloride threshold for the corrosion initiation of steel reinforcement (rebar). At a 2% air level, the generally recognised 'threshold' for corrosion is typically about 0.3% chloride by weight of cement. As air is eliminated and the solid alkali level at the rebar is increased (one is a partial consequence of the other) at the reinforcement, then that threshold may be dramatically increased (Fig. 7.8).[12]

This explains why voids, and particularly large voids, are usually the initiation sites of corrosion, leading to physical damage and the low durability of reinforced concrete. The move to self-compacting concrete and elimination of voids goes part of the way to increasing the chloride threshold. The second issue is the reserve of solid alkali at the rebar surface. It is worth noting that excessive use of pozzolans can actually reduce the level of corrosion protection.[13] Other work, however, indicates that in a very high alkalinity (high ordinary Portland cement)

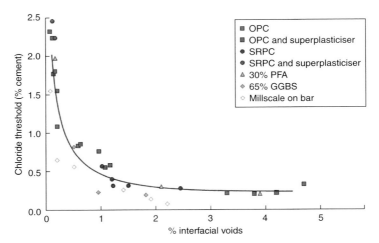

Fig. 7.8 Chloride threshold and voids (GGBS, ground granulated blast furnace slag; OPC, ordinary Portland cement; PFA, pulverised fuel ash; SRPC, sulfate-resisting Portland cement). Data courtesy of G. Glass, Imperial College of Science and Technology

environment, the use of certain pozzolans seems to present no problem with regard to corrosion.[14] Therefore, whilst the beneficial effects of pozzolans are increased use of waste, reduction of cement content, reduction of the alkali–aggregate reaction with reactive aggregates and, at modest levels, reduced permeability, there may be adverse effects, possibly due to loss of alkalinity.

7.1.3   Waterproofing admixtures

A porous structure such as concrete cannot properly be described as 'waterproof', but there are classes of admixtures that can contribute to lower permeability, lower water absorption and slower ingress of aggressive elements other than by reducing the water/cement ratio.

Sometimes these are also called 'corrosion inhibitors', but truly they may have no real effect upon chloride threshold for corrosion, rather delaying ingress (as can be determined by ASTM G109).

There are different proprietary approaches.[15] One uses aliphatic carboxylic acids as alkanolamine salts to provide a partial pore lining (and blocking) with increased hydrophobicity (Fig. 7.9). Such formulations can also be combined with a powerful superplasticiser for full effect.

Another approach is to effect crystalisation within the pore structure, the solid crystal material acting as a pore blocker.[16]

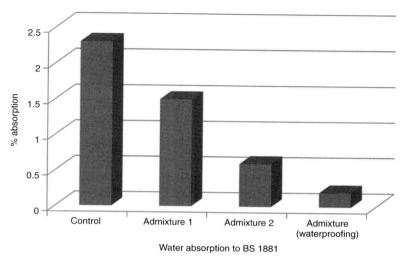

Water absorption to BS 1881

Fig. 7.9 The effectiveness of a typical waterproofing admixture

7.1.4 Shrinkage reducing admixtures

The phenomenon of shrinkage occurs as a result of several features of concrete. Firstly, water can be lost from the system. If that occurs in the early (plastic) stage, it can lead to what is known as 'plastic shrinkage' often displayed with attendant characteristic random cracking (Fig. 7.10). Note that 'map cracking could be due to other

Fig. 7.10 Plastic cracking. (Courtesy of NRMCA)

155

problems such as alkali–aggregate attack. For cracking to occur, it is simply necessary for the tensile stress to exceed the tensile strength.

Secondly, the volume of the hydrated cement is less than that of the unhydrated product and added water, although there can be early expansion reactions. Hydration shrinkage is sometimes called 'autogenous shrinkage', although that also includes the effect of capillary surface tension, which increases with the development of the capillary pore structure. Studies continue to investigate the mechanism in detail.[17,18] Admixtures to address shrinkage include polyhydroxy compounds that reportedly reduce the driving surface tension force.[19,20] In addition to reduction in surface tension, there are probably slight changes in hydrate morphology and a lower tendency to lose water. Such admixtures can be deployed to offset high shrinkage in high-performance concrete[21] or to simply reduce joints.

Best control is effected also by using expansive agents, which include graded calcium oxide and proprietary compounds comprising calcium sulfoaluminate or calcium aluminate.

A reduction in cement content and the water/cement ratio by using a good superplasticiser also contributes to lower shrinkage.

Rarely is concrete fully shrinkage compensated due to the cost of addition at effective levels. That results in most concrete having at least some cracks. Cracking can be exacerbated by a high exotherm, causing additional stress on cooling. Modern cements are designed to give high early and maximum strength up to 28 days. As such, that also leads to deleterious cracking, which, contrary to popular belief, does not necessarily become healed due to the subsequent later hydration of unhydrated cement. That may have been the case when cements had a higher (slower hydrating) dicalcium silicate content and were coarser ground.

The use of pozzolans is not always a solution for cracking. The use of high addition rates of silica fume will increase shrinkage.[22]

Cracks of 300 µm width are very porous to chloride ions – not surprising given that the hydrated radius of a chloride ion is about 3.2 Å or 3.2×10^{-4} µm. Research indicates that control of cracks to less than about 30 µm is required to prevent ready chloride passage.[23] Such cracks are difficult to see.

7.1.5   Corrosion inhibitors

Corrosion inhibitors may raise the concentration threshold for chloride attack or reduce corrosion rates. The best corrosion inhibitor is alkalinity,

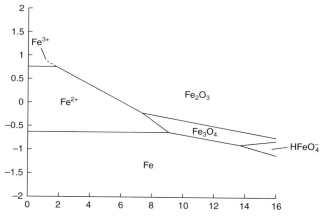

Fig. 7.11 A potential–pH (Pourbaix) diagram of iron in water

as described earlier and illustrated to some extent by the Pourbaix diagram (Fig. 7.11). After initiation, the pH in a corrosion pit can drop considerably compared with the bulk composition.

One long-proven cast-in-concrete inhibitor is calcium nitrite.[24]

Nitrite will inhibit corrosion due to chloride ingress until the molar concentration of nitrite is exceeded by nitrite. Practically, that means that about 30 litres/m^3 of concrete of a 35% solution of calcium nitrite is required to give good protection. Lower doses can be employed, but it has been found that, when corrosion initiation occurs, ensuing corrosion rates can be very high. That can partially be described as a large cathode driving a small anode, but there is also often a slight reduction in resistivity too. Despite claims to the contrary, the author has seen no evidence that calcium nitrate is a corrosion inhibitor. Lithium nitrite has also been proposed, but this is a potential hazardous material, being explosive when dry. Nitrites are toxic, requiring careful handling, but the evidence points to retention long term in concrete rather than being readily leached.

Alkanolamines are another group of corrosion inhibitors, and are often claimed to be effective at low addition rates – a few litres per cubic metre of concrete.[25] There is good evidence that corrosion rates can be reduced, but their effectiveness is often greater at higher addition levels than recommended, whilst too much attention is focused on in-vitro tests. There is poor evidence that alkanolamines can migrate through concrete to effect any significant reduction in corrosion rate when topically applied to a formed-concrete surface. Penetration levels can be detected, and are low (several ppm through 25 mm of concrete). Some inhibitors may at best be pore blockers.

The use of corrosion inhibitors for patch repair is not always successful, and may accelerate corrosion in untreated areas.[26]

7.1.6 Polymers and polymer dispersions

Whilst many admixtures are polymers, there are other polymer admixtures.

7.1.6.1 Thickeners/stabilisers

For underwater placement of concrete, high-molecular-weight polymers are used. A very good solution comprises small additions of polyethylene oxide as an anti-washout aid. Some cellulose ethers may also be suitable at slightly higher addition rates. The latter class also help water retention, thicken and slow down segregation. Polymers also act as pumping aids. Polyacrylamides have a very high activity, but care in choice is essential since they also tend to flocculate according to chemistry and molecular weight. They are usually added as powders.

7.1.6.2 Polymers and polymer dispersions

Dispersions, such as acrylic, styrene–butadiene rubber, vinyl acetate/ ethylene and other co- and ter-polymers have beneficial effects, reducing permeability, reducing brittleness, improving freeze–thaw resistance, improving tensile and flexural strength, and improving wear resistance. Cost limits their addition to most concrete mixes except for high-performance screeds and flooring. A reduction in modulus and an increase in creep also has implications for structural concrete. Use in mortars, tile adhesives/grouts and surface treatments is more common, where application factors (adhesion and cohesion) and robustness are important. Additions may be dispersible powders or liquid dispersions. Whilst it is argued that some polymers are susceptible to chemical attack, hydrolysis or environmental exposure breakdown, work has shown them to be quite durable.[27] Some 'polymer concretes' are not cementitious but are based upon epoxy binders and the like, not admixtures for concrete.

7.1.7   Air entrainment agents

The inclusion of air has benefits of adding to workability, but with some potential drawbacks as described earlier. Many plasticisers tend to increase air due to surface tension reduction. The inclusion of specifically designed air entrainment agents is designed to:

(*a*) increase workability, for example of mortar or 'semi-dry precast' mixes
(*b*) entrain air to provide freeze–thaw resistance
(*c*) produce foams for void fill or lightweight products.

An important property is to stabilise the air bubbles entrained. Many surfactants are capable of entraining air, but fewer give stability up until set. Synthetic surfactants are used, but some of the most stable formulations contain natural products such as vinsol resin (fine air for freeze–thaw resistance) and even keratin for foams.

Lightweight products provide enhanced insulation for building in addition to minimal use of primary raw materials and high use of wastes.

Void fill materials are used to provide stabilisation of ground with minimal strength materials. For example, in trench fill, high-strength is undesirable; in mines and tunnels, vast voids have been economically filled.

The improvement of freeze–thaw resistance is a result of providing space into which pore water may expand, thus obviating disruptive expansion (Fig. 7.12).

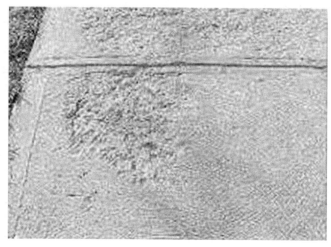

Fig. 7.12 Freeze spalling

The mechanism and other details have been described in several papers.[28]

7.1.8 Retarders and accelerators

Retardation can result in improved compaction and denser pore structure, which both improve durability. For bulk concrete, it would be wrong to assume that the heat output is reduced by the use of a retarder. It may simply delay heat output and may not even reduce the peak temperature. Excessive retardation can lead to segregation and even surface water bleed, leading to poor-quality surfaces.

Accelerators are often used to improve the build of shotcrete (Fig. 7.13). Early strength is improved, which may be critically important for, for example, ground stabilisation, but often later hydration, dimensional stability and strength may be reduced, particularly for some alkaline accelerators such as sodium aluminate.[29] Sodium aluminate is highly alkaline, and gives rise to skin burns. For that reason, accelerators based on non-alkaline materials such as aluminium sulfate are used.

Fig. 7.13 Shotcrete (Fosroc International Ltd)

Accelerators are often beneficial for cold weather working and hastening shuttering removal, but care should be taken on both the addition rate and type of admixture. These may comprise various salts, including nitrates, nitrites, thiocyanantes or formates. Alkanolamines can also accelerate set and increase early/late strength, depending upon the concentration deployed and the structure. Chloride is a very effective hydration accelerator, but chloride-containing products should never be used in reinforced concrete. The author is old enough to remember when there was a debate as to whether chloride was harmful or not. For unreinforced concrete, it is not a problem, and other effects can determine how much is needed to initiate corrosion. The lesson that the author learned about concrete is that what is 'good' under a selected set of parameters can be 'bad' when one or more parameters change – and there are many, many variables.

As a final comment, future admixtures may use greater biomass raw materials to contribute to their own sustainability and may have to work with different, low-carbon-footprint, cement chemistry.

References

1. Sakkai, K. (ed.) (1996) *Integrated Design and Environmental Issues in Concrete Technology.* Taylor and Francis, London.
2. Ramachandran, V.S. (1995) *Concrete Admixtures Handbook. Properties, Science and Technology (Building Materials Science).* Noyes Publications, New Ridge, NJ, 2nd edn.
3. Taylor, H.F.W. (1997) *Cement Chemistry.* Thomas Telford, London.
4. Yost, R.S. and Auten, R.W. (1956) *Resinous compositions and methods of making the same.* US patent 2407599.
5. Kim, B.-G., Jiang, S., Jolicoeur, C. and Atcin, P.-C. (2000) The adsorption behavior of PNS superplasticiser and its relation to fluidity of cement paste. *Cement and Concrete Research,* 30(6), 887–893.
6. Akao, K. *et al.* (1988) *Dispersant for Calcium carbonate.* Patent JP 1457558 (C). Nissan Kagaku Kogyo KK.
7. Margarotto, R. *et al.* (2003) Effect of alkaline sulphates on the performance of superplasticisers. In: *11th International Congress on the Chemistry of Cement.* Durban, 569–580.
8. Sobelov, K. and Soboleva, S. (2003) The effect of complex admixtures on strength of ultra high strength cement. In: *11th International Congress on the Chemistry of Cement.* Durban, 465–474.
9. Aitken, P.C. (2005) Admixtures and sustainability admixtures – enhancing concrete performance. In: *Proceedings of 6th International Conference.* Dundee, 122–128.

10. Concrete Society (1982) Technical Report No. 22. Concrete Society, 18–19.
11. Yang, C.C. and Cho, S.W. (2005) Approximate migration coefficient of percolated interfacial transition zone by using accelerated chloride diffusion test. *Cement and Concrete Research*, 35(2), 344–350.
12. Buenfeld *et al.* (2001) *Process for the protection of reinforcement in reinforced concrete*. International patent WO 2001/055056.
13. Manera *et al.* (2006) Effect of silica fume on the initiation of chloride induced corrosion in reinforced concrete. In: *8th Canmet/ACI International Conference on Superplasticisers and Other Chemical Admixtures in Concrete*. Sorrento, 187–200.
14. da Silva, G. and Liborio, J.B.L. (2006) A study of steel bar reinforcement corrosion in concretes with SF and SRH using electrochemical impedance spectroscopy. *Materials Research*, 9(2), 209–215.
15. Aldkiewiez, A.J. *et al.* (2006) Use of concrete admixtures to produce waterproof concrete. *8th Canmet/ACI International Conference on Superplasticisers and Other Chemical Admixtures in Concrete*. Sorrento, 53–61.
16. Iob, A., Saricimen, H., Narasimhan, S. and Abbas, N.M. (1993) Spectroscopic and microscopic studies of a commercial concrete waterproofing. Material. *Cement and Concrete Research*, 23(5), 1085–1094.
17. Barcelo, L., Moranville, M. and Clavaud, B. (2005) Autogenous shrinkage of concrete: a balance between autogenous swelling and self-desiccation. *Cement and Concrete Research*, 35(1), 177–183.
18. Mora-Ruacho, J., Gettu, R. and Aguado, A. (2009) Influence of shrinkage-reducing admixtures on the reduction of plastic shrinkage cracking in concrete. *Cement and Concrete Research*, 39(3), 141–146.
19. Rajabipour, F. *et al.* (2008) Interactions between shrinkage reducing admixtures (SRA) and cement paste's pore solution. *Cement and Concrete Research*, 38(5), 606–615.
20. Bettencourt Ribiero, A. *et al.* (2006) Behavior of mortars with different dosages of shrinkage reducing agents. In: *8th Canmet/ACI International Conference on Superplasticisers and Other Chemical Admixtures in Concrete*. Sorrento. SP-239-6, 77–92.
21. Sugamata, T. *et al.* (2006) Characteristics of concrete containing a shrinkage-reducing superplasticiser for ultra-high stength concrete. In: *8th Canmet/ACI International Conference on Superplasticisers and Other Chemical Admixtures in Concrete*. Sorrento, SP-239-4, 51–66.
22. Benz, D.P. (2008) A review of early-age properties of cement-based materials. *Cement and Concrete Research*, 38(2), 196–204.
23. Ismail, M., Toumi, A., François, R. and Gagné, R. (2008) Effect of crack opening on the local diffusion of chloride in cracked mortar samples. *Cement and Concrete Research*, 38(8–9), 1106–1111.
24. Berke, N.S. and Hicks, M.C. (2004) Predicting long-term durability of steel reinforced concrete with calcium nitrite corrosion inhibitor. *Cement and Concrete Composites*, 26(3), 191–198.

25. Munteanu, V.F. and Kinney, F.D. (2000) Corrosion inhibition properties of a complex inhibitor – mechanism of inhibition. In: *CANMET 2000*, 255–269.

26. Sprinkel, P.E. (2003) *Evaluation of corrosion inhibitors for concrete bridge deck patches and overlays*, VTRC 03-R14. Virginia Transportation Research Council.

27. Schulze, J. and Killermann, O. (2001) Long term performance of redispersible powders in mortars. *Cement and Concrete Research*, 3, 357–362.

28. Chatterji, S. (2003) Freezing of air-entrained cement-based materials and specific actions of air-entraining agents. *Cement and Concrete Composites*, 25(7), 759–765.

29. Paillère, A.M. (ed.) (1995) *Application of admixtures in concrete*, TC 84-AAC. Taylor and Francis, London.

7.2   Supplementary cementitious materials (SCMs)

Other names for these are:

- mineral admixtures
- pozzolans
- cement replacement materials.

A *pozzolan* is defined as: 'a siliceous or siliceous and aluminous material which in itself possesses little or no cementitious value but which will, in a finely divided form and in the presence of moisture, chemically react with calcium hydroxide at ordinary temperature to form compounds possessing cementitious properties'.[1]

The main ones used in concrete are:

- pulverised fly ash
- ground granulated blast furnace slag
- condensed silica fume (microsilica)
- metakaolin.

Ground granulated blast furnace slag is, strictly speaking, not a pozzolan but a hydraulic latent binder. The calcium oxide content is high, and if it is mixed with water it will set and harden. As such, an alternative term to 'pozzolan' has been sought to group all the above together. The one that appears to be the most preferred is 'supplementary cementitious materials' or SCMs. Metakaolin is the only one from this group of materials that is not a by-product of another industry. Their method of manufacture, chemical composition and their mode of action in concrete (fresh and early age properties and hardened properties, especially with respect to improving the resistance of concrete to degradation mechanisms) is discussed below.

Reference

1. ACI (2008) *201.2R-08: guide to durable concrete*. ACI Committee 201 on Durability of Concrete, Farmington Hills, MI.

7.2.1   Fly ash* for enhanced concrete durability

Lindon Sear, UK *Quality Ash Association, UK*

7.2.1.1 How pozzolanas work

Fly ash from the burning of pulverised coal in power stations has been used as a cementitious material for more than 50 years. Fly ashes can either by siliceous or calcareous, with the accepted difference being the proportion of reactive calcium oxide of $\leq 10\%$ in siliceous ash and $>10\%$ in calcareous fly ash. Siliceous fly ash is the predominant material used in concrete and cement production. Calcareous fly ash is usually self-cementing, as the high levels of calcium oxide are able to react pozzolanically, coupled with the high levels of sulfate usually contained in them forming ettringite on contact with water.

Fly ash is a beneficial addition to concrete because of the pozzolanic reactions that take place that are in addition to the normal hydration of the Portland cement component. The pozzolanic reaction was the basis for Roman concretes, produced prior to the invention of Portland cement in 1842, which relied on a pozzolanic material such as volcanic ash and crushed bricks or pottery being mixed with hydrated lime.

7.2.1.2 The reaction described

All pozzolanas consist of finely divided, silica-rich glassy particles. This silica reacts with alkalis in the presence of water, primarily with calcium hydroxide ($Ca(OH)_2$), to form additional hydration products to those resulting from the hydration of Portland cement with water (see equation (7.1)). It is these hydration products that give fly-ash-based concretes their enhanced low permeability and improved durability. This reaction is not exothermic, and rather temperature dependent. The finer the particle size of the ash, the greater the surface area exposed to alkali and the greater the reaction rate. Therefore, materials such as silica fume ($15\,000\ m^2/kg$) will react relatively quickly, fly ash ($3000\ m^2/kg$) will continue to react for many, years and with materials like furnace bottom ash and pumice, no reaction will be detected.

---

* We now use the term 'fly ash' rather than 'pulverised fuel ash' because the European Standards by which fly ash is supplied use this term in their titles. Also 'fly ash' is used throughout the rest of the world, and is more likely to be understood by a wider audience.

calcium hydroxide + silica = tri-calcium silicate + water

$$3Ca(OH)_2 + SiO_2 = 3CaO \cdot SiO_2 + 3H_2O \qquad (7.1)$$

7.2.1.3 Effects of curing conditions on fly ash

The pozzolanic reaction of lime with fly ash is generally a slow reaction, continuing over many years at normal ambient temperature. The reaction depends on there being alkalis (usually calcium hydroxide, though fly ash will react with sodium and potassium hydroxides) and water available. At 28 days in normal curing temperatures, a minimal pozzolanic contribution to strength will have taken place with materials such as fly ash, the benefits only being seen at later ages. This is normally seen as an increase in strength with time, resulting in higher strengths than would be achieved from a Portland cement mix of equal 28 day strength (Fig. 7.14). However, if the curing temperature is increased, the pozzolanic reaction rate will also increase. In thick sections, this is seen as very significantly higher in-situ strengths than would be expected from Portland cement, as in Fig. 7.15. It is for this reason that many precast concrete producers whom use accelerated curing techniques use fly ash as their preferred cementitious material.

Another benefit of using fly ash in concrete is the solid particulate effects on the water demand of the concrete produced. Finer fly ash tends to reduce the water requirement for a given workability, thereby

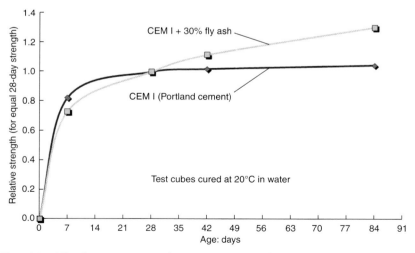

Fig. 7.14 The long-term strength gain properties of concretes designed to have equal 28 day strengths

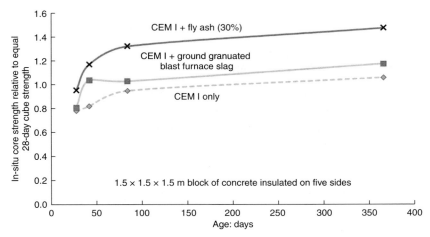

Fig. 7.15 The in-situ strength gain properties of various cements in thick sections

reducing the water/cement (w/c) ratio and permeability, and enhancing strength relatively, even at 28 days (Fig. 7.16). These effects are due to particle packing, and the rounded particle shape of fly ash particles acting as ball bearings within the mix (Fig. 7.17). It is for this reason that UK practice has been to classify fly ash to produce a finer material than is normally produced by UK power stations. Analysis of these data confirms this effect is primarily due to the water reducing properties of the finer fly ash, although there is an element of source dependency.

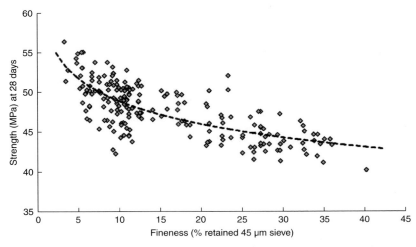

Fig. 7.16 Finer fly ash reduces the water demand

Fig. 7.17 Fly ash particles of <45 μm are generally spherical. (Reproduced with permission from the UK Quality Ash Association (UKQAA))

Loss on ignition (LOI) is usually cited as an issue with fly ash. LOI is partially burned carbon, which behaves as partially activated charcoal, exposing a large surface area to the concrete for a small mass. This basically dilutes the fly ash, reducing the pozzolanicity. However, the greater issue is the effect of LOI on the performance of admixtures, particularly air-entraining agents. As a result, LOIs are normally restricted to a maximum of 5% or 7% by the national concrete specification in force, depending on the country, though a limit of 9.0% is permitted within EN 450-1.[1]

The history and future of pozzolanic materials in concrete

Roman concrete

The Romans discovered in ~75 BC that a form of concrete could be created by mixing a red volcanic powder, found at Pozzuoli near Naples, and hence the name pozzolana, with lime. Similarly, Rivera-Villarreal and Cabrera[2] reported on the properties of a lightweight concrete developed around the same time in the ancient culture of Toto-nacas near the modern city of Veracruz, Mexico. Pozzolanic cements were first used at Pompeii in 55 BC. These concretes were only used as an infill material between brick and stone walls, but gradually this concrete became the exposed material, as in the arches of the Coliseum and the Pantheon, constructed in 115 AD. The dome of the Pantheon is an example of pozzolanic lightweight concrete, and is over 50 m in diameter, and made with a lightweight aggregate (pumice), with an air-entraining agent (animal blood) and a pozzolanic material (volcanic ash).

Early uses in the UK and how the market developed

It was in the late 1940s that research began into using fly ash in concrete. In 1949, the first use of fly ash as a cementitious material was in the USA, with the construction of the Hungry Horse Dam in Montana. Similarly, following research at the University of Glasgow,[3] fly ash, then known as pulverised fuel ash (PFA) in the UK, was first used in the construction of the Lednock,[4] Clatworthy and Lubreoch Dams. Thereafter, there followed a number of exemplary projects, but there was not a great acceptance of fly ash for use in concrete. The first UK standard for the use of fly ash as a mixer addition was BS 3892, which was published in 1965. Within this standard, fly ash was effectively treated as a fine aggregate, which resulted in little take up of the material.

In 1975, a new approach was taken by Pozzolan Ltd, which marketed a controlled-fineness, quality-assured fly ash. This was initially based on Agrément certificate, eventually being incorporated within BS 3892, Part 1 in 1982. Classification using air-swept classification or selection of fly ash became the norm for ~20 years for use in concrete. During this time onwards, the use of fly ash in concrete increased gradually as a mixer addition, even in the face of competition from other materials being available such as ground granulated blast furnace slag (GGBS). In 1995, the first European standard for fly ash in concrete was produced, EN 450. This allowed a wider range of ash fineness to be used than was normal UK practice. Eventually, the BS 3892 and EN 450 approaches were harmonised, with EN 450-1:2005[1] permitting both the basic fly ash as Category N and the processed finer ash as Category S, with BS 3892, Part 1 being finally withdrawn in January 2007.

Fly ash is permitted for use in concrete as a mixer addition, which is defined in BS EN 206-1 as a finely divided inorganic material used in concrete in order to improve certain properties or to achieve special properties. There are two types of additions:

- Type I – these are nearly inert additions
- Type II – these are pozzolanic or latent hydraulic additions.

While most fly ash is used as a Type II addition, there are no technical reasons why it cannot be used as a Type I addition replacing substantive proportions of the aggregates. Alternatively, fly ash may be used as a constituent of the cement. Due to increasing pressure on the Portland cement makers to reduce their overall CO_2 emissions, one relatively easy way is for them to add a pozzolana such as fly ash to the cement. This dilutes the emissions per tonne of product, and they also can chemically engineer an optimised Portland cement to improve the reactivity with the fly ash! In recent years, the cement industry has become one of the larger users of fly ash, both in blended cements and as a kiln feed material.

Recent developments and the future – ash processing
The coal-fired power generation industry became under increasing pressure to reduce emissions of NO_x and SO_x in the mid- to late-1990s. There are a number of ways of reducing such emissions, but the UK approach has been to retro-fit low-NO_x burners to the existing power plants and post-treat the flue gases to remove SO_x. While the latter does not affect the ash quality, low-NO_x burners have the effect of increasing LOI. As the UK limits LOI for use in concrete to a maximum of 7.0%, this has caused increasing problems in recent years.

Low-NO_x burners work by reducing the peak flame temperature within the power station furnace. Temperatures are such that the minerals within the coal are still turned into glassy spheres and therefore pozzolanic, but these lower temperatures are unable to burn out all the coal fully in the relatively short combustion period of 3–5 seconds. Overall, a more-even flame is produced within the furnace, but this leads to certain fractions within the coals, the so-called inertites, only being partially combusted. As a result, the LOI in the fly ash is increased, and the resulting char has a great affinity for surface-acting material such as admixtures. This is particularly noticeable with air-entraining agents, where dosages may have to increase by factors of 3–5 times that which would be used in a Portland cement concrete. A number of UK power stations are fitting ash-processing plants to electrostatically extract the carbon, producing a low-LOI fly ash which is very suitable for concrete. Other systems due for completion during 2008 use froth floatation and wet classification to process the fly ash.

One power station, at the time of writing, is able to reduce the carbon and classify the fly ash using air swept classifiers to remove the coarser fractions, further enhancing the reactivity and water-reducing properties of the ash. As a result, they can produce a low-LOI and more-reactive material eminently suitable for use in concrete.

Co-combustion, where small quantities of alternative fuels are burned in conjunction with the coal, is increasingly being carried out. By burning biomass and secondary fuels, the power generator can reduce the overall CO_2 emissions. With the advent of carbon trading, this has increasingly become an important part of coal-fired power station operation. The quantities of secondary fuel used in the UK are relatively small, and it has been shown to have had little effect on the performance of the resulting fly ash. The noticeable affects have been in changing the fineness, LOI and water demand within the fly ash. However, these are all factors that would be detected by routine quality monitoring. EN 450:2005 permits co-combustion fly ashes to be used within concrete, subject to some additional testing requirements.

In the future, many of the older UK stations will be replaced with new stations. Rather than opt for old technology, new coal-fired power stations will operate super critically, e.g. they use far bigger boilers allowing for residence times, therefore more efficient heat transfer and higher steam temperatures. This results in significant overall efficiency benefits to the power station. NO_x emissions will be controlled by selective catalytic reducers (SCRs), which, as in the motor car, convert the NO_x to nitrogen and oxygen using vanadium/titanium as catalysts. The result will be lower LOI in the resulting fly ash, making it suitable for use in concrete without any post-processing carbon reduction. However, the operation of SCRs requires the injection of ammonia into the gas stream to enhance precipitator efficiency. The SCR takes the oxygen from the NO_x in the presence of ammonia, and some by-products are formed as ammonium sulfate and bisulfate. If too much ammonia is injected into the gas stream, this can get into the fly ash, resulting in a material that when mixed into the alkali concrete releases ammonia gas. This is known as 'ammonia slip', and the smell of as little as 10 ppm of ammonia can be problematic, so it has to be carefully controlled. Limits of ~50 ppm by weight of ammonia have to be used when the fly ash is destined for use in concrete.

7.2.1.4 How fly ash enhances the durability of concrete

Fly ash has for many years been used to enhance the durability of concrete. It has been found through extensive research to have beneficial properties in respect of a number of potentially deleterious reactions that may occur in concrete. There are over 12,000 published papers on the production and use of fly ash, of which in excess of 4000 relate to concrete. The following is an overview of the use of fly ash in concrete.

Mechanical properties (strength, etc.)

Reduced permeability

The porosity and permeability of concrete containing fly ash will reduce with time as the pozzolanic hydration products are produced, presuming there is the presence of water and $Ca(OH)_2$. The relatively fine pore structure created in comparison with the Portland cement hydration products is able to block access of potentially deleterious compounds to the mass of the concrete.

It has been shown that pore size in fly ash concrete continues to reduce with time beyond that experienced with an equivalent Portland cement concrete.[5] The porosity has been shown to reduce by between 10 and 30% depending on curing regimes and fly ash fineness.[6] This reaction and the physical outcome reduce the porosity in fly ash concrete with time and increase the bond between the paste and particles.[7]

Studies of the oxygen permeability of fly ash concrete found it to be lower than a similar Portland cement concrete, even after only one day's curing.[8] Thomas *et al.*[9] concluded, for a given strength at the end of a curing period, that fly ash concrete is considerably less permeable than Portland cement concrete, especially at higher levels of replacement and higher strength. Permeability in concrete can be summarised to be dependent on the strength of the concrete, the curing regime and the proportion of fly ash within the mix. In general terms, the following rule of thumb is: the higher the strength and fly ash proportion (as a percentage of the total cementitious) the lower the permeability of the resulting concrete.

Alkali–silica reaction

The alkali–silica reaction (ASR) is an expansive reaction that occurs when some siliceous aggregates react with alkalis primarily from the Portland cement. It is dependent on the quantity of Portland cement, the sodium and potassium alkali content of the cement, and the presence of a reactive form of siliceous aggregate. These constituents form a silica gel that is able to absorb large quantities of water, producing a disruptive expansion. In order for this gel to form, sufficient highly alkali compounds must be present and available in the hardened concrete, such as sodium or potassium hydroxide with a source of silica. Fly ash when added to concrete reacts preferentially with any sodium or potassium alkalis before any significant setting and strength gain can take place within a concrete. The addition of fly ash quickly reduces the pH of the pore solution to below 13, at which point the ASR cannot occur. Calcium hydroxide is incapable of producing the very high pH required to form the disruptive gel.

Fly ash can contain quite high proportions of sodium/potassium oxide, and is theoretically capable of causing the ASR. Fortunately, these alkalis are held within the glassy phase of the fly ash particles and unable to react detrimentally if sufficient ash is used in conjunction with the Portland cement. Even when total alkalis, normally expressed as $Na_2Oeq.$, within the fly ash are as high as $5 \, kg/m^3$, fly ash has been found[10] able to prevent the ASR.

A pessimum ratio exists, where the greatest expansion will occur. However, by adding more reactive silica, e.g. fly ash, the dilution of the reaction with the alkalis, coupled with the low permeability of fly ash concrete, effectively mean that no disruptive reaction happens. The recommendations in BS 8500[11] within the UK require the producer to ensure a minimum of 25% fly ash or more, depending on the cement and aggregate combinations being used to prevent the ASR.

Chloride ingress

Chlorides can come from many sources. Firstly, it is important to restrict the chloride content in the constituents of the concrete to a minimum, and guidance is given within BS 8500. Chlorides from external sources are generally from seawater and de-icing salts used on roads. Tri-calcium aluminate (C_3A), a compound found in Portland cement, is able to bind chloride ions, forming calcium chloro-aluminate. Similarly, tetra-calcium aluminoferrite (C_4AF) can also reduce the mobility of chloride ions, forming calcium chloro-ferrite. Fly ash also contains oxides of alumina, which are able to bind chloride ions.

One highly effective way of reducing chloride ingress is to lower the chloride permeability of the concrete. Many researchers have shown that fly ash is capable of reducing the permeability to chloride ingress significantly, though this is dependent on the early age curing regime, the concrete strength and the proportion of fly ash within the concrete. The benefits of fly ash are reflected within BS 8500-1[11] durability tables for XS and XD exposure conditions, where chloride ingress is found. Mixes containing 30 and 50% fly ash benefit from lower required compressive strength and minimum cement contents and higher w/c ratios.

Sulfate resistance, including thaumasite
The ability of fly ash concrete to enhance the sulfate resistance of concrete has been recognised for many years within the UK. However, the discovery of the thaumasite form of sulfate attack some years ago in the M5 bridge foundations has led to a comprehensive review and more research.

Though much work has been done over the years on classified PFA and sulfate resistance, little had been done within the UK on unprocessed fly ash conforming to EN 450-1. During the research programme investigating the thaumasite form of attack carried out by both the Buildings Research Establishment (BRE)[12] and the University of Sheffield,[13] unprocessed fly ash and greater proportions of ash in the concrete were tried. The UKQAA has funded further work with BRE[14] that looked at the effect of the initial temperature and the resulting resistance to thaumasite attack. While no significant difference could be found between processed and unprocessed fly ashes or the effects that differing initial curing conditions had, this work did confirm that higher proportions of ash up to 55% of the cementitious content were more effective at resisting this form of sulfate attack.

The result of these projects has been to conclude that a 25% fly ash component, whether Category N or Category S, as a proportion of the total cementitious content does give a limited resistance to the thaumasite form of attack. However, higher proportions, >36% and up to the maximum of 55% fly ash content, give a far enhanced performance with no significant thaumasite attack being observed, as seen in Fig. 7.18.

Carbonation
Carbonation has often been cited as a weakness for fly-ash-based concretes. The pozzolanic reaction reduces the available lime, $Ca(OH)_2$, which is responsible for maintaining the alkali environment, preventing the corrosion of reinforcing. However, even with large proportions of fly ash the reaction rate is such that the available lime is never completely depleted. Additionally, the pozzolanic reaction produces more hydration products, reducing the permeability of the concrete, and reducing the ingress of CO_2 and oxygen significantly. Corrosion of the reinforcing cannot occur without oxygen and water. The overall effect is the carbonation front in properly compacted and cured fly ash concretes of equal 28 day strength does not advance significantly faster than would be found with Portland cement (CEM I) based concrete (Fig. 7.19). The problem arises in the manner in which assess-

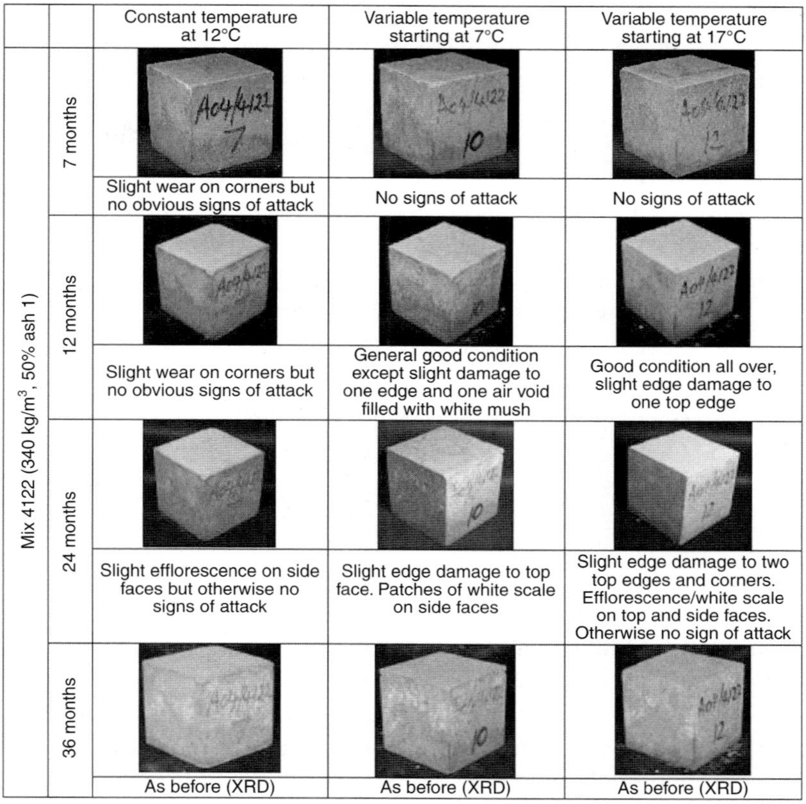

	Constant temperature at 12°C	Variable temperature starting at 7°C	Variable temperature starting at 17°C
7 months	Slight wear on corners but no obvious signs of attack	No signs of attack	No signs of attack
12 months	Slight wear on corners but no obvious signs of attack	General good condition except slight damage to one edge and one air void filled with white mush	Good condition all over, slight edge damage to one top edge
24 months	Slight efflorescence on side faces but otherwise no signs of attack	Slight edge damage to top face. Patches of white scale on side faces	Slight edge damage to two top edges and corners. Efflorescence/white scale on top and side faces. Otherwise no sign of attack
36 months	As before (XRD)	As before (XRD)	As before (XRD)

(Left axis label: Mix 4122 (340 kg/m³, 50% ash 1))

Fig. 7.18 Fifty per cent ash content concretes show no signs of thaumasite attack, irrespective of the initial curing temperature. (Reproduced with permission from BRE)

ment of carbonation performance is usually conducted. Most researchers used some form of accelerated testing involving higher concentrations of CO_2 and pressures than found naturally. While this accelerates the progress of the carbonation front, it generally does not accelerate the pozzolanic reaction, resulting in the apparent carbonation rate being far higher than found the CEM I concretes.

Heat of hydration and thermal cracking

Heat generated by the Portland cement in thick sections of concrete can lead to thermal cracking when the concrete subsequently cools down. These cracks allow the ingress of water and deleterious

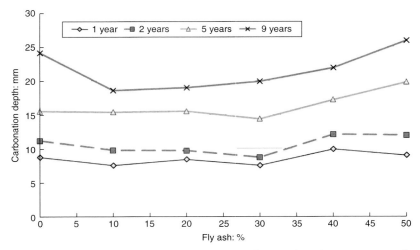

Fig. 7.19 Carbonation versus fly ash proportion for equal grade concrete – 3 day curing regime

compounds that can result in the premature failure of the concrete structure. Reducing the heat of hydration, thus limiting the peak temperature achieved in the concrete, can prevent such cracking. Research carried out by Dhir[15] has suggested that the fly ash has no effect on the maximum rate of heat production, irrespective of the fly ash source, at normal curing temperatures, e.g. the pozzolanic reaction is not exothermic. Even at higher temperatures the heat production due to the fly ash rises only marginally. These results are consistent with the normal understanding of the pozzolanic reaction.

CIRIA Report 660,[16] authored by Bamforth, gives guidance on when thermal cracking is likely to occur, and is based on the temperature differential of the core of the concrete to the exposed surface. The models have been refined, based on both Bamforth's own work and Dundee University's research, and contains more information about critical steel contents, restraint, and compliance with Eurocode 2 (EC2). The use of fly ash is strongly recommended to prevent disruptive thermal cracking, and it is clear that fly ash is the most effective material in reducing the risks of such cracking, particularly when higher proportions (>30%) are used.

Setting time and formwork striking times
Inclusion of fly ash in concrete will increase the setting time compared with an equivalent grade of Portland cement concrete. There is

176

undoubtedly a delay in the onset of the hydration of fly ash concrete, but is has been shown by Woolley[17] that the actual gain in strength once hydration has started is greater for fly ash concrete at normal temperature regimes. When 30% fly ash is used to replace Portland cement in a mix, the setting time may be increased by up to 2 hours. This increased setting time reduces the rate of workability loss. However, it may result in practical difficulties for finishing, particularly in periods of low temperature. In compensation, it will reduce the incidence of cold joints in the plastic concrete.

Formwork striking times at lower ambient temperatures normally will have to be extended in comparison with Portland cement concrete, especially with thin sections. In practice, vertical formwork striking times can be extended without this affecting the site routine, e.g. the formwork is struck the following day. For soffit formwork, greater care has to be taken. Reference should be made to BS 8110[18] for recommended striking times. Temperature-matched curing can be used to ensure that sufficient in-situ strength has been achieved while allowing for the concrete curing conditions.

7.2.1.5 The main standards for the use of fly ash in concrete

There are many ways in which fly ash may be incorporated within concrete, both as a cementitious component, a so-called Type II addition, or through the use of an EN 197-1 cement, and/or as an inert aggregate, as a Type I addition.

EN 450-1: Fly ash as a cementitious addition in concrete

Fly ash for use as a Type II concrete addition at the mixer is normally supplied complying with EN 450[1], where it can be considered as being part of the cement content of the concrete (see BS 8500[11]). EN 450 describes two basic types of fly ash: Category N and Category S. The differences are described below.

EN 450:2005 Category N fly ash

Category N (normal) is fly ash that is taken direct from the power station electrostatic precipitators. EN 450 imposes a series of quality control requirements on the ash, such as fineness, chemical properties, etc., but this material would normally be controlled by a process of selection and rejection based on the various control parameters. As a result, Category N fly ash is generally sourced straight from the power station silos without any processing. This could be considered as having zero environmental impact at the factory gate from production, and even possibly a positive benefit to the environment as the energy associated with conditioning and disposing of the ash may be more than that in transporting it to the concrete producer.

EN 450:2005 Category S fly ash

Category S, which we shall call 'special' fly ash, is again dry ash from the station. In the majority of cases this is processed to remove the coarser ash particles within the material. Typically, this is done with air-swept classifiers, and the process reduces the water demand and increases the strength of the resulting concrete by removing the misshapen and generally coarser fraction of material. This requires energy, typically 9.75 kW/h per tonne of Category S fly ash, which equates to ~4.2 kg of CO_2 per tonne of product. However, in relation to the improvements in reactivity and water demand of the resulting concretes, classification is a positive environmental benefit.

The main requirements of EN 450

EN 450 requires that fly ash is derived from the flue gases of the combustion of coal with or without co-combustion materials. It should contain at least 25% reactive silica and not less than 70% of the sum of the SiO_2, Al_2O_3 and Fe_2O_3 content. These clauses are intended to ensure the fly ash is pozzolanic. There is a table limiting the types and quantities of co-combustion materials permitted, and no ash shall be made from less than 80% coal by mass of fuel or 10% by mass of ash.

There are a series of limits on chemical parameters such as free calcium oxide, reactive calcium oxide, magnesium oxide, alkalis and soluble phosphate. These clauses are intended to prevent potentially expansive components occurring within the ash, with many only being applicable to ashes that contain co-combustion materials.

In order to control reactivity, fineness of the ash based on the percentage passing the 45 μm sieve when tested to EN 451-2 is required to be tested and reported. Category S ash has a maximum fineness of ≤12%. For Category N ash there are limits in fineness of ≤40% retained on the 45 μm sieve and a maximum variation ±10% of the manufacturer's declared value. The strength performance of the fly ash is assessed using the activity index test, which is where standard mortar prisms are prepared using test cement and a blend of the test cement with 25% fly ash using the cement test method, EN 196-1. The ratio of the fly ash mortar to the cement-only mortar prisms, the activity index, is measured at 28 days, where a minimum value of 75% is required, and at 90 days where a minimum value of 85% is required. Soundness, particle density, initial setting time and water requirement (for Category S ash only) are also required.

By a combination of physical and chemical requirements, fly ash produced to EN 450-1 is able to give the strength and durability characteristics required of such additions to concrete.

EN 197-1: fly ash in cement manufacture

This is the normal standard for common cements which permits a wide range of options for both siliceous and calcareous fly ashes. In practice, siliceous fly ashes are usually only regularly used in cement manufacture, and they are split into three basic categories: Portland fly ash cement (CEM II B-V), Portland composite cement (CEM V) and Pozzolanic cement (CEM IV B-V) as detailed in Table 7.1.

The main requirement on siliceous fly ash is that it shall contain >25% SiO_2 and have a maximum of 5.0% LOI, though this may be increased to 7.0%, as is used within the UK. There are few other

Table 7.1 Fly ash cements permitted through EN 197-1

Main cement type	Designation	Notation	Constituents Proportion by mass (%) based on the sum of the main and minor constituents			Minor additional constituents
			Clinker	Fly ash		
				Siliceous	Calcareous	
CEM II	Portland fly ash cement	CEM II/A-V	80–94	6–20		0–5
		CEM II/B-V	65–79	21–35		0–5
		CEM II/A-W	80–94		6–20	0–5
		CEM II/B-W	65–79		21–35	0–5
CEM IV	Pozzolanic cement	CEM IV/A	65–89	11–35[a]		0–5
		CEM IV/B	45–64	36–55[a]		0–5

[a] May also be natural pozzolana.

179

restrictions because the control is exercised through testing the final cement. When fly ash is being used in CEM IV formulations, the resulting cement shall comply with the pozzolanicity test described in EN 196-5.

EN 12620 and EN 13055-1: use as a filler aggregate in concrete

EN 12620, *Aggregates for Concrete*, and EN 13055-1, *Lightweight Aggregates for Concrete*, permit the use of fly ash as filler aggregates. Fly ash is easily capable of complying with both of these standards, and would be a Type I addition in concrete. There is no technical reason why fly ash cannot replace all the aggregate in a mixture, e.g. forming a grout, which is widely used for filling of mines and caverns and ground stabilisation. However, fly ash is also used as filler in self-compacting concrete, foamed concrete and low-strength pumpable concrete, to increase cohesiveness, etc., and yet still remains competitive with other fine aggregates.

EN 206: specifying fly ash in concrete

Fly ash is generally classed as an addition to concrete within BS EN 206,[19] which places few restrictions on its use, simply stating that additions of Type I and Type II may be used in concrete in quantities as used in the 'initial tests'. Initial tests are defined in an annex of BS EN 206, as those required for demonstrating that a mix satisfies all the specified requirements for the fresh and hardened concrete. These initial tests may be from laboratory work or from long-term experience. An addition is defined as a finely divided inorganic material used in concrete in order to improve certain properties or to achieve special properties. There are two types of additions:

- Type I – these are nearly inert additions
- Type II – these are pozzolanic or latent hydraulic additions.

The situation becomes more complex when additions are taken into account as part of the total cementitious content and when calculating the water/cementitious ratio. BS EN 206 contains specific rules for fly ash to BS EN 450, and these rules may be applied anywhere. It also permits the use of other rules, if the suitability of the rules is established.

The specific rules in EN 206 are based on the '*k* value' concept, which is:

- The term 'w/c ratio' is replaced by 'water/(cement + k × addition)'.
- The minimum cement content (MCC) can be reduced by k × (MCC − 200) kg/m^3. However, the amount of cement plus fly ash must never fall below the MCC value.
- Fly ash to BS EN 450 has a k value of 0.2 or 0.4, depending on the strength class of the Portland cement with which it is used. The k value does not vary with the quantity of ash being used. Up to a maximum of 25% fly ash by mass of the (cement + ash) is allowed to be counted as cementitious. Any additional ash within the mix is effectively a Type I addition, which is assumed to act as inert filler.

Other values of k or other k value concepts may be used if their suitability is established. One way of establishing suitability is via a national standard valid in the place of use of the concrete. BS 8500[11] utilises this option by including rules by which fly ash conforming to EN 450 may count fully towards the cement content and w/c ratio. These rules in BS 8500 are available for all concrete conforming to EN 206, which is to be used in the UK, but do not currently extend to concrete used in other countries unless permitted by their national standards.

BS EN 206 also contains an 'equivalent concrete performance concept' that may be applied to a combination of any specified cement with any specified addition if the suitability has been established. The test methods necessary for its implementation are not standardised, and the informative Annex C of EN 206 places limits on the application of the concept. However, it is used in some countries in which high levels of co-combustion materials are used in their power stations, and the resulting fly ash produced is tested in concretes to give the assurance of durability.

7.2.1.6 Examples of the use of fly ash in concrete

Fly ash has been used in a wide variety of concretes over the years. The first use in the UK was in the Lednock Dam, a structure that is still in excellent condition after more than 50 years. In recent years, fly ash as a concrete addition has been used in ~6 000 000 m^3 of concrete per annum, with notable contracts being the Channel Tunnel, the Thames Barrier (Fig. 7.20), Canary Wharf, Heathrow Terminal 5 and the Channel Tunnel Rail Link, to name but a few. Fly ash may be added to precast concrete products as filler aggregate, and, while still enjoying the technical and cementitious benefits of the material, be very economical. Additionally, fly ash is increasingly being added to cement by the cement manufacturers, particularly in bagged products,

Fig. 7.20 The Thames Barrier

meaning that even DIY enthusiasts are using fly ash within concrete and mortars.

7.2.1.7 Environmental considerations

Increasingly, fly ash is being seen as a way of offsetting the environmental impacts of producing concrete. When fly ash is used as a cementitious binder, greater environmental benefits are achievable than the simple displacement of virgin aggregates. Portland cement inherently produces a large quantity of CO_2 during its manufacture as it involves the calcination of calcium carbonate. This releases ~550 kg of CO_2 for each tonne of cement made. In addition to this chemical release, the raw materials and resulting cement clinker have to be ground to a fine powder, which in itself is an energy-intensive process. Even with the most efficient cement works, figures of 700 kg/tonne of CO_2 for CEM I are only just possible.

As well as a cementitious binder, fly ash can be used as a raw material within the cement manufacture process. It is used as a source of silica and alumina, replacing the clays and sands traditionally used. This market is increasingly significant to the fly ash marketing industry.

Replacing some of this Portland cement with pozzolanic materials such as fly ash has considerable environmental benefits without compromising technical aspects, e.g. strength and durability. In fact,

Table 7.2 Comparison of CO_2 emissions associated with some concrete mix types (excludes aggregates, which are considered constant for mixes)

Mix designation	Portland cement (CEM I)	Fly ash equivalent mix	Overall CO_2 savings
C30/37 design strength concrete	280 kg/m^3 of CEM I = 241 kg/m^3 CO_2	320 kg/m^3 of CEM I + 30% fly ash = 193 kg/m^3 CO_2	48 kg/m^3 (20%)
RC25/30 MCC260 w/c 0.65	270 kg/m^3 of CEM I = 232 kg/m^3 CO_2	290 kg/m^3 of CEM I + 30% fly ash = 175 kg/m^3 CO_2	57 kg/m^3 (25%)
XS1 50 mm cover	C40/50 MCC380 w/c 0.40 = 395 kg/m^3 of CEM I = 330 kg/m^3 CO_2	C25/30 MCC320 w/c 0.55 320 kg/m^3 of CEM I + 30% fly ash = 193 kg/m^3 CO_2	137 kg/m^3 (42%)
		C25/30 MCC260 w/c 0.65 = 330 kg/m^3 of CEM I + 50% fly ash = 142 kg/m^3 CO_2	188 kg/m^3 (57%)

fly ash enhances many durability aspects of the resulting concrete, e.g. improved sulfate resistance, prevention of the ASR, reduced permeability to chloride ions, etc.

The degree of benefit varies depending on the exact specification for the concrete mix and its application. Using BS 8500 criteria, Table 7.2 contains some estimates of the relative benefits of various mixes.

Currently, about 500 000 tonnes of fly ash is used in concrete, both as an addition and within blended cements. Additionally, cement companies take ~420 000 tonnes of fly ash as a raw material. A further 1 000 000 tonnes is used within the block and precast concrete sectors, both as aggregates and cementitious binders.[20]

The environmental benefits of using fly ash have been fully exploited by some, such as in airport runway construction, where the longer-term strength gain of the fly ash coupled with the extensive use of admixtures has been used to reduce runway thickness by 10%.[21] Another approach has been the use of the so-called high-volume fly ash (HVFA) concretes, which contain up to 70% fly ash by weight of cement. These have been shown to produce strong durable concretes when due allowance has been made for the extended setting times and slower strength gain properties.

7.2.1.8 Summary

Fly ash from coal-fired power stations has now been used in concrete for over 50 years. The pozzolanic reaction that occurs between lime and fly ash enhances the durability of the resulting concrete by reducing the permeability, preventing the ingress of harmful chemicals and reducing the risk of deleterious reactions. Additionally, there are a further series of beneficial effects, for example a reduced environmental impact in comparison with Portland cement, higher-quality finishes, a reduced water content for a given workability, and economic. This useful by-product material is readily available throughout the world, and using it, as well as enhancing the properties of concrete, prevents material being sent to landfill.

References

1. British Standards Institute, BS EN450 (2005) Fly ash for concrete. *Definition, specifications and conformity criteria*, BSI, London.
2. Rivera-Villarreal, R. and Cabrera, J.G. (1999) The Microstructure of Two-Thousand Year Old Lightweight Concrete, *Proceedings of the Int. Conf.*, Gramado, Brazil.
3. Fulton, A.A. and Marshall, W.T. (1956) The use of fly ash and similar materials in concrete, *Proc. Inst. Civ. Engrs.*, Part 1, Vol. 5, 714–730.
4. Allen, A.C. Features of Lednock Dam, including the use of fly ash, Paper No. 6326, Vol. 13, 179–196, *Proc. of Institute of Civil Engineers*, August 1958.
5. Cabrera, J.G. and Gray, M.N. (2007) Specific Surface, Pozzolanic activity and composition of pulverised fuel ash, *FUEL magazine*, 1973, Vol. 52, July.
6. Kandie, B. PhD thesis, University of Sheffield, UK.
7. Cabrera, J.G. and Plowman, C. (1981) Hydration and Microstructure of High fly ash content Concrete, *CIRIA*, Conf. on Concrete Dams, London.
8. Thomas, M.D.A., Matthews, J.D. and Haynes, C.A. (1989) *The effect of curing on the strength and permeability of fly ash concrete*, ACI SP-114, 191–217.
9. Thomas, M.D.A. and Matthews, J.D. (1994) *The Durability of pfa concrete*, BRE, Watford UK.
10. Alasali, M.M. and Malhotra, V.M. Role of concrete incorporating high volumes of fly ash in controlling expansion due to alkali–aggregate reaction, *ACI Materials Journal*, 88, No. 2, pp 159–163, 1991.
11. British Standards Institute, BS8500-1 (2006) *Concrete – complementary standard to BS EN206-1, Part 1: Method of specifying and guidance for the specifier*, BSI, London, UK. BS EN 8500-2. *Concrete – Complementary British standard to BS EN260-1, Part 2: Specification of constituent materials and concrete*, BSI, London.

12. The design of structural concrete to resist the thaumasite for of attack, Report 7 – 3 year confirmation report, DTI *Partners In Innovation project* no cc1879, September 2003. BRE, Watford, UK.
13. Hill, J., Byars, E.A., Cripps, J.C., Lynsdale, C.J. and Sharp, J.H. (2003) *An experimental study of combined acid and sulfate attack of concrete*, EPSRC Project Sheffield University.
14. Dunster, A. (2007) *Effects of temperature on thaumasite formation in PFA concretes: Final Report*, BRE, Watford, UK, UKQAA funded Project.
15. Dhir, R.K., Paine, K.A. and Zheng, L. (2006) *Design data for low heat and very low heat special cements, Stage 1*, Final Project Report, March 2006.
16. Bamforth, P. Publication C660 (2007) *Early-age thermal crack control in concrete*, CIRIA.
17. Woolley, G.R. and Cabrera, J.G. (1991) Early age in-situ strength development of fly ash concrete in thin shells, *Proceedings of the Int. Conf. On Blended cement*, Sheffield.
18. British Standards Institute, BS 8110 (1997) *Structural use of concrete. Code of practice for design and construction*, BSI, London.
19. British Standards Institute, BS EN206 (2001) *Concrete – Part 1: Specification, performance, production and conformity*, BSI, London.
20. UKQAA *Annual PFA statistics for 2006*, Unpublished data from members, UKQAA, Maple House, Kingswood Business Park, Holyhead Road, Albrighton, Wolverhampton, West Midlands, WV7 3AU.
21. NCE/BAA Environmental Construction Awards, March 2001 overall winner.

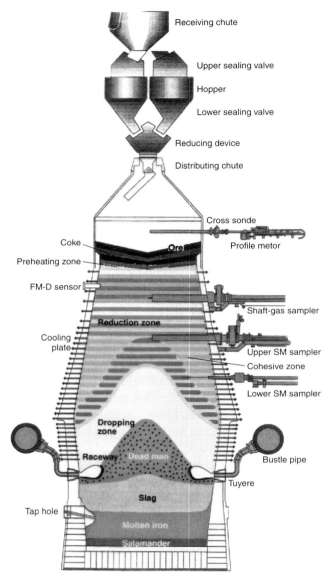

Fig. 7.21 Section through a typical blast furnace

its physical and chemical characteristics. There are two principle methods of cooling used. One is to pour the liquid slag into pits, and leave the material to gradually air-cool to ambient temperature over many hours. Such air-cooled material is of no use as a cementitious material and is generally used as road stone.

Fig. 7.22 Granulation in progress

The other method is to cool the blast furnace slag rapidly by quenching with water. There are two methods of achieving this granulation (Fig. 7.22) and pellitisation. Both methods use water: granulation relies on large volumes of water, while pelletisation uses water sprays, through which pellets of slag are propelled. Both methods result in a glassy homogeneous non-crystalline material that has latent hydraulic properties.

7.2.2.4 Ground granulated blast furnace slag (GGBS) production

The granulated slag is selected, stored and allowed to drain prior to processing. GGBS is produced by drying and reducing the practical size of the slag down to a similar size to that of CEM I Portland cement.

Types of mill

In the UK, much work has gone into refining the particle size reduction process, as this is highly energy intensive. There three main types of equipment in use throughout the cement industry:

(a) High-pressures rollers: two large-diameter rolls, one fixed, the other moving and held against the fixed roller under high pressure. Material is passed between the two rollers in a recirculation system, the finished-size material being extracted by separators and a bag house. High maintenance and low availability can be expected with this process.

(b) Vertical mills: similar in basic design to a flower mill, a rotating tablea number of rollers that apply high pressure. The material passes under the rollers and is crushed between the roller and table. The material is recirculated and separated in a similar way to the roller mills. The advantage of this system is that the table rollers and separator are housed in one chamber and material is recirculated within the chamber. Like the rollers, high maintenance and low availability can be expected with this process.

(c) Ball mills: these are the most common types used for final grinding in the cement industry. They consist of large-diameter steel tubes (typically 4 m diameter) which may have one or more chambers; the tubes are capped with slotted inlet and outlet diagrams. The tube contains steel balls of various sizes, from 50 to 20 mm diameter to approximately 30% of its volume. The tube is rotated at 95% of the critical centrifugal speed, so the balls cascade inside the tube, and material is circulated through the tube and is milled by the cascading balls. The final product is extracted by a separator and collected in a bag house. This process is more energy intensive than the other processes but has a large thermal mass which aids drying, and has high availability and low maintenance.

Hybrid or combi systems are also used, i.e. a ball mill with a roller press in the feed circuit pre-crushing the feed prior to the ball mill. These systems combine the energy efficiencies of the roller press with the high availability of a ball mill. This is considered by some to be most effective method milling.

Influence of fineness and particle size distribution
GGBS is milled to a finer particle size, typical 400–500 m^2/kg compared with 300–400 m^2/kg for CEM I. Generally, the finer the material is, the more reactive it is likely to be, although fine materials can have an adverse effect on water demand. This influence of fineness on reactivity is less significant than the influence of chemical composition. Much research has been conducted into the use of particle size distribution (PSD) as a control tool, but it has yet to be adopted. The PSD can

Table 7.3 Chemical compositions UK base blastfurnace slags

Parameter	Typical range: %	
CaO	42–38	
Al_2O_3	14–12	
SiO_2	36–34	
MgO	9–7	
Na_2Oeq	0.7–0.4	
V ratio	1.3–1.4	
Ih	1.5–1.8	
Glass content granulate	Granulate: 90–98%	Pellite: 80–90%[a]

[a] Generally only the fine fraction passing 20 mm is used for GGBS larger sizes can be used but the glass content will be lower.

have an influence on the water demand, and, as such, the performance in concrete. However, GGBS is used in combination with CEM I, typically in a 50% blend, and it is the overall PSD that is influential, and the combination tends to dilute any individual powder influences.

Chemical composition and glass content
The chemical composition in combination with the glass content has a significant influence on the performance of GGBS. The likely reactivity of a particular slag can be assessed by a number of chemical formulae, and the two most common are:

V ratio $\quad CaO + MgO/SiO_2$

Ih $\quad\quad CaO + MgO + Al_2O_3/SiO_2$

In both cases, the higher values indicate increased reactivity. Table 7.3 gives the range of compositions that are available in the UK.

The glass content is a measure of the granulation efficiency: higher values are indicative of improved reactivity for a given composition. The minimum specified in BS 6699[3] is 65%.

Additives and minor additions
Additives can be added, up to a maximum of 1%, of which 0.5% can be organic. Grinding aids are the most common additive, although other materials are used. These materials added to the grinding process that aid either grinding efficiencies and/or other performance aspects but

must not promote corrosion of reinforcement or impair other properties of the cement or concrete and mortar made from the cement. The Portland cement industry uses these materials extensively, and they are used to some extent by the GGBS industry elsewhere in the world. They are not used in the UK at this time, but were used for many years with great success with old open-circuit mills. Trials with more modern milling equipment indicate that improvement can be achieved but cost–benefit is not achieved.

Minor additions are inorganic materials that can be added up to a maximum of 5% but do not contribute to performance or grinding. These materials are used in Portland cement manufacture but not in the production of GGBS.

7.2.2.5 Certification and standards

- *BS 6699 and EN 15167*. The first UK standard was BS 669,[4] and this standard specified chemical composition with limiting values for loss on ignition, magnesia, chloride manganese and sulfur, expressed as both sulfide and sulfate. A number of physical and mechanical properties were also similar to those specified in the CEM I standards. The only additional requirement is a limiting value for glass content. This standard also included a method for determining conformity, where the performance of the GGBS is tested with individual CEM Is and a maximum replacement level that will satisfy the minimum strength requirements for the different strength grades. With the publication of BS 8500, the conformity procedure was removed from BS 6699 and included as an annex to that standard. BS EN 15167,[5] published in 2008, is a European equivalent to BS 6699 (the latter will be withdrawn). Similar tests are required in BS EN 15167, although the strength performance has limiting values for the activity index, the compressive strength of mortar containing 50% GGBS expressed as a proportion of a CEM I control. The limiting value for glass content has also been removed as a requirement, but a value can be requested by a customer.
- *EN 197:*[6] This is the European standard for cement, and includes CEM I and combinations of CEM I with GGBS, pozzolans, fly ash, silica fume and limestone fillers. The standard only refers to factory-produced cements and combinations.
- *BS 8500:*[1] This is the UK annex to EN 206,[7] and it includes the procedure to establish conformity of combinations that are made in the mixer. It is the procedure that was in BS 6699.

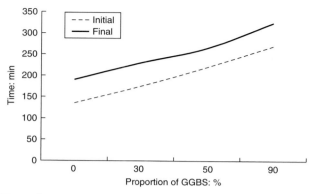

Fig. 7.23 Typical setting times for combinations of GGBS and CEM I

7.2.2.6 Influence on plastic properties

When tested as a paste, the stiffening rate of combinations of GGBS and CEM I will be slower than that of a similar paste containing CEM I alone: the difference increases with increasing proportions of GGBS (Fig. 7.23). The initial setting time will also be influenced by the curing temperature as well as the characteristics of the Portland cement component. This means that concrete containing increased amounts of GGBS will remain workable longer. This may be desirable in mass pour situation or hot weather conditions for the prevention of cold joints.

- *Ambient temperature.* All cement combinations containing GGBS are influenced by ambient conditions. Because GGBS is a latent hydraulic material, this influence is magnified depending on the proportion of GGBS. A high proportion of 70% and above in thin sections can have significant influence, and special precautions may be required. Generally, large sections, >600 mm minimum dimension, have sufficient thermal capacity not to be influenced even with high replacement levels.
- *Placing and finishing.* As indicated above, the time that concrete remains open is extended in line with increasing proportions of GGBS. This will be further influenced by ambient temperatures. In thin sections in cold conditions, e.g. flat slabs, this influence can be significant, and GGBS proportions are often reduced to 20% or not used at all. Because the concrete remains plastic for longer, formwork pressures are likely to be increased, and this may influence the pour rate. However, work in Dundee[8] indicates

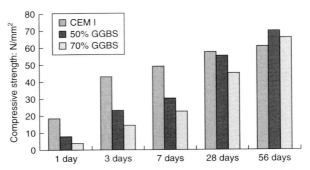

Fig. 7.24 Typical strength development in concrete containing different proportions of GGBS. Total cement content: 350 kg/m³. (Reproduced with permission of Trent Valley Aggregates)

that these increases are not significant and will be within the factor of safety in the formwork design code.

- *Bleeding, plastic shrinkage and settlement cracking.* As a further result of the extended plastic time, the potential bleed capacity may be increased when high percentages of GGBS are used. Bleeding may result in plastic settlement cracking if restraint is present, but this can be rectified by re-vibration. Plastic shrinkage occurs when the evaporation rate exceeds the bleed rate. As GGBS remains plastic for longer, it may be more vulnerable to this form of cracking. However, suitable curing systems will prevent this problem.

7.2.2.7 *Influence on mix design and hardened properties*

- *Rate of strength development.* As GGBS hydrates slower than CEM I, the rate of strength development will be slower. Increasing the proportion of GGBS increases the degree of this influence. As a result, the relationship between the water/cement (w/c) ratio and strength is different. The biggest difference is seen at early ages, with 28 day strengths being similar to that of a comparable CEM I concrete when normal addition rates up to 50% are in use. Figure 7.24 gives typical values. Typically to achieve parity at 28 days the cement content needs to be increased by between 0 and 20 kg/m³ this depends on aggregate type mix design and characteristic strength. As the GGBS content increases the strength difference at 28 days will increase.

- *Later age development.* As the hydration process for GGBS is largely pozzolanic, significant strength development is achieved at later

Table 7.4 Factors influencing sulfide colour and its rate of oxidation

Influencing factor	Influence on rate of colour change
Proportion of GGBS	Higher proportions of GGBS will give a deeper colour that take longer to oxidise
Concrete quality	High-performance concrete with a low w/c ratio and high cement content will take longer
Types of formwork	Low permeability surfaces such as treated plywood and polished steel are likely to produce a greater colour intensity
Release agents	Some types leave an impervious surface on the concrete which will inhibit oxidation
Curing and curing membranes	Effective curing such as under water or close contact with polythene will in essence slow down the rate oxidation. Sprayed on membranes will have a similar influence
Ambient condition	Cold, damp conditions will slow down the rate of oxidation. Warm, dry weather will have the reverse influence
Surface treatments	Lacquer treatments will seal in the colour. Acid treatment can expose the colour and seal the colour in

ages. Data are available that demonstrate development up to 5 years.

- *Sulfide blue.* GGBS contains small amounts of sulfides, which on hydration form complex polysulfide products that are blue-green in colour. This coloration is sometimes evident on the surface of concrete, and is indicative of the presence of GGBS in concrete when broken surfaces are exposed. Upon contact with air, these products oxidise, and the colours fade to the normal overall off-white appearance. The oxidation has no effect on the properties of the concrete other than its colour. The process of oxidation and the time taken to achieve a uniform colour is influenced by several interrelated factors, which are summarised in Table 7.4.

7.2.2.8 Influence on thermal properties

- *Thick sections.* GGBS combinations have long been recognised as a material with low heat-generating properties, and high proportions of GGBS are often specified to minimise temperature rise in thick sections. Increasing the GGBS content reduces the rate at which heat is generated as well as the rate of heat evolution, which will result in lower peak temperatures and related thermal gradients. There are a number of documents that give advice, such as CIRA 91,[9] 135[10] and C660,[11] all of which refer to cement type

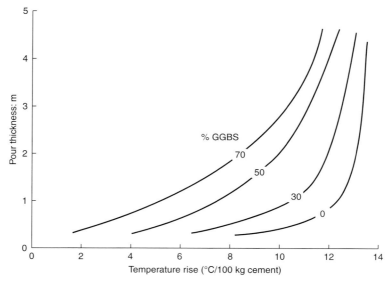

Fig. 7.25 *The influence of GGBS on the temperature rise in concrete elements of different thicknesses*[12]

as one of the means of influencing temperature rise. Figure 7.25 indicates the influence of increasing GGBS content.

- *Thin sections.* In thin sections the lower rate of heat evolution will influence the rate of in-situ strength development, particularly in low ambient temperature conditions. Depending on the proportion of GGBS, the mix design and the formwork type (in terms of insulating capacity and the strength required to release the formwork), the formwork striking times may be extended. In these conditions when high proportions of GGBS are used, extended curing times will also be required.

7.2.2.9 Influence on durability

The general term 'durability' is a measure of the ability of a concrete structure to survive in a given environment. GGBS concrete is inherently different in character to Portland cement concrete in that GGBS concrete has low permeability, a different hydrate structure and different cement particle composition. Concrete is subject to many different types of potential attack, which are considered separately in this section, with respect to the advantages of using GGBS in place of Portland cement.

Sulfate attack

Sulfate attack is a significant mechanism for failure in environments containing high-sulfate groundwater or in marine environments. Partial replacement of CEM I by GGBS generally increases the resistance of concrete to sulfate attack, and this is acknowledged in all major European codes of practice. The sulfate resistance of GGBS concrete is dependent on several parameters, principally the ratio of GGBS to Portland cement, the C_3A (tricalcium aluminate) content of the CEM I, the alumina content of the slag and the w/c ratio. Generally, the resistance increases with increasing replacement by GGBS, less C_3A in the CEM I, less alumina in the slag and a lower w/c ratio. When the GGBS content is 65% or greater, most combinations of GGBS and CEM I provide equivalent or better sulfate resistance than Sulfate-resisting Portland cement (SRPC). This is confirmed by BRE Special Digest 1,[13] whose recommendations have been incorporated in BS 8500.[5] These recommendations indicate that CIIIA cement (36–65% GGBS) is equivalent to SRPC while CIIIB (66–80% GGBS) is superior, as the recommendations allow lower cement contents and high w/c ratios for these cements compared with SRPC.

The mechanisms associated with sulfate attacks are discussed in detail in Chapter 5. The influence of GGBS on the mechanisms is debated by researchers. The principal process by which GGBS improves performance in sulfate conditions is by dilution of the vulnerable hydrates such as C_3A produced by the CEM I. This is further enhanced by the more refined pore structure that is less permeable to ingress by sulfate solutions. In-situ concrete foundations in sulfate-bearing ground are also less likely to suffer from thermal cracking when GGBS is used, further reducing possible paths for the ingress of sulfate-bearing solutions.

Thaumasite attack

Thaumasite requires a source of carbonate and sulfate to form, and these may be provided in many ways, as Table 7.5 indicates. The formation takes place rapidly in wet conditions at temperatures below 15°C but can occur at higher temperatures. Thaumasite forms as a result of the sulfates reacting with calcium carbonate and calcium silicates. This is distinct from conventional sulfate attack, where the calcium aluminate phases are attacked. At a Building Research Establishment (BRE) conference,[14] it was reported that after the initial diagnosis on the M5 bridges, a significant number of structures have been identified.

Table 7.5 Potential sources of carbonate and sulfate

Carbonate	Sulfate
Limestone aggregate	Lower Lias Clay (disturbed and used as a backfill)
Limestone filler	Kimmeridge Clay
Limestone fines	Rheatic Mudstone
Limestone-filled cements	Railway ash spoil from coal and oil shale
Groundwater high in carbon ions	Seawater
	De-icing salts
	Other sulfide-bearing soils
	Sulfate-bearing bricks, gypsum plasters

A 7 year research programme conducted at BRE indicated that concrete containing 70% GGBS was particularly resistant to this form of attack, and the following recommendations were developed for the following ground conditions as determined by BRE SD1[13] or BS 8500:[5]

- Design chemical class 3 (DC-3): use a cement type containing at least 70% GGBS and adopt a minimum cement content of $380\,kg/m^3$ and a maximum w/c ratio of 0.45.
- Design chemical class 4 (DC-4): use a cement type containing at least 70% GGBS and adopt a minimum cement content of $400\,kg/m^3$ and a maximum w/c ratio of 0.40.

Acid attack

In general, GGBS concrete is more resistant to dilute acids than Portland cement or SRPC concrete. This is largely due to the reduction in $Ca(OH)_2$ and the lower penetrability of well-cured GGBS concrete. Many European standards recommend the use of GGBS concrete for acid conditions. However, it should be noted that no concrete is completely immune to attack in acidic conditions, and the quality of the concrete (in particular the w/c ratio and compaction) can be more important than the type of cement.

Freeze–thaw resistance

The freeze–thaw resistance of CEM I concrete and GGBS concrete of similar strength and air content is essentially the same. There are no published reports of frost damage occurring in actual structures containing GGBS that are directly attributed to its presence. However,

Table 7.6 Diffusion of chloride ions at 25°C in cement pastes of w/c ratio 0.5[15]

Type of cement	Diffusivity: 10^{-8} cm^2/S
SRPC	100.0
OPC	44.7
70% OPC/30% PFA	14.7
35% OPC/65% GGBS	4.1

OPC, ordinary Portland cement; PFA, pulverised fuel ash.

because of their slower rate of strength gain, GGBS concrete may be more susceptible to frost attack at an early stage. There is some debate on the relationship between the indicated performance from laboratory test methods and field performance. Some tests that use freeze cycles down to −20°C can produce high degrees of scaling, but this is considered not to be representative of field performance.

Chloride attack
Chloride attack is a significant mechanism for failure in marine environments and structures subjected to de-icing salt application. Concrete containing GGBS provides high resistance to the penetration of chloride. Tables 7.6 and 7.7 (from Page et al.[15] and Bakker,[16] respectively) show typical diffusivities for various types of concrete where, at 65% GGBS, the diffusivity is reduced by an order of magnitude as compared with normal CEM I. This reduction in diffusivity has been confirmed by many investigators, and would appear to be due to two mechanisms. Firstly, the incorporation of GGBS reduces the permeability of the concrete; secondly, there is evidence which suggests

Table 7.7 Diffusion of coefficients in OPC and PBFC[16]

Diffused ion	Ionic diffusivity: 10^{-8} cm^2/S	
	OPC	PBFC
Sodium	2.38	0.10
Potassium	3.58	0.21
Chloride	4.47	0.41

OPC, ordinary Portland cement; PBFC, Portland blast furnace cement.

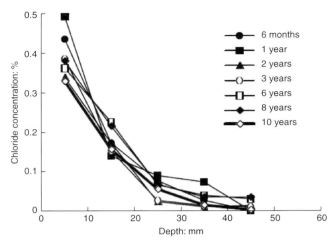

Fig. 7.26 Chloride profiles recorded over a 10 year period of splash zone exposure for 70% GGBS (w/c ratio 0.48)[17]

that the hardened paste of slag cement immobilises the chloride ions, removing the concentration gradient needed for diffusion.

A more recently published report by Bamforth[17] summarise the research, and concludes that significant benefits are achieved at addition levels in excess of 40%, with increasing resistance up to 65%.

In practice, it is common for chloride ingress in the near-surface zone to be rapid, but for further migration to occur very slowly with little change in the initial profile over many years (Fig. 7.26). In addition to substantially reducing the rate of chloride ingress by a combination of enhanced pore structure and chloride binding, GGBS also influences the pore solution chemistry. Although opinion is divided, some research suggested that GGBS may reduce the chloride threshold[18,19] by reducing the level of free chloride to a lesser extent than the reduction in hydroxide levels.

Resistance to carbonation

GGBS concretes will carbonate more rapidly than CEM I concrete. For concretes of a reasonable strength grade (C25/30) or greater and moderate levels of GGBS (50%), the difference will be marginal when concretes of equal grade are compared. Greater differences are reported for lower-strength grades with higher (<70%) GGBS additions. As a general rule for predictive purposes, a buffering capacity of 0.6 can be assumed. [17]

Alkali–silica reaction (ASR)

The ASR and the associated but rare alkali–carbonate reaction can occur in concrete, depending on the character and proportion of the reactive aggregate and the availability of mobile metal alkalis, details of which are discussed in Chapter 5.

Inclusion of GGBS in a concrete is accepted worldwide as a means of reducing the risk of damage due to the ASR. There have been many investigations, and without exception these have confirmed the ability of GGBS to reduce significantly the deleterious expansion caused by the ASR. Expansion due to the ASR reduces with increasing replacement of CEM I by GGBS. In the UK, it is generally agreed that use of a minimum of 50% GGBS can be a suitable preventative measure, provided that the total alkali content of the GGBS/CEM I blend does not exceed 1.1% Na_2O equivalent. At lower replacement levels, Concrete Society TR30[20] and BRE Digest 330[21] recommend that GGBS should be taken as contributing half of the total alkali. In a report that attempted to clarify this conflict, Connell/Higgins[22] found that at low replacement levels (25% GGBS) the half value was appropriate but at higher replacement levels (50% GGBS) there was zero contribution of alkalis from the GGBS. In practice, the alkali content of GGBS produced in the UK is such that concrete mixes can readily be designed which satisfy the requirements of both approaches. BS 8500[5] has incorporated these views, and recommends that the advice given in BRE Digest 330[21] be adopted.

There is some debate over the reasons why GGBS is effective in reducing deleterious expansion due to alkali reactions. The alkalis present in GGBS are predominantly acid-soluble, with a very small proportion water-soluble by comparison with those present in CEM I, which are predominantly water-soluble. The water-soluble alkali becomes available as soon as hydration commences, and is available to move to reaction sites. As such, GGBS would tend to act as a diluting influence on the water and total alkali available; however, as hydration progresses, the acid-soluble alkali in GGBS is likely to be released into solution. Also, GGBS remains effective at controlling expansion even when excess alkali up to $8\,kg/m^3$ Na_2Oeq are included in the concrete.[22] It is well reported that GGBS can be activated by alkalis, and it is thought that some of the released alkali is consumed in reaction with the GGBS. Additionally, as hydration progresses, the pore structure becomes increasingly refined, and diffusion coefficients reduce dramatically,[16] reducing the ability for alkali and water to move to reaction sites.

Acid attack

Acid attack may occur in industrial environments and in low-pH soil environments (e.g. peat). In general, GGBS concrete is more resistant to dilute acids than that made with Portland cement or SRPC. Many European standards recommend the use of GGBS concrete for acid conditions. However, it should be noted that no concrete is completely immune to attack in acidic conditions, and the quality of the concrete (in particular the w/c ratio and compaction) can be more important than the type of cement.

7.2.2.10 Environmental aspects

Environmental aspects are becoming increasingly important in construction, and the impact of construction materials are now considered by client's architects and designers. The manufacture of Portland cement is thought to contribute 5% of the world's CO_2. The use of additions can significantly reduce a number of environmental aspects, having a positive impact on sustainability. Table 7.8 gives values for some of these environmental aspects, and Table 7.9 shows the impact of similar grades of concrete with different cement types.

7.2.1.11 Specified application

Driven by the improved durability obtained by the use of GGBS, it is often specified for major contracts, particularly when extended design life is required and/or particularly aggressive environments are an issue. The following are some typical examples where GGBS has been used for these reasons, both internationally and in the UK:

- The Humber Bridge: for many years this structure was the longest single-span crossing in the world. GGBS was use at 70% replacement as an addition in the anchor blocks and foundations and 40% in the slip-formed towers.
- The Bahrain Causeway: all concrete elements were cast with blended cement imported to the gulf region, to produce concrete

Table 7.8 Environmental profiles

Issue	Measured as	Impact	
		Manufacture of 1 tonne of GGBS[a]	Manufacture of 1 tonne of CEM I Portland cement
Climate change	CO_2 equivalent	0.07 tonne[b]	0.88 tonne
Energy use	Primary energy[c]	1300 MJ	7000 MJ
Mineral extraction	Weight quarried	0	1.5 tonnes
Waste disposal	Weight to tip	1 tonne saved[d]	0.02 tonne

[a] The profile for GGBS is the impacts involved in processing granulated blast furnace slag to produce GGBS. No account has been taken of the impacts of iron-making because the slag evolves, irrespective of whether or not it can be used.
[b] Extracted from the BRE environmental profile.
[c] Includes losses involved in the generation and distribution of electricity.
[d] The use of slag for the manufacture of GGBS potentially saves it from having to be disposed of to tip.

capable of resisting the extremely aggressive conditions in the Gulf area.

- The QE II bridge across the Thames in London: all concrete was specified to contain 70% GGBS, mainly to give an extended design life, but also for the light colour of the concrete.
- Blackpool sea defence work constructed over 10 years up to 2008: GGBS was specified in much of the concrete, including precast step units to improve durability.
- Berwick-upon-Tweed sea defence cast in-situ wave wall: 70% GGBS was specified and used despite cold weather conditions, steel formwork and relatively thin sections.
- The new Tyne Tunnel under construction in 2008: this submerged tube construction was specified to contain 70% GGBS to insure water-tight durable construction.

Table 7.9 Calculated environmental impacts for 1 tonne of concrete

Impact	CEM I Portland cement reference concrete	GGBS concrete	Pulverised fuel ash concrete
Greenhouse gas	142 kg (100%)	85.4 kg (60%)	118 kg (83%)
Primary energy use	1070 MJ (100%)	760 MJ (71%)	925 MJ (86%)
Mineral extraction	1048 kg (100%)	965 kg (92%)	1007 kg (96%)

- The Tsing Ma Bridge, Hong Kong: the concrete for the towers of this suspension bridge was specified to have an extremely low coefficient of permeability, and GGBS exported from the UK was used to achieve the required permeability.

7.2.2.12 Special applications

GGBS is used mainly to produce cast in-situ concrete either by ready mix concrete companies or, on lager projects, produced on site by bespoke batching plants. Over the many years of use in the UK, a number of special applications have developed, partly driven by economy compared with CEM I but also by the specific properties attributed to GGBS.

- *Secant pilling.* A high proportions of GGBS, 80% and above, is used to produce concrete with very low early strength for cutting the secant but still retaining long-term durability, low permeability and strength. BRE IP17/05[23] gives details of the process.
- *Grout cut-off walls.* A number of specialist pilling contractors have developed a method of producing cut-off walls to satisfy the demand for impermeable structures around existing and new waste disposal sites. GGBS in combination with bentonite is used to produce slurry that is pumped into trenches under excavation, typically, 600 mm wide and deep enough to toe into a lower impermeable soil layer. The slurry provides support for the sides of the trench during excavation, and subsequently hardens to form a continuous wall that is stiff and impermeable to leachate and gas but still remains semi-plastic and able to take ground movement.
- *Soil stabilisation and hydraulically bound materials.* Partly driven by economic construction demands and the environmental aspects of construction, the use of soil stabilisation and hydraulically bound materials has grown considerably. A wide range of processes are now in common use in road and airfields, but also in building construction, reinstatement of brown field sites and stabilising contaminated ground. There are a number of codes of practice, advice documents and papers available[24–27] detailing the performance of GGBS in these applications.
- *Precast concrete applications.* Traditionally, the requirement for early formwork striking, lifting of units and transfer of pre-stressed loads has restricted the use of GGBS in these applications. However,

driven by environmental aspects, durability requirements and the colour of finished products, precast operators have developed mix designs and curing systems that have resulted in significant use of GGBS in many applications such as hollow core flooring, pigmented precast products and structural sea defence units.

7.2.2.13 Conclusion

Cement manufacturers worldwide are increasingly examining the use of alternative materials, partly to reduce emissions but also to reduce manufacturing costs. In the UK, the most common cement type is one containing an addition such as GGBS, fly ash or a limestone. The use of GGBS, most commonly as an addition at the mixer but also as a factory-produced CEM II or III, has resulted in significant reductions in the various environmental aspects of cement manufacture, including CO_2 emissions. Their use has also resulted in significant potential benefits in durability, leading increases in the design life and reductions in the whole-life cost of structures.

References

1. BSI (2006) *Concrete. Complementary British Standard to BS EN 206-1. Method of specifying and guidance for the specifier.* BS 8500-1:2006. British Standards Institution, London.
2. BSI (1996) *Specification for Portland blastfurnace cements*, BS 146:1996. British Standards Institution, London.
3. BSI (1996) *Specification for high slag blastfurnace cement*, BS 4246:1996. British Standards Institution, London.
4. BSI (1992) *Specification for ground granulated blastfurnace slag for use with Portland cement*, BS 6699:1992. British Standards Institution, London.
5. BSI (2006) *Ground granulated blastfurnace slag for use in concrete mortar and grout*, BS EN 15167-1:2006. British Standards Institution, London.
6. BSI (2000) *Composition, specifications and conformity criteria for common cements*, BS EN 197-1:2000. British Standards Institution, London.
7. BSI (2000) *Concrete, Part 1. Specification performance composition and conformity*, BS EN 206-1:2000. British Standards Institution, London.
8. McCarthy, M.J., Caliskan, S. and Dhir, R.K. (2004) Update on developing guidance for design formwork pressures. *Concrete Journal*, 38(5), 45–46.
9. Harrison, T. (1992) *Early-age thermal crack control in concrete*. CIRA, London, revised edn.
10. Bamforth, P.B. and Price, W.F. (1995) *Concreting deep lifts and large volume pours*. CIRA, London.

11. Bamforth, P.B. (2007) *Early-age crack control in concrete*, C660. CIRA, London.
12. Bamforth, P.B. (1982) *Early age thermal cracking in concrete*. Technical note TN/2. Institute of Concrete Technology, Camberley.
13. BRE (2005) *Concrete in aggressive ground*. Special Digest SD1:2005. Building Research Establishment, Garston, Watford.
14. BRE (2002) *Proceedings of the 1st International Conference on Thaumasite in Cementitious Materials*. Building Research Establishment, Garston, Watford.
15. Page, C.L., Short, N.R. and El Tarras, A. (1981) Diffusion of chloride ions into hardened cement paste. *Cement and Concrete Research*, 11(3), 395–406.
16. Bakker, R.F.M. (1981) About the causes of the resistance of blastfurnace cement to the alkali silica reaction. In: *5th International Conference on Alkali Aggregate Reactions in Concrete*. Cape Town, Paper 252/29.
17. Bamforth, P.B. (2004) *Enhancing reinforced concrete durability*. Technical report 61, Concrete Society, Camberley.
18. Bamforth, P.B. (1998) *Guide to the pretension of corrosion in reinforced concrete exposed to salt*, Part 1. *Performance of r.c blocks exposed for 10 years in marine splash zone*. DETR PIT project CI39/3/231. Report No. 1303/98/10248. Taywood Engineering, Watford.
19. Tuutti, K. (1982) *Corrosion of steel in concrete*. Report No. 4.48. Swedish Cement and Concrete Association, Stockholm.
20. Concrete Society (1987) *Alkali–silica reaction: minimising the risk of damage to concrete*. Technical Report No. 30. Concrete Society, Camberley.
21. BRE (2004) *Alkali aggregate reactions in concrete*, Part 2. BRE Digest 330. Building Research Establishment, Garston, Watford.
22. Higgins, D. and Connell, M.D. (1992) Effectiveness of granulated blastfurnace slag in preventing alkali–silica reaction. In: *9th International Conference on Alkali–Aggregate Reactions*. London, 175.
23. BRE (2005) *Concrete with high ggbs for use in hard/firm secant piling*. BRE Information paper IP 17/05. Building Research Establishment, Garston, Watford.
24. BSI (2004) *Hydraulically bound mixtures*, Parts 1–5. *Specifications*, BS EN 14227. British Standards Institution, London.
25. Britpave (2005) *The immediate trafficking of cement-bound materials*. Technical Report. BP/14. Britpave, Camberley.
26. Britpave (2005) *Stabilisation of sulfate-bearing soils*. Technical Guidelines BP/16. Britpave, Camberley.
27. Kennedy, J. (2006) *Hydraulically bound mixtures for pavements*, CCIP-009. Concrete Centre, Camberley.

7.2.3   Silica fume for concrete

Robert Lewis, Elkem Limited, UK

7.2.3.1 The material

The terms 'condensed silica fume', 'microsilica', 'silica fume' and 'volatilised silica' are often used to describe the by-products extracted from the exhaust gases of silicon, ferrosilicon and other metal alloy smelting furnaces. However, the terms 'microsilica' and 'silica fume' are the ones used to describe those condensed silica fumes that are of high quality, for use in the cement and concrete industry. The latter term, 'silica fume', is the one used in the European standard – EN 13263.05.

Silica fume was first 'obtained' in Norway, in 1947, when environmental restraints made the filtering of the exhaust gases from the furnaces compulsory. The major portion of these fumes was a very fine powder composed of a high percentage of silicon dioxide. As the pozzolanic reactivity of silicon dioxide was well known, extensive research was undertaken, principally at the Norwegian Institute of Technology. There are over 7000 papers now available that detail work on silica fume and silica fume concrete.

Large-scale filtering began in the 1970s, and the first standard, NS 3050, for use in a factory-produced cement, was granted in 1976.

Production and extraction

Silica fume is produced during the high-temperature reduction of quartz in an electric arc furnace, where the main product is silicon or ferro-silicon (Fig. 7.27). Due to the vast amounts of electricity needed, these furnaces are located in countries with abundant electrical capacity: China, Scandinavia, Europe, the Americas, Canada, South Africa and Australia, amongst others.

High-purity quartz is heated to 2000°C with coal, coke or wood chips as fuel, and an electric arc introduced to separate out the metal. As the quartz is reduced, it releases silicon monoxide gas. This mixes with oxygen in the upper parts of the furnace, where it oxidises and condenses into microspheres of amorphous silicon dioxide.

The fumes are drawn out of the furnace through a pre-collector and a cyclone, which remove the larger coarse particles of unburnt wood or carbon, and then blown into a series of special filter bags.

207

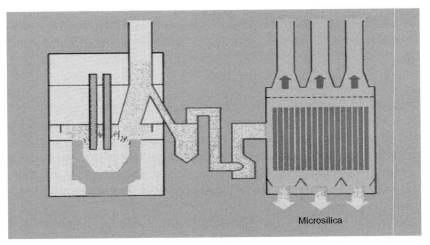

Fig. 7.27 Typical plant layout (left to right): furnace, main stack, pre-collector, cooling pipes and fan, and baghouse

The chemistry of the process is very complex and temperature-dependent. The SiC formed initially plays an important intermediate role, as does the unstable SiO gas which forms the silica fume. The following summarises the process:

$$t > 1520°C$$

$$SiO_2 + 3C = SiC + 2CO$$

$$\downarrow$$

$$t > 1800°C$$

$$3SiO_2 + 2SiC = Si + 4SiO + 2CO$$

The unstable gas travels up in the furnace, where it reacts with oxygen to give the silicon dioxide (Fig. 7.28):

$$4SiO + 2O_2 = 4SiO_2$$

The silica fume is not collected as purely this material. Other particles and chemicals make up the powder that is collected in the filter bags (Table 7.10).

Characteristics
Silica fume is, when collected, an ultrafine powder having the following basic properties:

- at least 85% SiO_2 content.
- mean particle size between 0.1 and 0.2 µm.

208

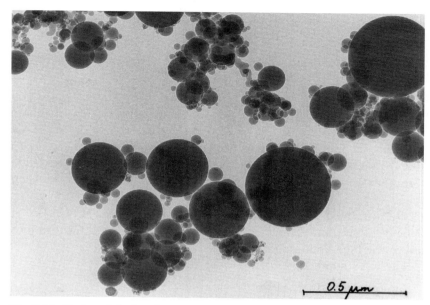

Fig. 7.28 Undensified silica fume (scanning electron micrograph)

- minimum specific surface area of $15\,000\,\text{m}^2/\text{kg}$.
- spherical particle shape.

The powder is normally grey in colour, but this can vary according to the source. Operation of the furnace, the raw materials and the quality of

Table 7.10 A comparison of cementitious materials available in the UK

	Portland cement	Pulverised fuel ash	Ground granulated blast furnace slag	Microsilica
Physical data for cementitious materials				
Surface area (m^2/kg)	350–500	300–600	300–500	15 000–20 000
Bulk density (kg/m^3)	1300–1400	1000	1000–1200	200–300
Specific gravity	3.12	2.30	2.90	2.20
Chemical data for cementitious materials				
SiO_2	20	50	38	92
Fe_2O_3	3.5	10.4	0.3	1.2
Al_2O_3	5.0	28	11	0.7
CaO	65	3	40	0.2
MgO	0.1	2	7.5	0.2
$Na_2O + K_2O$	0.8	3.2	1.2	2.0

209

metal produced, will all have an effect on the colour of the powder. This variation can show up in the material from different furnaces; thus, it is advisable to ensure a consistency of supply from one source.

Health and safety

As with all fine powders, there are potential health risks, particularly when relating silicon dioxide to the lung disease silicosis. In studies of the material and of workers in the ferrosilicon industry[1,2] it has been found that the silicon dioxide that causes silicosis is the crystalline form. Silica fume, produced as described, is amorphous, and has been found to be non-hazardous.

The threshold limit values for the respirable dust have been set at $3 \, \text{mg/m}^3$ (low) and $5 \, \text{mg/m}^3$ (high), and exposure to concentrations above the high limit are not advised.

The CAS number for both powder and slurry is:

amorphous SiO_2: 7631-86-9.

For full health and safety information it is always advised to contact the producers.

Available forms of silica fume

As the powder is 100 times finer than ordinary cement, there are transportation, storage and dispensing considerations to be taken into account. To accommodate some of these difficulties, the material is commercially available in various forms. The differences between these forms are related to the shape and size of the particles, and do not greatly affect the chemical make-up or reaction of the material. These differences will influence the areas of use, and careful thought should be given to the type of silica fume chosen for a specific application.

The main forms are:

- Undensified: bulk density $200–350 \, \text{kg/m}^3$. Due to the very low bulk density, this form is considered impractical to use in normal concrete production. The main areas of use are in refractory products and formulated bagged materials such as grouts, mortars, concrete repair systems and protective coatings.
- Densified: bulk density $500–650 \, \text{kg/m}^3$. In the densification process the ultrafine particles become loosely agglomerated, making the particulate size larger. This makes the powder easier to handle, with less dust, than the undensified form. Areas where this material

is used are in those processes that utilise high-shear mixing facilities such as pre-cast works, concrete roof tile works or ready mixed concrete plants with 'wet' mixing units.

- Micropellitised: bulk density 700–1000 kg/m³. Micropellitisation involves forming the powder into small spheres about 0.5–1 mm in diameter. The material in this form does not readily break down in conventional concrete mixing, and is best suited to inter-grinding with cement clinker to produce a composite cement. Icelandic cement is made in this fashion, and contains 7.5% silica fume to combat the potential alkali–silica reaction (ASR) from the local aggregates.
- Slurry: specific gravity 1400 kg/m³. This material is produced by mixing the undensified powder and water in equal proportions, by weight, to produce a stable slurry.

Mixing and maintaining a stable slurry requires expensive hi-tech equipment, and cannot be done easily; therefore, all slurries should be obtained from one specific supplier to maintain quality. In this form, the material is easily introduced into the concrete mix. It is also the most practical form for dispensing by weight or volume. It can be used for virtually all forms of concrete, from semi-dry mixes for pre-cast products to self-compacting concrete, and is ideally suited to the ready mixed industry.

7.2.3.2 Inclusion in concrete

Standards and specifications
Various national standards, codes of practice and recommendations have been published since the use of silica fume has become globally accepted.
The ones in major use are:

- America (global use): ASTM C1240
- European (global use): EN 13263
- Canada: CAN/CSA-A23.5-M98
- China: GB/T18736
- India: IS 15388
- Japan: JIS A6207.

Reference should be made to the relative issuing body – or a supplier – to ensure that the latest version of a standard is being used.

The main ones that are accepted on a global basis are ATSM C1240 and CSA A23.5. Only the latest versions of any of these standards should be accepted for specification use. Table 7.11 gives the mandatory

Table 7.11 Comparison of standards

	USA ASTM C1240	Norway NS 3045	Canada CSA, A23.5-98	Australia AS 3582	Europe EN 13263 -1	France NF P 18-502 1992	Japan JIS A6207: 2000
SiO$_2$: %	85.0	85	85	85	85	85	85
SO$_3$: %			1.0	3	2.0	2.5	3.0
Cl: %		Report if >0.10		Report	0.3	0.2	0.1
CaO: %		2			1.0	1.2	1.0
MgO							5.0
Si (free): %					0.4		0.4
Total alkalis						4	
Free carbon						4	
Moisture content: %	3.0			2		1	3.0
LOI: %	6.0	5	6.0	6	4.0		5.0
Specific surface: m^2/g		>12			15–35	20–35	>15
Bulk density	Report			Report			
Pozzolanic activity index: %	85% – 7 days accelerated curing, w/cm = variable	95% – 28 days Normal curing w/cm = 0.5	85% – 7 days accelerated curing	Report	100% – 28 days normal curing w/p ratio = 0.5		95% – 7 days 105% – 28 days w/cm = 0.5
Retained on 45 µm sieve: %	10		10				
Density: kg/m^3						2100–2300	
Autoclave expansion: %			0.2%				Report
Canadian foaming test			No visible foam				
Notes	Characteristic values	Characteristic values			Characteristic values. Not an official standard, in the approval process	Material type A	

chemical and physical requirements; several of the standards also contain optional requirements. Where there are blanks in Table 7.11, no mandatory requirements exist. Note that the Norwegian and French standards are now superseded by the European standard – they have been left in the table for comparison.

Effects on fresh concrete

Due to the nature and size of the silica fume, a small addition to a concrete mix will produce marked changes in both the physical and chemical properties.

The primary physical effect is that of adding, at the typical dosage of 8–10% by cement weight, between 50 000 and 100 000 microspheres per grain of cement. This means that the mix will be suffused with fine material, causing an increase in the cohesiveness of the concrete. When using a powder form of silica fume, this will mean an increased water demand to maintain mixing and workability, and therefore powders are most often used with plasticisers or superplasticisers.

Regarding workability, it should be noted that a fresh silica fume concrete will have a lower slump than a similar ordinary concrete, due to the greater cohesion. When the mix is supplied with energy, as in pumping, vibrating or tamping, the silica fume particles, being spherical, will act as ball bearings and lubricate the mix, giving it a greater mobility than the similar ordinary concrete. Silica fume concrete is often referred to as being thixotropic in nature to describe this. Thus, when measuring the slump of a silica fume concrete, it must be remembered that the figure will only show the consistency of the concrete and will not relate to its workability. The most favourable test for such concrete is the DIN-standard flow table, or similar, which gives a reaction to an energy input and thus gives a better visual appraisal of the workability of the mix.

In mixes using the slurry material as an addition, there can be a slight increase in water demand to maintain a given slump. This demand is normally offset by the use of a standard plasticiser. Another way of negating the effect is to modify the fines in the mix, reducing the sand content by a small percentage.

In the ready mix industry, use of the slurry material is nearly always enhanced by a nominal dosage of a plasticiser, to ensure full dispersion and maintain the water/cement (w/c) ratio.

As the concrete is more cohesive, it is less susceptible to segregation, even at very high workabilities such as in flowing or self-compacting concretes. This lack of segregation also makes it ideal for incorporation into high-fluidity grout.

The ultrafine nature of the particles will mean a much greater contact surface area, and thus will improve the bond between the fresh concrete and the substrate or reinforcement.

Another aspect of this non-segregation, and the filling of the major voids in the fresh concrete, is that a silica fume concrete will produce virtually no bleed water. The concrete must therefore be cured, in accordance with good site practice, as soon as it has been placed, compacted and finished. The lack of bleed water means that processes such as powerfloating can be commenced much sooner than with ordinary concretes.

However, this lower slump, lack of bleed water and 'gelling' (stiffening when not agitated) does not indicate a rapid set. Silica fume is a pozzolana, and therefore requires the presence of calcium hydroxide to activate it. The calcium hydroxide is produced by the cement hydrating, and thus the silica fume can only work after the cement has started reacting. The setting times for microsilica concretes should be similar to those of ordinary concretes, except when specifically designed for such features as ultra-high strength or low heat.

Mix design criteria

It is considered by most of the producers of silica fume that the powder forms are best suited to specialised production or mixing facilities or those precasters and ready mix producers using high-power, forced-action mixers. This is inherently due to the water demand of the powders and the subsequent difficulty in dispersion through a drier mix.

In most ready mix or precast operations, it is normal for the slurry product to be used, dosing at the same time as the normal water addition. Although the slurry disperses more readily, care is needed to ensure a uniform quality of mix before allowing the concrete to be used.

There are variations of dosage for given types of application which serve as guidelines for initial trial mixes. It should be stressed that, even with the precedent of past work, trial mixes should always be conducted before acceptance of a mix design.

As silica fume is most often used to produce high-performance concrete, the dosage is always given as an addition to the total cementitious content of the mix. The following are often used as starting points:

- pumping aid: 2–4%
- normal concretes: 4–7%
- high strength: 8–10% (higher values may be used in special circumstances)
- high chemical resistance: 10–12% (higher values may be used in special circumstances)
- underwater concretes: 10–15%.

The fine filler effect caused by adding a material with a bulk density one-fifth of the cement will nearly always require a modification to the mix constituents, particularly the coarse/fine aggregate ratio, to achieve optimum rheology. The use of particle packing software programs are an advantage in blending the mix to give the best balance of coarse to fine products – and thus the optimum mix rheology. In nearly all ready mix production, a plasticiser – or, more often, a superplasticiser – is used to give optimum dispersion, while maintaining the w/c ratio.

In some countries, microsilica is added as a large-scale cement replacement across the board in ready mix production, but this practice is decreasing as quality control restrictions become tighter.

Mixing times will need to be adjusted to allow for the maximum dispersion of the silica fume. This is most important when using any of the powder forms, to prevent agglomerations within the mix. The slurry form, when used in the ready mix environment, will not need excessive mixing times, and the 'mass action' of a truck mixer has been known to produce better results than laboratory trials for the same mix.

Like most cementitious materials, silica fume will function more efficiently with some types of chemical admixtures than with others, and trial work is advised. Most of the producers of silica fume will have basic mix designs for various situations and aggregate types.

Effects on setting and hardening concrete
As the concrete sets and hardens, the pozzolanic action of the silica fume takes over from the physical effects. The silica fume reacts with the liberated calcium hydroxide to produce calcium silicate hydrates. These both increase the strength and reduce the permeability by densifying the matrix of the concrete.

Silica fume, having a greater surface area and higher silicon dioxide content, has been found to be much more reactive than pulverised fuel ash (PFA) or ground granulated blast furnace slag (GGBS).[3] This increased reactivity appears to push the hydration of the C_3S fraction of the cement in the first instance,[4] thus creating more calcium hydroxide, but settles down to more normal rates beyond 2 days.

The high reactivity and consumption of the calcium hydroxide has prompted questions relating to the pH levels of the concrete and the related effects on steel passivity and carbonation rates. Studies have shown that the effect on the carbonation rate depends greatly on the quality of the mix produced. Good-quality, well-proportioned silica fume concrete does not exhibit any greater carbonation than an ordinary Portland cement concrete. The reduction of pH in a concrete mix is usually of the order of 13.5–13.0, and this is well above the level for steel passivity. It has been established that a 25% addition of silica fume[4] would be required to use up all the calcium hydroxide being produced in a concrete, and studies have shown that the pH still does not drop below 12.0. In normal practice, the highest dosage advised for concrete is 15%, and this should have no deleterious effects.

This reduction in the alkalinity and the binding up of the K^+ and Na^+ ions in the pore solution[5] is one of the ways in which the addition of silica fume decreases the risk of the ASR in concrete.[6]

As the silica fume reacts, and produces the calcium silicate hydrates, the voids and pores within the concrete are filled, as the crystals formed bridge the gaps between cement grains and aggregate particles. Coupling this with the physical filling effect, it can be seen that the matrix of the concrete will be very homogeneous and dense, giving improved strength and impermeability.[7] It has been found that the relatively porous section that surrounds the aggregate grains in normal concrete – the transition zone – is virtually absent in high-quality silica fume concrete.

The measure of reactivity, the cementing efficiency, of silica fume has been found to be between four and five times that of ordinary Portland cement.[8] This implies that large amounts of Portland cement could be replaced by small dosages of silica fume and a concrete would still achieve the required strength. While this is possible, it is not considered to be an 'ethical' usage. Reducing the cementitious content to very low levels, though still achieving strength, will have adverse effects on the durability of the concrete despite the benefits imparted by incorporating the silica fume.

Even though the rate of reaction is very high in the initial stages, not all the silica fume is used up, and studies have shown that unreacted material is still present at later ages.[9]

In general, the heat evolution of a microsilica mix depends on the design. If a high early strength is needed, addition of silica fume to a high-cement content mix will produce a higher heat of hydration for the initial stages, lessening as the reactivity slows. Additions to 'normal' concretes do not usually produce significant changes in heat evolution.

Silica fume concrete is very susceptible to temperature variations while hardening. The rate of strength gain can be reduced with low temperatures, below the 20°C optimum, and accelerated with increased temperatures.[10] This relates to concrete and not to some of the specialised repair materials available, which can be used at very low temperatures.

Blended cement mixes

While silica fume is compatible with both PFA and GGBS, it is a pozzolanic material and hence will give differing results depending on the mix designs used. High PFA percentages will be affected by the ability of the microsilica to rapidly take up the calcium hydroxide. This may give good early strengths but a reduced rate of long-term gain. The GGBS blends are less affected because of the latent hydraulic nature of the GGBS. When high replacement levels of GGBS are used, 50–70%, silica fume can be added to improve the early age strength, as it reacts faster in the first 3 days, or to improve the consistency of the fresh concrete. High levels of GGBS can cause problems with high water contents, leading to segregation and bleeding, not only on the surface of the concrete but also within the matrix itself. The silica fume will virtually eliminate this bleeding, and hence maintain the integrity of the concrete. With more standard levels of PFA and GGBS, 25–40%, silica fume can be added to give enhanced performance. In such cases where there would be a minor reduction in strength due to using these materials, this is offset by the silica fume, and high early, and ultimate strength, can be achieved without an excessive increase in the cost of the concrete. These triple-blend cements employ the beneficial characteristics of both pozzolanic materials in producing an optimally durable concrete. This type of concrete is being specified where concrete structures are expected to last for upwards of 100 years such as the Storebaelt in Denmark, the

217

Bonding

Silica fume concrete has a much finer paste phase, and the bond to substrates, old concrete, reinforcement, fibres and aggregates will be improved. Investigation has shown[19] that the aggregate cement interface is altered when silica fume is present, and pull-out tests[21,22] show improved strength. Bonding to fibres is greatly improved.[22-24] This is particularly beneficial in steel fibre/silica fume-modified shotcrete. It negates the use of mesh reinforcement, which would have to be fixed to the substrate. The high reactivity and extreme cohesiveness of the silica fume shotcrete or gunite also reduces the need for an accelerator. Such use of silica fume has resulted in rebound figures of less than 5%.

Shrinkage

Shrinkage in cement pastes has been found to be increased when using silica fume,[25] and so, when paste, mortar or grout systems are used, a compensator is usually added.

In concretes, the shrinkage is related to the aggregate volume and aggregate quality, and many reports are available[25,26] that show that the addition of silica fume will reduce shrinkage in concrete when close control is exercised over the mix design and aggregate selection. Curing is again stressed in these papers, due to the marked effect that curing can have on this concrete.

Creep

There is little information about the effect of silica fume on the creep of concrete. What is available refers to high-strength, high-dosage mixes. In general, it appears that silica fume concrete exhibits less creep in comparison with ordinary concrete.[26,27]

Fire resistance

Fire resistance is a property not often considered in normal concretes, though still of great importance. It is known that high-strength concretes may explode when exposed to fire. Several tests[28-30] have shown that under normal fire conditions, silica fume concrete does not behave any differently to normal concretes. Ultra-high-strength silica fume concrete may be susceptible to this type of failure due to increased brittleness, and consideration should be given to the mix design (lowest free water content) in this instance. ACI Report 216 is good a source for data on this aspect of silica fume concrete.

Abrasion/erosion

Low w/c ratio, high-strength silica fume concrete shows greatly improved resistance to abrasion and erosion. A large amount of silica fume concrete has been produced to specifically utilise this quality. A major repair project on the Kinzua Dam, USA, has been studied,[31] and results show good performance of the concrete used. A high number of hydropower projects in India are utilising silica fume concrete for this performance.

7.2.3.4 Hardened concrete: durability related properties

The use of a silica fume concrete, with its potential for greater strengths, both compressive and tensile, its more refined pore structure and its lower permeability, gives the opportunity of providing a more durable concrete with a longer working lifespan than a conventional concrete in the same environment.

Permeability

The two main methodologies of measuring permeability are either statically, such as allowing a concrete to dry out and noting the weight loss, or actively, by subjecting the material to a liquid or gas under pressure, and measuring the depth of penetration.

In studies using drying methods,[16,32] the efficiency factor for silica fume concrete was between 6 and 8. This indicates that the physical size and high reactivity of the silica fume have more influence on permeability than on compressive strength.

In actively testing the permeability, with water under pressure, early tests[32] showed results ranging from very little permeability to impermeable. Since these tests (1960s), several comprehensive evaluations have been made[33–35] which back up the previous results. Examinations on mature specimens[36] gave reduced permeability, and microscopic study revealed a very dense microstructure and the virtual absence of the weak layer normally surrounding the aggregate grains.

In all these tests, close attention was paid to the curing regimes used, and it was found that the curing time had a very marked influence on the results.

221

Permeability should not be confused with porosity, as the pore structure is modified but not decreased.[16]

The permeability of concretes, particularly to chemicals such as chlorides and sulfates, is a great concern around the world with regard to durability. Research has been ongoing, and results are frequently added to the literature.

Sulfate resistance

A major study was initiated in Oslo in the first years of testing silica fume concretes. Various specimens were buried in the acidic, sulfate-rich ground in Oslo. The 20 year results are available for this trial.[37,38] These results indicate that the silica fume concretes performed as well as the sulfate-resistant ones, and this is confirmed in the 40 year report and in further laboratory tests.[39,40]

When testing in conjunction with GGBS or PFA,[41] silica fume mixes were found to show greater resistance to sulfate attack than special sulfate-resisting cements. This has resulted in silica fume concrete being specified in areas such as the Arabian Gulf to combat severe deterioration in the concrete.[42]

Such performance of the concrete can be attributed to:

- the refined pore structure and thus the reduced passage of harmful ions[43]
- the increased amount of aluminium incorporated in the microsilica, thus reducing the amount of alumina available for ettringite formation
- the lower calcium hydroxide content.

Chloride resistance

Chlorides do not have to solely penetrate into the concrete to attack the reinforcement. Some chlorides may well be present in the aggregates or even the mixing water. Therefore, the ability of the concrete to withstand chloride penetration, from seawater or de-icing salts, is important, as is the reactivity of the mix in binding up the aggressive fraction of the chlorides that becomes present in the pore water. Studies have been made[5,44–46] which discuss the varying effects of the lower permeability and the reduction of pH in the pore water and how this relates to the presence of chlorides and the state of passivity of embedded steel.

It is considered that the lower the pH, then the lower the threshold limit for depassivation of the steel by chlorides. In studies of the penetration co-efficient of chlorides,[47] this was shown to be much lower in the silica fume concretes, thus negating the effect of the lower threshold value.

In general for equivalent strengths, initiation of chloride attack will be delayed in a concrete containing silica fume.

The initial curing and correct amount of cover are still important – reducing the penetration rate is no good if the distance to be travelled is also reduced.

In the USA a special rapid test – ASTM C1202[48] – has been applied to compare the chloride penetration of high-strength silica fume concrete with latex-modified and low w/c ratio concrete. The test is based on the resistivity of the concrete to the passage of an electrical charge. Properly designed, produced and finished silica fume concrete normally achieves a very low rating.

Carbonation
Results for tests on carbonation and carbonation rates are somewhat varied and contradictory depending on the viewpoint taken when analysing the findings. In a study[49] into the effect of silica fume on carbonation and transport of oxygen, adding up to 20% silica fume caused a slight reduction in these two actions in water-saturated concrete. In essence, the conclusions shown by those reports available[14,49,50] are that for equal strengths and any concretes below 40 MPa, carbonation is higher in silica fume concrete. Concretes above 40 MPa show reductions in the carbonation rate, and it is only these concretes that are deemed susceptible to attack and damage if there is reinforcement present.

As silica fume concrete is normally used where the average strengths are above 40 MPa, it is a moot point as to whether carbonation is a serious risk. Correct curing procedures are essential to ensure optimum performance of the silica fume concrete.

Leaching and efflorescence
This problem mainly occurs where one surface of the concrete is subjected to either continuous water contact or intermittent wetting and drying. The excess calcium hydroxide is leached through the concrete to the surface, where it carbonates, giving a white powdery deposit. Although a problem of aesthetics more than anything, if sufficient leaching occurs, it will result in increased porosity and permeability, and ultimately a weaker concrete. It has been found in studies[51] that the addition of silica

fume will reduce leaching. This is due to the refined pore structure and increased consumption of the calcium hydroxide. The results indicated that the more efficient the curing and the longer curing time before exposure, the more resistant the concrete became.

Frost resistance

It will be appreciated that as the major producers and users in the early years were the Scandinavian countries, this particular property of silica fume concrete has been well scrutinised. Investigations have included using the silica fume on its own, with superplasticisers, with varying dosages of air entrainers, and with different aggregates, and using various curing regimes.

In entraining air into the concrete, it is necessary to achieve the correct amount of air, and the right dispersion of the bubbles, so that the mix is stable enough to withstand compaction methods and will reach the required strength.

Many different concretes have been compared[52,53] to determine the effect of silica fume addition. It was found to be difficult to entrain air in a silica fume mix that did not use a plasticiser, but that by increasing the dosage of air entrainer and adding a plasticiser, it was easy to achieve the desired levels. There is some speculation as to the reason for the increased dosage of air entrainer, the most likely one being that the air and the silica fume are competing for the same space in the mix. Once in the mix, it was noted that the bubble spacing and stability was greatly improved. The variations in air content for given dosages, as sometimes happens when using PFA, were not noticed in the silica fume concretes.

Use of silica fume and air entrainment is put forward as the best option of all, with the microsilica maintaining good stability and uniform bubble spacing of the air, which gives maximum frost protection, based around a mix design guideline for a concrete of 30–50 MPa, utilising 8% microsilica and 5% air entrainment as an optimum. In all cases, the concrete should be cured for the longest allowable time before exposure to the working environment.

Alkali–silica reaction

To proceed into a lengthy discourse on this particular problem, its causes and prevalence would be inappropriate here, and the subject is well covered in the available literature.

To view the effect of silica fume on this form of attack, it is necessary to remember the three main factors that are required for potential reaction:

(*a*) a high alkali content in the mix
(*b*) reactive aggregates
(*c*) available water.

Silica fume reacts with the liberated calcium hydroxide to form calcium silicate hydrates, and this reduction leads to a lowering of the pH and a lower risk of reaction due to high alkali. In the formation of the calcium silicate hydrates, the K^+ and Na^+ ions are bound into the matrix, and cannot react with any potentially siliceous aggregates.

The minute size and the pozzolanic reactivity of the silica fume greatly refines the pore structure of the concrete, reducing the permeability, and this means less water can enter the concrete.

Thus, a normal dosage of microsilica, 10% by cement weight, can negate the main factors that could lead to the ASR in a concrete. Many reports are available[61–63] that document numerous studies on this.

7.2.3.5 Summary

The addition of silica fume to concrete will produce significant changes in the structure of the matrix, giving a densified, refined pore system and greater strength through both physical action and a pozzolanic reaction. In most cases, it is the refinement of the pore system, reducing permeability, that has the greatest effect on the performance of the concrete, rather than the increased strength.

Usage can be made of these improved qualities in designing concretes to comply with onerous requirements or greater resistance to certain hostile environments.

Silica fume should be considered as an addition to a mix rather than a replacement for cementitious content, and sensible mix design is essential.

Silica fume concrete is susceptible to poor curing conditions, and the effects of such are more pronounced than in ordinary concrete. Close attention to curing methods and times is important to ensure optimum performance.

Tailoring the design of a silica fume concrete to the specific requirements should always be a matter of consultation between the client, the contractor, the ready mix or precast supplier, and the technical section of the silica fume supplier. Reference should be made to any previous project or use, and trial work is 'compulsory' to ensure correct usage of the material and an appreciation of the characteristics of this type of concrete.

7.2.3.6 Acknowledgements

Background information for this chapter, as well as that released by Elkem Materials Ltd, was obtained from the following sources:

- FIP (1988) *Condensed silica fume in concrete.* Thomas Telford, London.
- Concrete Society (1993) *Microsilica in concrete.* Concrete Society Technical Report 41. Concrete Society, Camberley.
- Fidjestøl, P. and Lewis, R. (2003) Microsilica as an addition. In: P. Hewlett (ed.), *Lea's chemistry of cement and concrete.* Butterworth-Heinemann, Oxford, 4th edn, ch. 12.

Further reading

ACI (2006) *Guide to the use of silica fume.* Report 234R-06. American Concrete Institute, Farmington Hills, MI.
Elkem Materials: www.concrete.elkem.com.
Fidjestøl, P. and Lewis, R. (2003) Microsilica as an addition. In: P. 4th edn, ch. 12.
Silica Fume Association CD-Rom: www.silicafume.org.
Silica Fume Association User's Manual CD-Rom: www.silicafume.org.

References

1. Aitcin, P.C. and Regourd, M. (1985) The use of condensed silica fume to control alkali silica reaction – a field case study. *Cement and Concrete Research*, 15, 711–719.
2. Jorgen, J.A.H.R. (1980) *Possible health hazards from different types of amorphous silicas – suggested threshold limit values.* Revised report HD806/79. Institute of Occupational Health, Oslo.
3. Regourd, M. (1983) In: P.C. Aitcin (ed.), *Condensed silica fume.* Universite de Sherbrooke, Quebec, 20–24.
4. Andrija, D. (1986) In: *8th International Congress on Chemistry of Cement.* Rio de Janeiro, vol. 4, 279–285.
5. Page, C.L. and Vennesland, O. (1983) Pore solution composition and chloride binding capacity of silica fume cement pastes. *Materials and Structures*, 16, 19–25.
6. Parker, D.G. (1986) *Alkali aggregate reactivity and condensed silica fume*, SA 854/3C. Elkem Materials, Kristiansand.

7. Diamond, S. (1986) In: *8th International Congress on Chemistry of Cement.* Rio de Janeiro, vol. 1, 122–147.
8. Page, C.L. (1983) *Influence of microsilica on compressive strength of concrete made from British cement and aggregates.* Elkem Materials, Kristiansand.
9. Li, S., Roy, D.M. and Kumar, A. (1985) Quantitative determination of pozzolanas in hydrated systems of cement or $Ca(OH)_2$ with fly ash or silica fume. *Cement and Concrete Research*, 15, 1079–1086.
10. Sandvik, M. (1981) *Fasthetsutvikling for silicabetong ved ulike temperaturniva.* Report STF65 F81016. FCB/SINTEF, Norwegian Institute of Technology, Trondheim (in Norwegian).
11. Loland, K.E. (1983) Fasthets-og deformasjonsegenskaper i herdnet tilstand – herdebetingelser. In: *Bruk av silika i betong, Norsk Sivilingeniorers Forening* (seminar). Oslo (in Norwegian).
12. Loland, K.E. and Hustad, T. (1981) *Report 2: mechanical properties.* Report STF65 A81031. Norwegian Institute of Technology, Trondheim.
13. Sellevold, E.J. and Radjy, F.F. (1983) *Condensed silica fume (microsilica) in concrete: water demand and strength development*, SP-79. American Concrete Institute, Farmington Hills, MI.
14. Johansen, R. (1981) *Report 6: long term effects.* Report STF65 A81031 FCB/SINTEF, Norwegian Institute of Technology, Trondheim.
15. Maage, M. and Hammer, T.A. (1985) *Modifisert Portlandsement. Delrapport 3, Fasthetsutvikling og E-modul.* Report STF65 A85041. FCB/SINTEF, Norwegian Institute of Technology, Trondheim (in Norwegian).
16. Sellevold, E.J., Bager, D.H., Klitgaaxd Jensen, E. and Knudsen, T. (1982) *Silica fume-cement pastes: hydration and pore structure.* Report BML 82.610 Norwegian Institute of Technology, Trondheim.
17. Justesen, C.F. (1981) *Performance of densit injection grout for prestressing tendons.* Technical Advisory Service, Aalborg Portland, Aalborg.
18. de Larrard, F., Boulay, C. and Rossi, P. (1987) Fracture toughness of high strength concrete. In: *Proceedings of the Symposium on the Utilisation of High Strength Concrete.* Stavanger, 215–223.
19. Carles-Gibergues, A., Grandet, J. and Ollivier, J.P. (1982) Contact zone between cement paste and aggregate. In: P. Bartos (ed.), *Proceedings of the International Conference on Bond in Concrete.* Applied Science, London, 24–33.
20. Gjorv, O.E., Monteiro, P.J.M. and Mehta, P.K. (1986) *Effect of condensed silica fume on the steel-concrete bond.* Report BML 86.201. Norwegian Institute of Technology, Trondheim.
21. Monteiro, P.J., Gjorv, O.E. and Mehta, P.K. (1986) *Effect of condensed silica fume on the steel–cement paste transition zone.* Report BML 86.205. Norwegian Institute of Technology, Trondheim.
22. Bache, H.H. (1981) Densified cement/ultrafine particle-based materials. In: *2nd International Conference on Superplasticisers in Concrete.* Ottawa.

23. Krenchel, H. and Shah, S. (1985) Applications of polypropylene fibres in Scandinavia. *Concrete International: Design and Construction*, 2(3), 32–34.

24. Ramakrishnan, V. and Srinivasan, V. (1983) Performance characteristics of fibre reinforced condensed silica fume concrete. *American Concrete Institute*, 11, 797–812.

25. Traetteberg, A. and Alstad, R. (1981) *Volumstabilitet i blandingssementer med rajernslagg og silikastov*. Report STF65 A81034. FCB/SINTEF, Norwegian Institute of Technology, Trondheim (in Norwegian).

26. Johansen, R. (1979) *Silicastov i fabrikksbetong. Langtidseffekter*. Report STF65 F79019. FCB/SINTEF, Norwegian Institute of Technology, Trondheim (in Norwegian).

27. Buil, M. and Acker, P. (1985) Creep of silica fume concrete. *Cement and Concrete Research*, 15, 463–466.

28. Wolsiefer, J. (1982) Ultra high strength field placeable concrete in the range 10,000 to 18,000 psi (69 to 124 MPa). In: *American Concrete Institute Annual Conference*. Atlanta.

29. Maage, M. and Rueslatten, H. (1987) *Trykkfasthet og blaeredannelse pa brannpakjent hoyfastbetong*. Report STF65 A87006. FCB/SINTEF, Norwegian Institute of Technology, Trondheim (in Norwegian).

30. Shirley, S.T., Burg, R.G. and Fiorato, A.E. (1988) Fire endurance of high strength concrete slabs. *ACI Materials Journal*, 85, 102–108.

31. Holland, T.C. (1983) *Abrasion-erosion evaluation of concrete admixtures for stilling basin repairs, Kinzua Dam, Pennsylvania*. Miscellaneous paper SL-83-16. US Army Engineer Waterways Experiment Station, Structures Laboratory, Vicksburg, MS.

32. Sorensen, E.V. (1982) Concrete with condensed silica fume. *A preliminary study of strength and permeability*. Report BML 82.610. Norwegian Institute of Technology, Trondheim (in Danish).

33. Markestad, S.A. (1977) *An investigation of concrete in regard to permeability problems and factors influencing the results of permeability tests*. Report STF65 A77027. FCB/SINTEF, Norwegian Institute of Technology, Trondheim.

34. Hustad, T. and Loland, K.E. (1981) *Report 4: permeability*. Report STF65 A81031. FCB/SINTEF, Norwegian Institute of Technology, Trondheim.

35. Sandvik, M. (1983) *Silicabetong: herdevarme, egenkapsutvikling*. Report STF65 A83063. FCB/SINTEF, Norwegian Institute of Technology, Trondheim (in Norwegian).

36. Skurdal, S. (1982) *Egenskapsutvikling for silicabetong ved forskjellige herdetemperaturer*. Report BML 82.416. Norwegian Institute of Technology, Trondheim (in Norwegian).

37. Maage, M. (1984) *Effect of microsilica on the durability of concrete structures*. Report STF65 A84019. FCB/SINTEF, Norwegian Institute of Technology, Trondheim.

38. Fiskaa, O. *et al.* (1971) *Betong i Alunskifier.* Publication 86. Norwegian Geotechnical Institute, Oslo (in Norwegian).

39. Fiskaa, O.M. (1973) *Betong i Alkunskifier.* Publication 101. Norwegian Geotechnical Institute, Oslo (in Norwegian).

40. Mather, K. (1980) Factors affecting the sulphate resistance of mortars. In: *Proceedings of the 7th International Conference on Chemistry of Cements.* Paris, vol. 4, 580–585.

41. Carlsen, R. and Vennesland, O. (1982) *Sementers sulfat- og sjovannsbestandighet.* Report STF65 F82010.FCB/SINTEF, Norwegian Institute of Technology, Trondheim (in Norwegian).

42. Rasheeduzzafar, *et al.* Proposal for a code of practice for durability of concrete in the Arabian Gulf environment.

43. Popovic, K., Ukraincik, V. and Djurekovic, A. (1984) Improvement of mortar and concrete durability by the use of condensed silica fume. *Durability of Building Materials,* 2, 171–186.

44. Mehta, P.K. (1981) Sulphate resistance of blended Portland cements containing pozzolans and granulated blastfurnace slag. In: *Proceedings of the 5th International Symposium on Concrete Technology.* Monterey, CA.

45. Page, C.L. and Havdahl, J. (1985) Electrochemical monitoring of corrosion of steel in microsilica cement pastes. *Materials and Structures,* 18(103), 41–47.

46. Monteiro, P.J.M., Gjorv, O.E. and Mehta, P.K. (1985) Microstructure of the steel–cement paste interface in the presence of chloride. *Cement and Concrete Research,* 15, 781–784.

47. Fisher, K.P. *et al.* (1982) *Corrosion of steel in concrete: some fundamental aspects of concrete with added silica.* Norwegian Geotechnical Institute, Oslo, Report 51304–06.

48. Christensen, D.W., Sorensen, E.V. and Radjy, F.F. (1984) Rockbond: a new microsilica concrete bridge deck overlay material. In: Proceedings of the International Bridge Conference. Pittsburgh, PA, 151–160.

49. Vennesland, O. and Gjorv, O.E. (1983) *Silica concrete – protection against corrosion of embedded steel,* SP-79. American Concrete Institute, Farmington Hills, MI.

50. Vennesland, O. (1981) *Report 3: corrosion properties.* Report STF65 A81031. FCB/SINTEF, Norwegian Institute of Technology, Trondheim.

51. Samuelsson, P. (1982) *The influence of silica fume on the risk of efflorescence on concrete surfaces.* Report BML 82.610. Norwegian Institute of Technology, Trondheim (in Swedish).

52. Okkenhaug, K. and Gjorv, O.E. (1982) *Influence of condensed silica fume on the air-void system in concrete.* Report STF65 A82044. FCB/SINTEF, Norwegian Institute of Technology, Trondheim.

53. Okkenhaug, K. (1983) *Silikastovets innvirkning pa luftens stabilitet i betong med L-stoff og med L-stoff i kombinasjon med P-stoff.* CBI Report 2:83.

Fig. 7.30 High-resolution scanning electron micrograph (SEM) of kaolinite

of their essential constituents, and by the nature and amount of accessory, associated minerals. The precise selection of the calcination temperature is essential, as the reactivity of the calcined product increases with increasing calcination temperature to an optimum, and declines again as the temperature increases further. Under heating, three main processes – dehydration, dehydroxylation and recrystallisation – take place, depending on the temperature:

(a) low temperature (below ~400°C), involving dehydration
(b) intermediate temperature (400–750°C), involving dehydroxylation
(c) high-temperature reactions (above 750°C), resulting mostly in recrystallisation of new phases.

The thermal reactions of kaolinite are as follows:

$$\text{kaolinite} \rightarrow \text{MK} (\sim750°C) \rightarrow \text{spinel} (\sim900°C)$$

$$\rightarrow \text{mullite} (\sim1000°C)$$

The optimum calcination temperature, from the point of view of maximum strength of MK–portlandite paste, varies slightly. For example, Ambroise *et al.*[5] determined the effect of kaolinite calcining temperature (600–800°C) on the strength development of 1:1 MK–portlandite

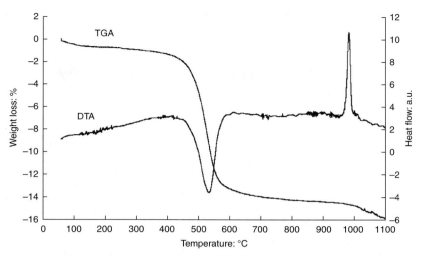

Fig. 7.31 Differential thermal analysis (DTA) and thermogravimetric analysis (TGA) curves for pure kaolinite

pastes of water to a solid ratio of 0.72 at different curing times, and found an optimal calcination temperature of 700°C. Fialips *et al.*[3] and Odler[4] found that the optimum calcination temperature of kaolinite was 650°C.

On heating, kaolinite undergoes both chemical and structural transformations. The thermal transformations of kaolinite have been the subject of a large number of investigations, which showed that the temperature of heating, the rate and the time, as well as the cooling parameters at the end of the production cycle, significantly influence the dexydroxylation process.[6,7] It should be noted that heating may have both positive and negative effects on the reactivity of pozzolana.

The major quantitative criterion for the evaluation of the effectiveness of the heat treatment of kaolinite is the degree of its dehydroxylation. The dehydroxylation of kaolinites in a normal atmosphere results in a mass loss of 13.95%, which corresponds to the transition of $SiO_2 \cdot 2Al_2O_3 \cdot 2H_2O$ to $SiO_2 \cdot 2Al_2O_3$. A strong endothermic effect in the range 450–600°C (the exact temperature interval depends on the crystallinity and particle size) is typical for most kaolinites. The well-defined endothermic peak (Fig. 7.31) associated with this effect corresponds to the formation of a new disordered phase – MK ($Al_2Si_2O_7$) – via the dehydroxylation route followed by effluent of water, as shown by mass spectrometry (Fig. 7.32). The dehydroxylation of kaolinite is preceded by a 'predehydroxylation' stage in the range 80–150°C, where the evolution of a small amount of water is observed

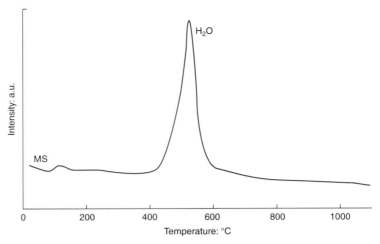

Fig. 7.32 MS pattern of effluent water for pure kaolinite

(Fig. 7.32). Between 900 and 1100°C, the exothermal formation of crystalline phases, such as spinel and mullite ($Al_6Si_2O_{13}$), where silica is totally or partly segregated, is characterised by an exothermic peak on the DTA curve and insignificant weight loss on the TGA curve (see Fig. 7.31).

According to Shvarzman et al.,[8] the dehydroxylation is accompanied by kaolinite amorphisation, which affects the pozzolanic activity of MK.

Mechanochemical method

Kaolinite may be also activated by a mechanochemical method,[9-12] which is usually based on combined grinding (either dry or wet) and intercalating kaolinite with molecules which otherwise would not readily insert between the kaolinite layers.

It was observed that the mechanochemical treatment of kaolinite results in partial dehydroxylation of kaolinite.[1] New additional phases of kaolinite were found, and modification of the hydroxyl surfaces was extensive even with mild heating.

The mechanochemical modification of kaolinite through grinding results in a high-surface-area material. Intense local heating can occur, thus resulting in the formation of new surfaces.

Mechanochemical activation causes significant changes in the kaolinite structure (increasing the number of lattice defects), in surface energy and in chemical reactivity. Mako et al.[10] concluded that the destruction of the crystal structure of kaolinite (by the rupture of the O–H, Al–OH, Al–O–Si, and Si–O bonds) was significantly influenced by the kaolinite/quartz ratio and by grinding time.

7.2.4.3 Hydration of the cement–MK system

MK–portlandite reactions

Many researchers have studied the kinetics of the MK–portlandite reactions, detailing the nature of the hydration products formed. Hydrates formed are essentially calcium silicate hydrate (C–S–H), gehlenite hydrate (C_2ASH_8), C_3AH_6 and C_4AH_{13}. The MK content, curing conditions and the presence of chemical activators are usually the main influencing factors on the hydration reaction and new-formed hydrated products.[13–18]

Murat[19] studied MK–lime mixtures. For a mass ratio of 1:1, hydration products such as C–S–H and C_2ASH_8 were observed.

De Silva and Glasser[20] reported that in an MK–lime mixture of the same composition at 55°C, C–S–H developed after 24 hours of hydration, and that after 90 days the hydration products were C_2ASH_8 and C_4AH_{13}, hydrogarnet and C–S–H. They indicated that C_2ASH_8 and C_4AH_{13} are probably present as metastable phases, which after long curing periods convert to hydrogarnet with variable composition, $C_3AS_zH_{6-2z}$. This phenomenon may lead to a negative influence on the performance of MK-blended matrixes, mainly on durability.

Gabrera and Rojas[21] investigated the reaction kinetics of MK–lime mixtures in water at 60°C. They reported that the first product of hydration is C–S–H, which is detected clearly at 6 hours. At 12 hours of hydration, C–S–H, C_2ASH_8 and C_4AH_{13} are clearly detected. After 21 hours of hydration, hydrogarnet appears. According to Cabrera et al.,[13] hydrogarnet is not formed from a transformation reaction but is the result of a direct reaction between MK and lime.

Wild and Khatib[22] have suggested that in Portland cement (PC)–MK–water systems at ambient temperatures, the most likely phase to coexist with C–S–H at short curing times and low MK levels is C_4AH_{13}, whereas at long curing times and high levels of MK it is C_2ASH_8.

Rojas and Cabrera[23] related the absence of C_4AH_{13} in the reaction kinetics of the MK–PC system to the presence of SO_4^{2-} ions in the matrix.

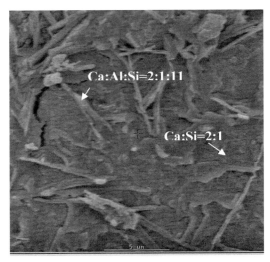

Fig. 7.33 SEM of MK–portlandite paste (at age 90 days)[41]

Shi and Day[15] concluded that the increase in the curing temperature accelerated the reaction rate but did not have an effect on the hydration products in the pastes.

According to Murat[19] and Wild *et al.*,[31] the hydration of MK may occur in the following form:

$$AS_2 + 7CH + 9H \rightarrow C_4AH_{13} + C_3S_2H_3$$

$$AS_2 + 5CH + 3H \rightarrow C_3AH_6 + 2CSH$$

$$AS_2 + 3.5CH + 6H \rightarrow C_2ASH_8 + 0.5C_3S_2H_3$$

Microstructure of MK–portlandite pastes

Figure 7.33 shows an SEM of hydrated MK–portlandite paste at age 90 days. Thermally treated (under 600°C) standard pure kaolin was used for the preparation of MK–portlandite paste. It can be seen that the paste with the treated standard kaolin is a dense, well-formed crystalline structure containing a large amount of gehlenite hydrate (fibre-like plates) and some nodular spongy areas, corresponding to the C–S–H microstructure. Ca:Si and Ca:Si:Al ratios were measured by energy-dispersive spectrometry (EDS) as the average of three measurements at the same zone. According to the EDS analysis, the Ca:Si ratio was equal to 2 for C–S–H; the Ca:Al:Si ratio was 2:1:1 for gehlenite hydrate.

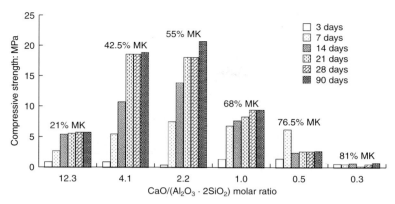

Fig. 7.34 Compressive strength versus CaO/(Al$_2$O$_3$·2SiO$_2$) molar ratio[41]

Chemical activity of MK

The reactivity of a calcined clay such as MK with calcium hydroxide is often determined by a so-called 'Chappelle test', in which a dilute slurry of the calcined clay is reacted with excess calcium hydroxide at 95°C for 18 hours. At the end of this period, the amount of unreacted calcium hydroxide is determined by titration. The results are expressed as the weight of Ca(OH)$_2$ in milligrams absorbed by 1 g of calcined clay, e.g. MK.[24]

The compressive strength test is another method to determine chemical activity of MK. The compressive strength development of pastes with different lime–MK, CaO/(Al$_2$O$_3$·2SiO$_2$), molar ratios is presented in Fig. 7.34. As can be seen, the molar ratio influences compressive strength development. There is an optimal CaO/(Al$_2$O$_3$·2SiO$_2$) molar ratio of 2.2–4.1, which corresponds to an MK content of 42–55%, from the viewpoint of compressive strength (chemical activity).[25] In the molar ratio range 0.2–4.1, a drastic increase in compressive strength was observed. As can be seen from Fig. 7.34, such intensive strength growth was observed within a very short curing time (within the first 3 weeks). However, at molar ratios >4.1 a severe decrease in compressive strength occurred at all the ages tested.

Effect of the portlandite/MK ratio on phase formation and volumetric quantities

The DTA and TGA curves for portlandite–MK paste with different molar ratios tested at age 90 days are shown in Figs 7.35 and 7.36,

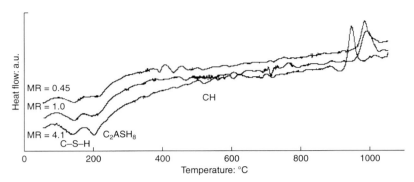

Fig. 7.35 DTA curves for portlandite–MK paste with different CaO/(Al₂O₃·2SiO₂) molar ratios (MRs)[41]

respectively. Figure 7.37 shows the weight loss associated with decomposition of hydration products as a function of the portlandite/MK molar ratio. Figure 7.38 shows the content of hydrated products as a function of portlandite/MK molar ratio at age 90 days.

As can be seen from Fig. 7.38, the main phases formed at hydration are C_2ASH_8 (gehlenite hydrate) and C–S–H.

From Figs 7.35 and 7.36 the following is apparent:

- Dehydration of calcium silicate hydrates, which is manifested by endotherms at ~115–170°C, increased in size with increasing molar ratio. The small endotherm, below 100°C, present in all samples, is most likely due to moisture adsorbed during sample preparation for thermal analysis.

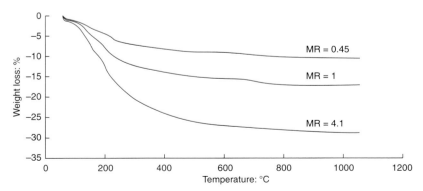

Fig. 7.36 TGA curves for portlandite–MK paste with different CaO/(Al₂O₃·2SiO₂) molar ratios[41]

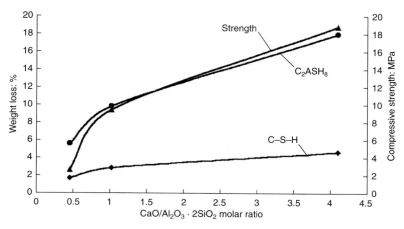

Fig. 7.37 *Influence of $CaO/(Al_2O_3 \cdot 2SiO_2)$ molar ratio on weight loss and compressive strength (age 90 days)* [41]

- The endotherm centred at ~200°C, due to the dehydration of gehlenite hydrate, C_2ASH_8, increased in size with increasing molar ratio.

7.2.4.4 *Properties of cement–MK systems at early ages*

The properties of MK-based cementitious systems at early ages often have a primary influence on their durability. This is why it is important

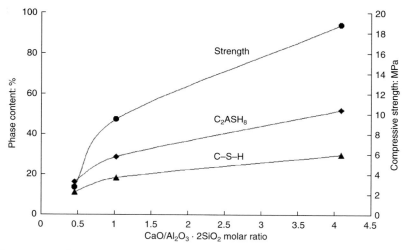

Fig. 7.38 *Influence of $CaO/(Al_2O_3 \cdot 2SiO_2)$ molar ratio on the content of hydrated products (age 90 days)* [41]

to address at least some of these properties before discussing durability aspects.

Water demand

The mean diameter of commercial MK particles is about 1–5 μm. The small size of the MK particle means that the material has a large specific surface area, about 12 000–18 000 m²/kg. The high specific surface area of MK leads to an increased water demand, which should be taken into account in concrete mix design.

Workability

MK adversely affects the workability of concrete.[26] This is the consequence of high water demand, one of the disadvantages of MK. Bai et al.[27] reported a decrease in both slump and compacting factor and an increase in the Vebe time as the content of MK in concrete mixtures increased from 0 to 15%. Rols et al.[28] reported that the utilisation of MK requires higher quantities of superplasticiser compared with that in a control concrete. Wild et al.[40] found it necessary to employ up to 3% superplasticiser to produce moderate slumps (75 mm) in MK concrete (10% MK) with a water/binder ratio of 0.45.

Heat of hydration

Recent work indicates that MK–PC blends produce greater heats of hydration and temperature rise in concrete and mortar than does PC alone. For example, Zhang and Malhotra[29] reported a temperature rise 7°C greater than PC concrete when 10% of PC was replaced with MK, whereas 10% silica fume produced a temperature rise only slightly above (0.5°C) that of the PC concrete. Ambroise et al.,[5] in similar work on MK–PC mortar using semiadiabatic calorimetry, reported temperature rises above that of the control PC mortar by 8, 6 and 1°C, at PC replacement levels by MK of, respectively, of 10, 20 and 30%. This enhanced temperature rise could lead to severe problems of thermal stress and cracking, particularly in large-volume mass concrete.

Early age strength

The effect of MK on compressive strength development has been the subject of several studies. Recent work has shown that when MK is

240

used as a partial cement replacement in concrete, it gives significantly enhanced early strength and increased long-term strength. This is because it acts as a filler, it accelerates initial cement hydration and, in the early stages of curing, it rapidly consumes the portlandite produced by cement hydration to produce additional cementitious reaction products.

Ambroise *et al.*,[5] in an investigation of the properties of PC–MK pastes (at water/binder ratios between 0.25 and 0.54 and MK contents of 0, 10, 20, 30, 40 and 50%), reported maximum strength at 10% replacement with MK. A similar influence of MK on the strength of paste and mortars has been reported by several authors.[30,36]

Chemical shrinkage

Wild *et al.*[31] investigated the autogenous and chemical shrinkage and swelling of MK–PC pastes with different contents of MK (5–25%), with a water/binder ratio of 0.55. The specimens were monitored for periods of between one and 45 days. Chemical shrinkage was also found to reach a maximum between 10 and 15% MK, and then decrease sharply for higher MK contents.

Autogenous shrinkage

Autogenous shrinkage can be a major problem for high-performance concrete. As mentioned before, the use of certain pozzolanas, such as silica fume and fly ash, results in increased autogenous shrinkage.

The effect of MK on the autogenous shrinkage of cement pastes can be the consequence of four phenomena: (*a*) cement dilution by MK, less cement generating less shrinkage; (*b*) heterogeneous nucleation of hydrates on the surface of MK particles, accelerating cement hydration and, consequently, increasing shrinkage; (*c*) the pozzolanic reaction of MK with CH; (*d*) increase of capillary tension, due to the refinement of pore size distribution, leading to an increase in autogenous shrinkage.

Using MK as a partial cement replacement may lead to reduced autogenous shrinkage. For example, Rols *et al.*[28] measured autogenous shrinkage in concrete containing 9% MK. At all times the autogenous shrinkage was found to be lower than that in the control specimens.

Brooks and Johari[32] found that at very young ages, from the initial setting up to 20 hours, the autogenous shrinkage of concrete with a low water/binder ratio of 0.28 decreased with increasing of MK content.

Wild *et al.*[33] showed that, for cement–MK pastes (water/binder ratio of 0.55), autogenous shrinkage increased for an MK content up to a maximum of 10%, then it appeared to be comparable to that of the control cement paste for an MK content above 15%. An expansion up to 14 days for all compositions (0–25% MK) was also observed, except for 10% MK. Kinuthia *et al.*[34] observed in cement–MK pastes (water/binder ratio of 0.50), that 5 and 10% MK increased the autogenous shrinkage of cement pastes, while at 15 and 20% MK, they observed a significant decrease.

According to Gleize *et al.*,[35] a significant part of the reduction of long-term autogenous shrinkage in cement–MK pastes is due to the MK pozzolanic activity. A long-term autogenous shrinkage of cement–MK paste, with water/binder ratios of 0.3 and 0.5, decreased as the cement replacement level by MK increased (5, 10, 15 and 20%).

7.2.4.5 *Properties of hardened cement–MK systems*

Porosity

In most cases, mortar and concrete containing material with pozzolanic characteristics have porosity values equal to or superior to that of PC concrete. This evaluation of the porosity depends on the characteristics of pozzolanic materials, such as fineness, specific surface area, mineralogical and chemical composition, content and quality of the amorphous phase, and an important aspect to point out is the water/binder ratio. It is well known that the critical factor affecting the performance and durability of concrete structure is the pore size distribution, rather than the total porosity.

The main physical effect of MK on the microstructure of hardened concrete is a refinement of the pore structure.

Khatib and Wild[36] reported a refinement of the pore structure and a total intruded pore volume increase between 14 and 28 days for pastes with 5, 10 and 15% MK.

Rojas and Cabrera[23] studied the effect of MK content on the pore size distribution and the degree of hydration of cement–MK paste. They observed that the total porosity decreases up to 28–56 days of curing time. Above this age, the total porosity of MK mixes increases with respect to the PC paste. A similar phenomenon for the capillary porosity was also observed. In this case, the values obtained for MK mixes decreased with respect to PC paste. The best evidence for the effect of MK on the refining of the pore structure was detected in pores with a radius smaller than 100 Å. Between 7–90 days, the gel porosity of MK mixes increased, while the porosity in the PC mix was practically constant.

Curing conditions influence the porosity of the MK system.[37] The pore volume of autoclaved plain and MK concrete was found to be less than that of moist plain and MK concrete.

The improvement in pore structure is reflected in the higher compressive strength, but it is also of great importance to the permeability of the material.

Another important factor is the improvement in quality of the interfacial transition zone. The partial replacement of cement with ultra-fine particles of MK plays an important role in densifying the interfacial transition zone because of the microfiller effect due to the relatively fine particles of MK.[38]

Compressive strength
The development of strength in concrete containing MK is similar but not identical to that observed in mortars and pastes. Factors such as the amorphous phase content, the type of cement, the type and content of the chemical admixture, aggregate properties, and the type of mix influence the strength development of MK concrete.

Caldarone et al.,[39] Wild et al.[40] and Shvarzman[41] reported that MK–PC concretes exhibited strengths which were slightly greater than silica fume–PC concretes at the same levels of cement replacement by pozzolanas, which is shown in Fig. 7.39.

Wild et al.[40] and Sabir et al.[26] studied the influence of two factors, the filler effect and curing temperature, on the compressive strength. The filler effect, which is the immediate cause of the acceleration of hydration during the first 24 hours, and the pozzolanic reaction, has its maximum effect within the first 7–14 days for all MK levels between 5 and 30%. The degree to which the strength is enhanced declines beyond 14 days, although strength gains relative to the control are still present after 90 days. Little strength advantage is gained for MK levels in excess of 15%. Similar findings were observed by Curcio et al.[42] in mortars containing 15% MK.

The influence of curing temperatures on the strength development in concretes containing up to 15% MK was studied by Sabir et al.[26] They reported increased early strength (7 days) of concrete at curing temperature of 50°C compared with the strength of specimens cured

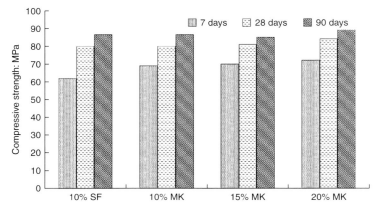

Fig. 7.39 Compressive strength development of concrete with different additives (SF, silica fume; water/binder ratio of 0.40)[41]

at 20°C. The acceleration in strength development due to the high temperature diminishes in the long term.

Khatib and Hibbert[43] investigated the influence of incorporating ground granulated blast furnace slag (GGBS) and MK on the strength of concrete. The incorporation of MK caused an increase in compressive strength, especially during the early ages of curing. The decrease in compressive strength during the early ages of curing of GGBS concrete is compensated by the inclusion of MK. The presence of MK in concrete containing GGBS causes an increase in flexural strength. The effect was particularly beneficial at 20% MK and 60% GGBS, where the flexural strength exceeded that of the control at 90 days of curing.

Flexural strength
The relationships between flexural and compressive strength in MK concrete are similar to those of ordinary concrete. The flexural strength of MK concrete is higher that that of silica fume concrete (Fig. 7.40).

Elasticity
The modulus of elasticity of concretes at age 28 days is illustrated in Fig. 7.41. At 28 days, the concrete containing silica fume had a higher modulus of elasticity compared with MK concrete. The modulus of elasticity increases with the increase in MK content.

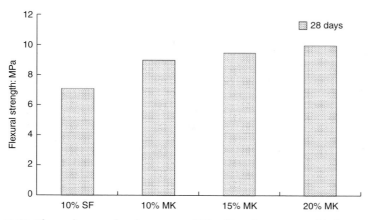

Fig. 7.40 Flexural strength of concrete (SF, silica fume; water/binder ratio of 0.40)[41]

Drying shrinkage

MK concrete has lower drying shrinkage compared with that of silica fume concrete. Figure 7.42 shows the dry shrinkage strain of concretes with different contents of MK and silica fume at age 28 and 90 days. At age 90 days the MK concrete had a drying shrinkage strain of 376×10^{-6} compared with 636×10^{-6} for silica fume concrete. An increase of MK content from 10 to 20% results in an insignificant decrease in drying shrinkage.

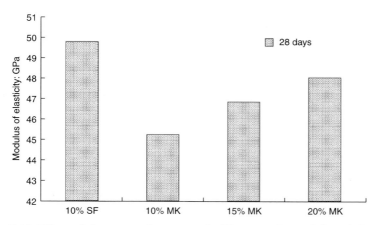

Fig. 7.41 Effect of the type and content of additive on the modulus of elasticity concrete (SF, silica fume; water/binder ratio of 0.40)[41]

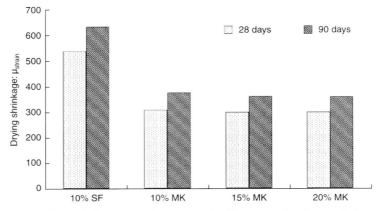

Fig. 7.42 Effect of the type and content of additive on the drying shrinkage of concrete (SF, silica fume; water/binder ratio of 0.40)[41]

7.2.4.6 Durability aspects

Portlandite consumption

The amount of MK required for complete elimination of calcium hydroxide depends on a number of factors such as the purity of the MK, the Portland cement composition, the wate/binder ratio and the curing conditions. The reduction in the calcium hydroxide content results in superior strength and durability performance.[22,26]

Oriol and Pera[44] determined the level of portlandite consumption in MK–PC pastes under microwave curing conditions. They reported that total elimination of calcium hydroxide is achievable in binders containing 15% MK. They also reported that between 30 and 40% MK is required to remove all the calcium hydroxide in PC–MK pastes when cured in lime-saturated water at room temperature for 28 days.

Resistance to chloride ion penetration

Ingress of chloride ions into reinforced concrete and the corresponding corrosion behaviour is one of the major problems in the cement and building materials industry.

The ability of concrete to bind chloride ions and retard their rate of penetration is an important factor in determining the service life of concrete structures with regard to chloride-induced corrosion.

In hardened cement, chloride may be bound in the C–S–H gel or as a result of the formation of complex calcium oxychloride, Friedel's salt

($Ca_2Al(OH)_6Cl \cdot 2H_2O$). MK–Portland cement paste can bind considerable amounts of dissolved chloride present in pore water with a relatively low reduction in pH due to the formation of stable Friedel's salt.

The increase in chloride-binding capacity observed for MK-blended paste samples could be attributed to the participation of calcium aluminate species in the formation of Friedel's salt which would otherwise be engaged in the formation of C_2ASH_8 and C_4AH_{13}.[45] The mechanism of chloride ion binding in lime–MK mixtures was studied by Zibara *et al.*[46] and Saikia *et al.*[47] The influence of the calcium-to-silica (C/S) and calcium-to-alumina (C/A) ratios of hydration products on the chloride-ion-binding capacity of lime–MK systems were investigated. The alumina content of the mixture has an important influence on the chloride-ion-binding capacity. At a high C/A ratio, the formation of monocarboaluminate is favoured, which has a high ability to form Friedel's salt, and does so at low chloride concentrations (less than 0.1 M). With a low C/A ratio, the formation of stratlingite is favoured, with little formation of monocarboaluminate. Stratlingite does convert to Freidel's salt, but at a much lower rate, and much of the stratlingite persists at higher chloride concentrations.

Boddy *et al.*[48] studied the chloride penetration resistance of concrete containing MK. The results showed that a higher MK content and a lower water/binder ratio decreased diffusion, permeability and conductivity but increased resistivity. Resistance to chloride migration increased with increasing MK content and decreasing water/binder ratio. Similar results were obtained by other authors.[41,49–52]

The resistance of the MK concrete to chloride ion penetration was significantly higher than that of silica fume concrete (Fig. 7.43). At 28 days, the charge passed through the MK concrete (365 coulombs) was very low in comparison with the SF concrete (750 coulombs). The charge passed through the MK concrete decrease with increased curing time, and at 90 days the charge passed through the 10% MK concretes was only 160 coulombs, and 86 coulombs for concrete with 20% MK. According to ASTM C1202, when the charge passing through concrete is below 100 coulombs, the concrete is inert and can be considered as impermeable to chlorides.

Resistance to freezing and thawing
Both scaling resistance of concrete surfaces exposed to freezing–thawing cycles in the presence of de-icing chemicals (ASTM C672)

247

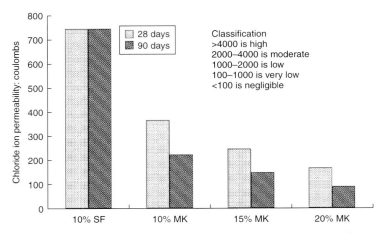

Fig. 7.43 Resistance of concrete to chloride ion penetration (SF, silica fume; water/binder ratio of 0.40)[41]

and the direct freeze–thaw test under water-saturated conditions (ASTM C666) showed that MK–cement mixes exhibited high freeze–thaw durability, similarly to that of silica fume concrete.[39]

Carbonation

The carbonation of concretes mixes with MK is characterised by a significant increase in the carbonation depth according to the increase in the replacement ratio.[41] It can be explained by the fact that the replacement of cement by MK decreases the content of portlandite in hydrate products due to the pozzolanic reaction.

Sulfate resistance

The reported properties of MK suggest that it should be, in general, highly effective in promoting sulfate resistance.[53,54] Sulfate attack on concrete is primarily attributed to sodium, magnesium and calcium salts. However, the samples exposed to sodium sulfate and magnesium sulfate environments show different behaviours, which indicates that the mechanisms of deterioration in these environments are different.

Khatib and Wild[55] have investigated the effect of cement type with different C_3A content on the resistance of MK mortar to sulfate attack under moist curing conditions. The sulfate expansion results demonstrated that the sulfate resistance was increased as the replacement

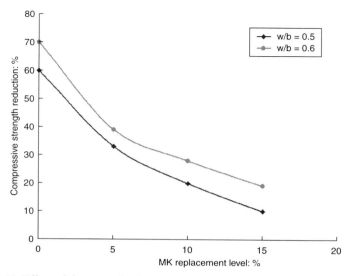

Fig. 7.44 Effect of the water/binder ratio on the compressive strength reduction of MK concrete[51]

level of cement with MK increases, up to at least 25% replacement. After prolonged periods of exposure to Na_2SO_4 solution, there was significant strength loss of PC mortars and mortars with a low level of MK (5 and 10%); for mortars with a high level of MK (15–25%) there was a consistent strength gain.

Al-Akhras[37] studied the effect of MK replacement of cement on the resistance of concrete to sodium sulfate attack under different parameters, including MK replacement level, water/binder ratio, curing type, initial moist curing and air content. The sulfate resistance of MK concrete increased with increasing MK replacement level. Autoclaved MK concrete showed superior sulfate resistance compared with moist-cured MK concrete. Increasing the air content (from 1.5 to 5%) improved the sulfate resistance of MK concrete (Figs 7.44 and 7.45).

Limestone cement mortar is susceptible to the thaumasite-kind of sulfate attack at low temperature. Incorporation of MK improved the resistance of the limestone cements against sulfate attack.[56] MK had been found effective in reducing the thaumasite type of sulfate attack in concrete containing limestone filler.[57]

Lee et al.[58] found that MK negatively influences the performance of mortar or paste specimens when exposed to magnesium sulfate solution.

249

A short list of recent applications of MK–cement concrete reported by the BASF company follows:

- High-reactivity MK was specified for the new US Embassy in Beijing. The $50\,000\,m^2$ building was completed in time for the 2008 Olympic Games.
- MK with high-range water-reducing admixture has been used to produce very high strength concrete. For example, the New York department of transportation approved MK for use in high-performance concrete jobs. As a result, the concrete admixture was used in high-visibility work on the Erie Canal, along with new Interstate 87 bridges over the Hudson River, and other projects, around the state.
- Departments of Transportation from Florida to Wyoming, the Port Authority of New York and New Jersey, and the Pennsylvania Turnpike Commission have approved high-reactivity MK for use in high-performance concrete.
- MK has been used in a 23-storey condominium in a seismically active zone (Figs 7.46 and 7.47). The high early strength of the

Fig. 7.46 A 23-storey condominium being built with 80 MPa concrete containing high-reactivity MK. The large columns and beams of the moment frame, exposed at the centre of the east and west façades, is in counterpoint to the minimal columns required around the rest of the building (Courtesy of Michael Chusid, www.chusid. com.)

Fig. 7.47 This condominium in Los Angeles is believed to be the first high-rise structure to use high-reactivity MK in concrete with a strength of 80 MPa (Courtesy of Michael Chusid, www.chusid.com.)

MK-enriched concrete prevented thermal cracking despite concerns about the heat of hydration in the thick columns.

- The resistance of MK to the penetration of chloride ions is also attracting interest for marine applications. MK concrete was used for a wharf at Shakespeare Bay in Queen Charlotte Sound, New Zealand. The wharf was built on concrete piles up to 30 m long, and has a 200 m-long deck.
- Underwater cellular MK concrete was used to repair and protect piles and pile caps on a 183 m long, 50-year-old dock on the East River, New York.

253

7.2.4.8 Summary

- MK can be considered as a highly pozzolanic material, and appears to have excellent potential as a supplementary cementing material for high-performance cementitious materials.
- The main phases that form in the hydrated MK–portlandite system are C_2ASH_8 and C–S–H.
- The hydration process of the MK–portlandite system is very intensive at an early age. The rate of both strength development and the formation of hydration products achieves the maximum at age 2 weeks.
- The portlandite/MK molar ratio influences the hydration rate, type, content and volume of the hydration phases and, consequently, strength development.
- The type, content and volume of hydrated products, as well as their morphology, significantly influence the development of compressive strength.
- The content of MK influences the compressive strength of high-performance concrete.
- The optimum level of cement replacement seems to be around 10–20%, where maximum strength is observed. The compressive strength of concrete made with 10–20% of MK was 20–25% higher than that of reference concrete at age 90 days.
- MK concrete has lower drying shrinkage compared with silica fume concrete.
- MK–cement mixes exhibit high freeze–thaw durability, similar silica fume concrete.
- The addition of MK reduces the ingress of chloride by improving the microstructure and chloride-binding behaviour. The resistance of MK concrete to the penetration of chloride ions is higher than that of silica fume concrete.
- MK enhances both the mechanical properties and durability of glass-fibre-reinforced cementitious materials, and reduces the alkali–silica reaction and efflorescence.
- MK has a beneficial effect on the resistance to sodium sulfate attack, and reduces the thaumasite type of sulfate attack, but negatively influences the resistance to magnesium sulfate attack.

References

1. Mendelovici, E. (1997) Comparative study of the effects of thermal and mechanical treatments on the structures of clay minerals. *Journal of Thermal Analysis*, 49, 1385–1397.

2. Onina, M. (1993) New method of activation. *Beton i Zhelezobeton*, 6, 12–14 (in Russian).

3. Fialips, C., Petit, S. and Decarreau, A. (2000) Hydrothermal formation of kaolinite from various metakaolins. *Clay Minerals*, 35, 559–572.

4. Odler, I. (2000) *Special inorganic cements*. Spon, London.

5. Ambroise, J., Maximilien, S. and Pera, J. (1994) Properties of metakaolin blended cements. *Advanced Cement Based Materials*, 1, 161–168.

6. Akolekar, D., Chaffee, A. and Howe, R.F. (1997) The transformation of kaolin to low-silica X zeolite. *Zeolites*, 19, 359–365.

7. He, C., Osbaeck, B. and Makovicky, E. (1995) Pozzolanic reactions of six principal clay minerals: activation, reactivity assessments and technological effects *Cement and Concrete Research*, 25(8), 1691–1702.

8. Shvarzman, A., Kovler, K., Grader, G. and Shter, G. (2003) The effect of dehydroxylation/amorphization degree on pozzolanic activity of kaolinite. *Cement and Concrete Research*, 33, 405–416.

9. Sánchez, R.T., Basaldella, E.I. and Marco, J.F. (1999) The effect of thermal and mechanical treatments on kaolinite: characterisation by XPS and IEP measurements. *Journal of Colloid and Interface Science*, 215, 339–344.

10. Mako, E., Frost, R., Kristof, J. and Horvath, E. (2001) The effect of quartz content on the mechanochemical activation of metakaolinite. *Journal of Colloid and Interface Science*, 244, 359–364.

11. Frost, R., Mako, E., Kristof, J., Horvath, E. and Kloprogge, J. (2001) Mechanochemical treatment of kaolinite. *Journal of Colloid and Interface Science*, 239, 458–466.

12. Grigorieva, T., Vorsina, I., Barinova, A. and Boldyrev, V. (1996) Solid-state interaction of kaolinite and acids during joint mechanical activation. *Journal of Materials Synthesis and Processing*, 4, 299–305.

13. Rojas, M.F. and Cabrera, J. (2002) The effect of temperature on the hydration rate and stability of the hydration phases of metakaolin–lime–water systems. *Cement and Concrete Research*, 32, 133–138.

14. Coleman, N. and Mcwhinnie, W. (2000) The solid state chemistry of metakaolin-blended ordinary Portland cement. *Journal of Materials Science*, 35, 2701–2710.

15. Shi, C. and Day, R. (2000) Pozzolanic reaction in presence of chemical activators. Part 2: reaction products and mechanism. *Cement and Concrete Research*, 30, 607–613.

16. Love, C., Richardson, I. and Brough, A. (2007) Composition and structure of C–S–H in white Portland cement – 20% metakaolin pastes hydrated at 25 °C. *Cement and Concrete Research*, 37, 109–117.

17. Rojas, M. (2006) Study of hydrated phases present in a MK–lime system cured at 60°C and 60 month of reaction. *Cement and Concrete Research*, 36, 827–831.

18. Lagier, F. and Kurtis, K. (2007) Influence of Portland cement composition on early age reactions with metakaolin. *Cement and Concrete Research*, 37, 1411–1417.

19. Murat, M. (1983) Hydration reaction and hardening of calcined clays and related minerals. 1. Preliminary investigation of metakaolinite. *Cement and Concrete Research*, 13, 259–266.

20. De Silva, P. and Glasser, F. (1992) Pozzolanic activation of metakaolin. *Advances in Cement Research*, 4(16), 167–178.

21. Gabrera, J. and Rojas, M.F. (2001) Mechanism of hydration of the metakaolin–lime–water system. *Cement and Concrete Research*, 31, 177–182.

22. Wild, S. and Khatib, J. (1997) Portlandite consumption in metakaolin cement pastes and mortars. *Cement and Concrete Research*, 27(1), 137–146.

23. Rojas, M.F. and Cabrera, J. (2001) Influence of MK on the reaction kinetics in MK/lime and MK-blended cement systems at 20°C. *Cement and Concrete Research*, 31, 519–527.

24. Largent, R. (1998) Estimation de l'activite pouzzolanique – recherche d'un essai. *Bulletin de Liaison des Laboratories des Ponts et Chaussées*, 93, 63.

25. Shvarzman, A., Kovler, K., Grader, G. and Shter, G. (2003) Hydration of portlandite–metakaolin and Portland cement–metakaolin systems, In: *11th International Congress on the Chemistry of Cement*, Durban, 843–852.

26. Sabir, B., Wild, S. and Baj, J. (2001) Metakaolin and calcined clays as pozzolans for concrete: a review. *Cement and Concrete Composites*, 23, 441–454.

27. Bai, J., Sabir, B., Wild, S. and Kinuthia, J. (2000) Kinuthia, strength development in concrete incorporating fly ash and metakaolin. *Magazine of Concrete Research*, 52(3), 153–162.

28. Rols, S., Mbessa, M., Ambroise, J. and Pera, J. (1999) Influence of ultra-fine particle type on properties of very high strength concrete. In: *Proceedings of the 2nd CANMET/ACI International Conference on High Performance Concrete and Performance and Quality of Concrete Structures*, Brazil, 671–686.

29. Zhang, M. and Malhotra (1995) Characteristics of a thermally activated alumino-silicate pozzolanic material and its use in concrete. *Cement and Concrete Research*, 25(8), 1713–1725.

30. Kim, H., Lee, S. and Moon, H. (2007) Strength properties and durability aspect of high strength concrete using Korean metakaolin. *Construction and Building Materials*, 21, 1229–1237.

31. Wild, S., Khatib, J. and Roose, L. (1998) Chemical shrinkage and auto-genous shrinkage of Portland cement–metakaolin pastes. *Advances in Cement Research*, 10(3), 109–119.

32. Brooks, J., and Johari, M.A.M. (2001) Effect of metakaolin on creep and shrinkage of concrete. *Cement and Concrete Composites*, 23(6), 495–502.

33. Wild, S., Khatib, J. and Roose, L. (1998) Chemical shrinkage and autogenous shrinkage of Portland cement–metakaolin pastes. *Advances in Cement Research*, 10(3), 109–119.
34. Kinuthia, J., Wild, S., Sabir, B. and Bai, J. (2000) Self-compensating autogeneous shrinkage in Portland cement–metakaolin–fly ash pastes. *Advanced Cement and Research*, 12(1), 35–43.
35. Gleize, P., Cyr, M. and Escadeillas, G. (2007) Effect of metakaolin on autogenous shrinkage of cement pastes. *Cement and Composites*, 29, 80–87.
36. Khatib, J. and Wild, S. (1996) Pore size distribution of metakaolin paste. *Cement and Concrete Research*, 26(10), 1545–1553.
37. Al-Akhras, N.M. (2006) Durability of metakaolin concrete to sulfate attack. *Cement and Concrete Research*, 36, 1727–1734.
38. Asbridge, A.H., Page, C.L. and Page, M.M. (2002) Effects of metakaolin, water/binder ratio and interfacial transition zones on the microhardness of cement mortars. *Cement and Concrete Research*, 32, 1365–1369.
39. Caldarone, M., Gruber, K. and Burg, R. (1994) High reactivity metakaolin: a new generation mineral admixture. *Concrete International*, 37–40.
40. Wild, S., Khatib, J. and Jones, A. (1996) Relative strength, pozzolanic activity, and cement hydration in superplasticised metakaolin concrete. *Cement and Concrete Research*, 26(10), 1537–1544.
41. Shvarzman, A. (2003) *Properties and microstructure of high performance cementitious systems with metakaolin*. Ph.D thesis, Faculty of Civil Engineering, Technion – Israel Institute of Technology, Haifa.
42. Curcio, F., DeAngelis, B.A. and Pagliolico, S. (1998) Metakaolin as a pozzolanic microfiller for high-performance mortars. *Cement and Concrete Research*, 26(6), 803–809.
43. Khatib, J. and Hibbert, J. (2005) Selected engineering properties of concrete incorporating slag and metakaolin. *Construction and Building Materials*, 19, 460–472.
44. Oriol, M. and Pera, J. (1995) Pozzolanic activity of metakaolin under microwave treatment. *Cement and Concrete Research*, 25(2), 265–270.
45. Gruber, K., Ramlochan, T., Boddy, A., Hooton, R. and Thomas, M. (2001) Increasing concrete durability with high reactivity metakaolin. *Cement and Concrete Composites*, 23, 479–484.
46. Zibara, H., Hooton, R., Thomas, M. and Stanish, K. (2007) Influence of the C/S and C/A ratios of hydration products on the chloride ion binding capacity of lime–SF and lime–MK mixtures. *Cement and Concrete Research*, 38, 422–426.
47. Saikia, N., Kato, S. and Kojima, T. (2006) Thermogravimetric investigation on the chloride binding behaviour of MK–lime paste. *Thermochimica Acta*, 444, 16–25.
48. Boddy, A., Hooton, R. and Gruber, K. (2001) Long-term testing of the chloride penetration resistance of concrete containing high reactivity metakaolin. *Cement and Concrete Research*, 31, 759–765.

49. Hooton, R., Gruber, K. and Boddy, A. (1997) The chloride penetration resistance of concrete containing high reactivity metakaolin. In: *Proceedings of the PCI/FHWA International Symposium on High Performance Concrete*. New Orleans, 172–183.

50. Caldarone, M. and Gruber, K. (1995) High reactivity metakaolin – a mineral admixture for high-performance concrete. *Concrete Under Severe Conditions*, 1015–1023.

51. Asbridge, A., Chadbourn, G. and Page, C. (2001) Effects of metakaolin and the interfacial transition zone on the diffusion of chloride ions through cement mortars. *Cement and Concrete Research*, 31, 1567–1572.

52. Basheer, P., Gilleece, P., Loong, A. and McCarter, W. (2002) Monitoring electrical resistance of concretes containing alternative cementitious materials to assess their resistance to chloride penetration. *Cement and Concrete Composites*, 24, 437–449.

53. Vu, D., Stroeven, P. and Bui, V. (2001) Strength and durability aspects of calcined kaolin-blended Portland cement and mortar. *Cement and Concrete Composites*, 23, 471–478.

54. Courard, L., Darimont, A., Schouterden, M., Ferauche, F., Willem, X. and Degeimbre, R. (2003) Durability of mortars modified with metakaolin. *Cement and Concrete Research*, 33, 1473–1479.

55. Khatib, J. and Wild, S. (1998) Sulphate resistance of metakaolin mortar. *Cement and Concrete Research*, 28(1), 83–92.

56. Tsivilis, S., Kakali, G., Skaropoulou, A., Sharp, J. and Swamy, R. (2003) Use of mineral admixtures to prevent thaumasite formation in lime stone cement mortar. *Cement and Concrete Composites*, 25, 969–976.

57. Smallwood, I., Wild, S. and Morgan, E. (2003) The resistance of metakaolin (MK)–Portland cement (PC) concrete to the thaumasite-type of sulfate attack (TSA) – program of research and preliminary results. *Cement and Concrete Composites*, 25, 931–938.

58. Lee, S., Moon, H., Hooton, R. and Kim, J. (2005) Effect of solution concentrations and replacement levels of metakaolin on the resistance of mortars exposed to magnesium sulfate solutions. *Cement and Concrete Research*, 35, 1314–1323.

59. Poon, C., Azhar, S., Anson, M. and Wong, Y. (2003) Performance of metakaolin concrete at elevated temperatures. *Cement and Concrete Composites*, 25(1), 83–89.

60. Gutierrez, R., Diaz, L. and Delvasto, S. (2005) Effect of pozzolanas on the performance of fiber reinforced mortars. *Cement and Concrete Composites*, 27, 593–598.

61. Thiery, J. and Soukatchoff, P. (1991) *Method for selecting a pozzolan, intended to be incorporated into a composite materials comprising cement and glass*. US Patent 4994114.

62. Barger, G.S., Hansen, E.R., Wood, M.R., Neary, T., Beech, D.J. and Jaquier, D. (2001) Production and use of calcined natural pozzolans in concrete. *Cement, Concrete and Aggregates*, 23(2), 73–80.

7.3 Alternative reinforcing materials to conventional steel reinforcement (black bars)

Concrete, if properly designed, will offer adequate protection for steel reinforcement in the majority of environments, but this is assuming that an appropriate cover to the reinforcement has been provided. There are environments in which, however, improving the concrete mix may not be sufficient to protect the steel reinforcement. Alternative reinforcing materials to conventional steel reinforcement (black bars) may therefore be considered to ensure a durable reinforced concrete structure. These are:

- epoxy-coated steel reinforcement
- galvanised steel reinforcement
- stainless steel reinforcement
- glass-fibre-reinforced polymers.

The first two of these are still black bars, but they have a protective coating, while the third one offers improved protection by a change in the composition of the metal. The last one is the only one that is not metallic. Their methods of manufacture, how they provide improved durability, their properties that may affect design procedures that have been developed for conventional reinforcement, and the reputation that they have based on past applications and performance in real structures is discussed in this section.

7.3.1 Considerations for the design, specification and use of epoxy-coated reinforcing steel bars

Scott Humphreys, Concrete Reinforcing Steel Institute, USA

7.3.1.1 Introduction

In North America, the first in-depth corrosion research project that identified epoxy-coated reinforcing steel as an option for improved corrosion resistance was an early 1970s National Bureau of Standards corrosion research project.[1] This research and others[2] was initiated in response to widespread premature deterioration of reinforced-concrete bridge decks due to the dramatic increased use of de-icing material in the late 1960s and early 1970s. Today, the use of epoxy as a protective coating on reinforcing steel is widespread within North America. A recent survey of North American transportation agencies documented that that the only option used more frequently than epoxy-coated reinforcing is increased concrete cover.[3]

The intent of this chapter is to provide the designer of reinforced-concrete structures located in corrosive environments with an overview of epoxy-coated reinforcing steel and the related technology. Designing and specifying reinforced-concrete structures for extended service life in corrosive environments is unique in its set of challenges while balancing the ever-increasing expectations of society. Despite the ubiquitous use of epoxy-coated reinforcing steel bars and the deceptively simple approach to improved corrosion resistance, the designer will benefit from having an understanding of the information provided in this chapter.

The topics addressed in this chapter are:

- epoxy-coated reinforcing steel: applications and history
- epoxy: material and history
- North American epoxy-coated steel standards
- corrosion resistance
- performance of epoxy-coated reinforcing steel: accelerated tests
- performance of epoxy-coated reinforcing steel: service life extension
- field condition assessments and service life modelling
- best practices.

Fig. 7.48 Epoxy-coated reinforcing steel bundled – waiting for placement

7.3.1.2 Epoxy-coated reinforcing steel: applications and history

Epoxy-coated deformed-steel reinforcing bars are the most common type of corrosion resistant steel bars by tonnage when compared with all the other types of corrosion resistant steel bars. Currently, it is estimated that 600 000 tons are produced annually in North America. Epoxy-coated steel reinforcing bars are used for bridge deck reinforcement, bridge substructures, barrier walls for highways, retaining walls, tunnels, environmental structures, parking garages, foundations, exterior walls and balconies of high-rise buildings (Figs 7.48–7.55).

Fig. 7.49 Epoxy-coated reinforcing bars for continuously reinforced-concrete pavement

261

Fig. 7.50 Reconstruction of the Dan Ryan Expressway, Chicago

Fig. 7.51 Reconstruction of I-294 western suburbs, Chicago

262

Fig. 7.52 Coiled epoxy-coated reinforcing bar

According to the Federal Highway Administration's National Bridge Inventory database, there has been over 63 000 bridges that have used epoxy-coated reinforcing steel bar in the deck. Today, there are approximately 37 coating plants in North America,[4] and there are reports of similar numbers of coating plants in the Middle-East, China, India, Korea and Japan.

There is also a relatively small, but significant, quantity of other reinforcing steel products and accessories that are coated with epoxy. These include welded wire reinforcment, post-tensioning strands, load transfer devices (or concrete pavement dowels) and wire bar supports.

7.3.1.3 Epoxy: material and history

Epoxy or polyepoxide is a polymer that is polymerised or cured by two different techniques. One technique is when the epoxy is mixed with a hardener.[5] The other technique of curing (or cross-linking) uses elevated temperatures.\The ability of epoxy to cross-link at elevated temperatures is what defines epoxy as a thermosetting plastic. Another characteristic of epoxy, as with all thermosetting plastics, is that once it is cross-linked it will not turn into a liquid when heated.[6] Current

Fig. 7.53 Epoxy-coated bar balcony of Aqua Chicago

industry practice for the production of epoxy-coated reinforcing steel is to blast clean the bars and then heat them to an elevated temperature, apply the powder so that it cross-links and then to use epoxy with a hardener for patching material.

Epoxy is not new. The material was first patented over 100 years ago in the 1890s. However, the first practical uses of epoxy did not occur until the 1930s in Germany and Switzerland. Reportedly, after World

Fig. 7.54 Epoxy-coated reinforcing steel box girder of the I-35 Bridge, Minneapolis

Fig. 7.55 People mover at Dulles International Airport, Washington, DC

War II the use of epoxy in the laboratory became more common, as the material was found to be useful for its inert and adhesive qualities. Production on a commercial basis did not start until the 1950s.[5,7]

From the 1960s to the present, epoxy is widely used as a protective coating on petroleum pipe lines. Today, there are thousands of miles of epoxy-coated pipelines. Many of these are located off the coast of Texas and Louisiana, placed on the bottom of the Caribbean ocean and embedded in concrete, serving oil wells.

Currently, resins of epoxy or epoxy resin flakes are produced in large quantities by a few chemical companies, and then others, called formulators, utilise the epoxy resin flakes as a basic ingredient of epoxy powder. The formulators modify the epoxy resin flakes depending on the final application. This modification essentially means that the various epoxy coatings are tailor made specifically for the application. Some of the typical modifications include the addition of colorisers and mineral fillers – talc, silica and alumina. Also, flexibilisers, viscosity reducers, colorants, thickeners, accelerants and adhesion promoters are used.[5,7]

Overall, epoxy is known for excellent adhesion, chemical and heat resistance, good-to-excellent mechanical properties, and very good electrical insulating properties.[5] Epoxy resins are also noted for being an ideal binder, and can be used in various forms.[7] The American Concrete Institute's document ACI 503, *Use of Epoxy Compounds with Concrete*, notes that there are many characteristics of epoxies that make them desirable for use with concrete. These qualities include adhesion, versatility, chemical resistance, low shrinkage, rapid hardening, moisture resistance and strain compatibility.

7.3.1.4 *North American epoxy-coated steel standards*

The North American material standards for epoxy-coated steel reinforcing consist of:

- ASTM A775/A775M, *Standard Specification for Epoxy-Coated Steel Reinforcing Bars*. The majority of epoxy-coated bars in North America are coated to this standard. Currently, there are 36 Concrete Reinforcing Steel Reinforcing Institute certified plants in North America.
- ASTM A882/A882M, *Standard Specification for Filled Epoxy-Coated Seven-Wire Prestressing Steel Strand*.
- ASTM A884/A884M, *Standard Specification for Epoxy-Coated Steel Wire and Welded Wire Reinforcement*. This standard is for

266

epoxy-coated welded wire reinforcement, and allows for the use of epoxy coating qualified under ASTM A775 (flexible) or ASTM A934 (rigid) to be used.

- ASTM A933/A933, *Standard Specification for Vinyl (PVC) Coated Steel Wire and Welded Wire Fabric for Reinforcement.* This standard is for vinyl-coated welded wire reinforcement – not epoxy – but is otherwise comparable. Currently, there is no known producer of this in North America.
- ASTM A934/A934M, *Standard Specification for Epoxy-Coated Pre-fabricated Steel Reinforcing Bars.* The majority of this material is used in California and for US Navy projects.
- ASTM A1055, *Standard Specification for Zinc and Epoxy Dual Coated Steel Reinforcing Bars.* This standard is the most recent epoxy-related ASTM standard issued. This specification covers plain and deformed steel reinforcing bars that are produced by thermal spraying (metalising) of a zinc/zinc alloy after the bars are blast cleaned and prior to coating with an epoxy coating. The zinc layer is required to be 35 μms or 1.4 mil in thickness, and the bars can be conventional ASTM A615 or A706 bars. The zinc coating is intended to serve sacrificially at holidays and coating damage, thereby providing a more durable coated bar.
- ASTM D3963/D3963M, *Standard Specification for Fabrication and Jobsite Handling of Epoxy-Coated Steel Reinforcing Bars.*

7.3.1.5 Corrosion resistance

An important distinction related to the coating of epoxy is that the epoxy serves as a protective coating. A protective coating is fundamentally different from what is generally called 'paint'. A protective coating is required to provide certain protective characteristics, whereas paint is essentially for aesthetics. A properly applied coating of epoxy serves to protect the reinforcing steel. To provide corrosion protection, a protective coating must (*a*) resist the transfer or penetration of salt ions through the coating; (*b*) resist the action of osmosis; (*c*) expand and contract with the underlying surface; and (*d*) maintain a good appearance.[8] Epoxy powders are qualified for use as protective coating by the applicable ASTM standard for epoxy coating steel, based on the ability of the particular epoxy powder to pass the specified test criteria. In order for an epoxy powder to be qualified for use as a coating for reinforcing bars meeting ASTM A775 *Standard specification for epoxy-coated steel reinforcing bars.* The standard requires that the powder provide acceptable performance for chemical resistance, cathodic disbondment, chloride permeability, coating flexibility, relative bond strength in concrete, abrasion resistance and impact resistance.

In concrete, when the epoxy coating is intact, the ability of the coating to provide corrosion resistance is widely recognised. In this state, chloride ions that may initiate continued corrosion cannot reach the steel surface. However, it is widely recognised through the coating process, fabrication and placing operations that damage to the coating does occur. At these damaged locations, when chloride ions are present in sufficient quantities, corrosion is initiated. However, since the large cathodic area is electrically isolated, the corrosion process is limited to the damage site.[9]

Fundamentally, the epoxy coating provides increased resistance to corrosion in three basic ways. The first is that the coating itself acts as a barrier to oxygen. The second is that the coating acts as a barrier to chlorides ions. Third is that the coating limits the conductivity or the rate at which the cathodic process can occur, as it provides electrical isolation. There is much conjecture on the exact nature and importance of various aspects of the deterioration process of epoxy-coated bars. Later in this chapter, more details related to this process (i.e. accelerated corrosion tests and condition assessments) will be reviewed.

A central issue that one needs to understand for a greater understanding of the phenomena of corrosion and the selection of corrosion resistant systems is that a single accelerated corrosion test that replicates corrosion resistance and deterioration over the actual life of a reinforced-concrete structure does not exist. Currently, the practice is to use an accelerated corrosion test that is more severe than the actual application. The assumption is that exposing the material in question to a more aggressive environment will correspond to the long-term deterioration process. The other assumption is that the accelerated tests will provide an indication of corrosion resistance and a relative performance between the tested specimens. However, the reality is that there are a large number of factors that will affect the actual performance, and it remains uncertain if the results of the tests will actually correspond to long-term performance.

The most common accelerated test regimes that have been used to evaluate the corrosion resistance of epoxy-coated reinforcing steel are the solution immersion, salt spray (fog) test, rapid macrocell, southern exposure, cracked beam, and in-situ tests.

7.3.1.6 *Performance of epoxy-coated reinforcing steel: accelerated tests*

In spite of the prior discussion concerning some of the deficiencies of accelerated corrosion tests, there is much that can be learned from the results of accelerated corrosion tests of reinforced concrete. Indeed, over the last 35 years there have been countless accelerated corrosion tests, many of which have used epoxy-coated reinforcing steel. These tests have been conducted by the industry, the US Federal Highway Administration, state transportation agencies and university researchers. However, just three will be discussed, as they quantify some of the key parameters necessary to understand the overall performance of epoxy-coated reinforcing.

The first study by Al-Amoudi included coated and uncoated steel reinforcing bars, and was unique in the number of variables tested.[9] For the epoxy-coated steel, the test combined the corrosive effect of varying chloride levels along with various coating defect sizes for an extended period of time, 2600 days. The test results provided a comparison of epoxy-coated steel reinforcing verses uncoated steel reinforcing, varying amounts of coating damage, pinholes ('holidays') and varying amounts of chlorides. This accelerated testing regime utilised chloride that were mixed into the concrete and thereby did not include what is called time-to-initiation. This testing regime provides insight into the chloride threshold and propagation time by varying the amount of coating damage, and varying the amount of chlorides.

Specimens varied in the amount of chloride that were mixed with concrete (0.4, 1 and 2%). Specimens that had epoxy-coated reinforcing bar varied in the amount of coating damage (0, 0.5, 1 and 1.5%) as well as bars with a varying number of holidays (1, 2 and 3). As one might expect, specimens with the most coating damage in the highest contaminated concrete corroded the most.[9]

The following conclusions can be drawn from Table 7.12 summarised from that report. For uncoated bars and epoxy-coated bars, the tendency for embedded bar corrosion is higher where higher amounts of chlorides exist. In all cases, epoxy-coated bars exhibited significantly less corrosion than the uncoated bars. For epoxy-coated bars with no damage sites, the condition of the specimens indicated that they were essentially impervious to chlorides. Another interesting conclusion for epoxy-coated bars is that coating damage and quantity of chlorides are significant, whereas the number of holidays appears not to be significant.

Table 7.12 Corrosion rating at 2600 days: uncoated versus epoxy-coated steel reinforcement[9]

	Corrosion rating		
	0.4% chlorides	1% chlorides	2% chlorides
Uncoated	3	4	5
0% coating damage	0	0	0
0.5% coating damage	0	1	1
1.0% coating damage	1	1	2
1.5% coating damage	2	2	3
1 holiday	0	0	0
2 holidays	0	0	0
3 holidays	0	0	0

The report also documented the current density (listed in Table 7.13), to assess corrosion. For reinforcing steel to be considered in a passive condition (i.e. not corroding), the current density is less than $0.1\,\mu A/cm^2$. If the current density is greater than $0.3\,\mu A/cm^2$, the steel is considered to be actively corroding. Likewise, it has been proposed that if the current density is less than $0.01\,\mu A/cm^2$, then an indefinite service life can be achieved. From Table 7.13, it can be seen that at all levels of coating damage, the coated bars resulted in significantly lower rates of corrosion when compared with the uncoated bars. One other significant item reported was that the concrete specimens with uncoated bars cracked after 5 years but the specimens with coated bars did not. In all cases, from Table 7.13 it can be observed that, regardless of the amount of coating damage, the epoxy-coated bars had dramatic decreases in current density regardless of the coating damage or chloride level.

Table 7.13 Corrosion current density at 2600 days: uncoated versus epoxy-coated steel reinforcement[9]

Chlorides: %	Current density: $\mu A/cm^2$				
	Uncoated	0% coating damage	0.5% coating damage	1% coating damage	1.5% coating damage
0.4	6.12	0.015	0.149	0.223	0.319
1	11.3	0.016	0.198	0.295	0.395
2	17.83	0.0191	0.231	0.32	0.46

The second study by Krondratova *et al.* at the University of New Brunswick investigated the various aspects of coating and the resulting corrosion resistance in both a simulated marine environmental simulated set-up tank (MESS) and a natural marine exposure site at Treat Island, Maine.[10] The MESS exposure was described as a set of tanks in which the specimens are exposed to a 2 hour wet cycle and 4 hours of drying for a total of four cycles a day. In total, four different coatings were evaluated. Two of the coatings were currently available, and the other two were experimental. However, each was prepared in a different way for the corrosion tests. The ways in which the specimens were prepared were described as grit blasted, polished, salt contaminated, cleaner for chlorides, primer, under cured, and thin coating. Undamaged and straight bars performed the best in the MESS over 4 years, with nearly negligible corrosion rates. Specimens that were coated and then bent had a measureable corrosion rate of 1/10th that of the uncoated specimens, even though the uncoated specimens were exposed for 2 years while the coated specimens were exposed for 4 years. The other important conclusion was that the surface treatment or type of coating had no long-term effect on corrosion performance. However, the exception to this was the specimens that were prepared with a surface contaminated with chlorides. The chloride-contaminated specimens had a rate of corrosion that was over 2 MPY, while the damaged but unbent specimens had a corrosion rate of 0.4 MPY or a difference of five times indicating the importance of under coating chloride on corrosion performances.

The third study which provides many insights into the performance of epoxy-coated reinforcing steel and relative performance to other materials was a comprehensive multi-year programme funded by the US Federal Highway Administration.[11,12,13] The goal of this 5 year study was to identify and develop new types of organic, inorganic, ceramic and metallic coatings, as well as metallic alloys that would serve as corrosion protection in corrosive applications. More specifically, it was anticipated that the investigation would lead to more cost-effective types of coating or cladding with better performance when exposed to adverse environments than the typical fusion-bonded epoxy coatings. The last noteworthy objective of this program was to identify cost-effective reinforcement alloys that had all the inherent properties of black steel but whose corrosion resistance characteristics when exposed to adverse environments exceeded those of the fusion-bonded epoxy coating.

To evaluate the various types of bars (and coating), a series of corrosion tests were utilised, and described as prescreening tests, screening

Table 7.14 1993–1998 US Federal Highway Administration: 96 weeks of accelerated exposure to 15% chloride solution

Test No.	Description of test specimen				Voltage: μV
	Straight (S) or bent (B)	Bottom mat coated?	Pre-formed concrete crack?	Damaged coating: %	Average of (5) proprietary epoxy coatings
1	S	No	No	0.5	356
2	S	Yes	No	0.5	9
3	B	No	No	0.5	362
4	S	No	Yes	0.5	782
5	S	No	No	0.004	37
6	S	Yes	No	0.004	5
7	B	No	No	0.004	22
8	S	No	Yes	0.004	280
Control group – uncoated					
9	S	No	No	NA	3525
10	B	No	No	NA	2142
11	S	No	Yes	NA	4053

From US FHWA Report FHWA RD-98-153 (*Corrosion Evaluation of Epoxy-Coated, Metallic Clad and Solid Metallic Reinforcing Bars in Concrete*): see Tables 26, 28, 31, 34, 37, 40, 43, 46 and 49.

tests and in-concrete tests. The prescreening tests consisted of corrosion tests that used various solutions, as well as cathodic disbondment tests. In the next phase, called screening tests, a series of solution tests were again used, but this time with varying amounts of chlorides in both neutral solutions and solutions that replicated the concrete pore water pH. The final test was an in-concrete test. This test is frequently referred to as the 'southern exposure' test, and replicates a bridge deck, since the bars are placed over each other, in a manner similar to a bridge deck. The concrete specimens are then ponded with a saturated chloride solution (15%) in a cyclic manner for a total of 96 weeks. In total, 33 different organic coated bars, 14 ceramic, inorganic and metallic-clad bars, and 10 solid metallic bars were tested.

To understand the significance of the voltages listed in Table 7.14, consider that a uniform thickness of corrosion 1 mil in depth can theoretically cause the concrete to crack, and that the desired service life of the structure is 100 years. From 1 mil and 100 years, one can determine that the approximate maximum macrocell voltage is 27 μV. Similarly, for an indefinite service life, a current of 10 μV is appropriate.

This testing programme led to many insights related to corrosion and the use of epoxy-coated reinforcing steel. The first test results showed significantly less corrosion when both the top and bottom bars were coated, and the bars were less sensitive to coating damage. Thus, epoxy-coated steel placed over uncoated steel should not be expected to perform as well as when epoxy-coated bars are used throughout. There is also a significant detrimental effect due to the cracking of concrete and, as such, concrete cracks should be repaired. Other conclusions included that specimens with either high or low macrocell voltages correlated well to performance. Finally, of the 29 specimens that reduced the macrocell corrosion by 99%, none cracked, and of the 48 specimens that reduced the macrocell corrosion by 90%, only three specimens cracked.

Other significant conclusions were that low corrosion rates were found correlate with specimens that had high mat-to-mat resistance, and that the results from the first series of tests that used solution immersion and cathodic disbondment were poor predictors of performance in concrete. Also noteworthy was the fact that epoxy-coated bars with pretreatments did not perform significantly better than bars that were not.

McDonald, the primary author for this study, concluded the 5 year study substantiated the need for using epoxy-coated reinforcing bar for both top and bottom mat in bridge decks (and similarly throughout the structure when subjected to a corrosive environment). He also concluded that it is important to minimise coating damage during shipping and placement, and that the repair of coating at the job-site and the repair of concrete cracks in the bridge deck would also be a benefit.

7.3.1.7 Performance of epoxy-coated reinforcing steel: service life extension

There are two primary aspects to the service life of reinforced concrete. One is the initiation phase, and the other is the propagation phase. Combining the time period during the initiation phase and the time period during the propagation phase provides the total time that the structure is usable. This time period is also called the service life. During the initiation phase, chloride ions accumulate in the concrete and migrate from higher levels of concentration at the surface to lower levels of concentration down through the concrete. Given a sufficient period of time, chloride ions migrate down to where the reinforcing steel is located – presuming the concrete is not cracked. Once sufficient chloride ions have accumulated at the bar level, corrosion is initiated. The initiation of corrosion ends the first phase, and the propagation phase is initiated. The propagation phase is from the point at which corrosion is initiated to the point at which the products of corrosion accumulate such that the concrete cracks. As compared with uncoated carbon steel, epoxy-coated reinforcing steel extends the time-to-initiation and extends the time of propagation. The time-to-initiation has been established through numerous laboratory corrosion tests and through field assessments where the chlorides have been shown to be well above the quantity of chlorides to initiate corrosion of uncoated carbon steel. While more difficult to verify, the propagation phase can be seen in the test results documented by Al-Amoudi and discussed earlier.

Another noteworthy aspect of the corrosion of epoxy-coated reinforcing steel and the issue of the propagation phase is that epoxy-coated steel corrodes at discrete locations – coating damage sites. This deterioration at discrete sites is fundamentally different from uncoated carbon steel bar where corrosion tends to occur uniformly along the bar. Darwin[14] made note of this in a presentation and that a model had been developed by Torres-Acosta.[15] Based on this model it was shown that there would have to be as much as 2500 μm to cause cracking of concrete at a discrete location, whereas 25 μm (1 mil) is considered enough to cause cracking.

The concept of service life, service and repairability is also evolving. Historically, a bridge deck would be considered to be serviceable until a certain damaged percentage of the surface is deemed to be unacceptable. This is normally between 10 and 12%. However, one of the newer concepts, reportedly borrowed from seismic design, is a concept called damage avoidance design, which attempts to ensure that the structure can provide full service and functionality for a prescribed limit state. The other is an approach for the control and repairability of damage, in which damage is allowed but in a controlled manner.[16] Clearly, different structures have different service lives, different expectations and represent different risks for the owner and users when they need repair. Many transportation structures have service lives of the order of 50–100 years, while others have indefinite service lives.[17]

7.3.1.8 Field condition assessments and service life modelling

Samples and Ramirez investigated the condition of (123) bridges containing epoxy-coated reinforcement, uncoated reinforcement, and several other corrosion resistant systems. On behalf of the Indiana Department of Transportation.[18] Of these (123) bridges 44 percent had signs of distress, and six decks were selected for additional investigation. Three and contained epoxy-coated reinforcement. Corrosion of epoxy-coated reinforcement was reported in areas of concrete cracking and shallow cover on two of these three bridge decks. Samples and Ramirez reported that the use of steel surface pretreatment did not appear to increase corrosion performance of epoxy-coated reinforcement. Also, adhesion damage site did not always correlate to corrosion performance. It was concluded that, if mat-to-mat resistance was greater than 5000 Ω, there was no corrosion, and that the uncoated black bar cathode had a detrimental effect on epoxy-coated reinforcement specimens with 0.5% coating damage.

Other noteworthy conclusions of this study was that thicker coated bars tended to reduce coating damage during construction, and that, when placing concrete, reducing the vertical drop reduces the number of holidays by 50%.

Sohanghpurwala and Scannell investigated over 80 bridges and removed three core specimens from each bridge, a total of 240, which resulted in 473 bar specimens on behalf of New York Department of Transportation and the Pennsylvania Department of Transportation.[19] Chloride and condition analysis of the epoxy-coated bar specimens were undertaken. Of the 240 core specimens 49 were determined to have exceeded the corrosion threshold of 1.2 pcy of chloride for a period of five years, which was expected to be long enough to initiate corrosion. Only two bars were found to be exhibiting progressive corrosion. The report concluded that adhesion reduction was not a good predictor of corrosion.

Fanous *et al.* examined 81 bridges in Iowa in order to estimate the service life of bridge decks with epoxy-coated reinforcing steel on behalf of the Iowa Department of Transportation.[20] Concrete cores were taken at cracked as well as uncracked locations. No signs of corrosion were present on any of the epoxy-coated bar specimens removed from the solid concrete. The chloride concentration found 0.5 in. below the specimen was found to be 14 lb/yd^3 and a diffusion constant of 0.05 in^2/year. The corrosion threshold was determined to be between 3.6 and 7.2 lb/yard^3.

In 1976–1986, the Iowa Department of Transportation used epoxy in the top mat only, and then switched to both mats.

Fanous *et al.* note that corrosion is an electrochemical process. This means that there has to be a current flow in order for the process to proceed. They also note that the corrosion process is dynamic: it occurs at different rates throughout the year. Both wet and dry conditions as well as temperature affect the rate of corrosion. Similarly, the degree to which the reinforcing bars are interconnected affects the rate of corrosion. Even the quantity of corrosion by products needed to produce cracks varies: some reports have stated that as little as 0.1 mm is required, but it has been shown that in some situations, less than 0.1 mm is necessary.

In another study on service life extensions eleven bridge decks on phase I and six Iowa bridge decks in phase II were investigated.[21] Phase I bridge decks contained epoxy-coated reinforcement in the top mat only, while phase II decks contained epoxy-coated reinforcement in both mats. Phase I decks varied in age from 19 to 27 years, while the phase II age range was 9–15 years. Phase I decks averaged 1.3% damaged surface area but ranged from 0 to 8.2%. All six phase II decks had no surface damage, and four of the phase I decks did. The average chloride concentration at the top reinforcing mat in cracked delaminated cores was found to be twice as high as in intact cores.

A service life model was used by Lee and Krause to project the service life of the uncoated bars, based on a corrosion threshold of 300 ppm or (1.2 lb/yard³).[21] For decks with epoxy-coated reinforcement in the top mat only, a threshold of 1185 ppm or 4.7 lb/yard³ was used. For decks with epoxy-coated reinforcement in both mats, 3750 ppm or 15 lb/yard³ was used. A propagation period of 5 years was assumed for uncoated bars, epoxy-coated reinforcement in the top mat, plus 15 years for epoxy-coated reinforcement in both mats. The conclusion was that decks with uncoated bars would have a service life extension of the order of 10.5–30 years (average of 16–19 years) while the service life for the top-mat-only decks would be anywhere from 24 to 100 years (average of 40 years).[21] Decks with epoxy-coated reinforcement in both mats had a projected service life extension of 82 years.

7.3.1.9 Best practices

For epoxy-coated reinforcing steel there are three distinct areas in which industry-recommended practices have to be followed in order for the desired service life to be achieved. These areas are production, fabrication and placing.

The Concrete Reinforcing Steel Institute developed a quality control programme[22] for the production of epoxy-coated reinforcing bars in the early 1990s because it had become clear that the quality of the coating could have a significant impact on the service life of reinforced concrete structures. The certification programme has two main objectives. First, it verifies that a plant has the appropriate capabilities in both staff and equipment as well as the procedures in place to produce epoxy-coated bars in accordance with current standards and related industry recommendations. Second, it provides recognition of those plants that successfully implement and follow the requirements of the programme.

The first and foremost requirement of the certification programme is that the plant's quality control personnel must be knowledgeable with respect to the required quality control procedures. The personnel are required to understand the requirements of the programme and be capable of correctly performing the various tests specified by the programme. In addition, the programme requires there always to be enough trained personnel that quality control tests can be performed in a timely manner and documented. Also, the quality control personnel must have the authority to make changes that are necessary to ensure that the requirements of the programme are met.

During the inspections required for certification, the independent third-party inspector will observe the quality control personnel conduct the required tests. In addition, the inspector will review the documentation that the plant is required to keep on file as a record of its quality control procedures.

Certification indicates that the plant's procedures comply with the programme requirements for:

- examination of the uncoated bars
- surface preparation
- testing for contamination and gradation of the blast abrasive
- cleanliness of bar (chlorides, mill scale and dust)
- measuring of bar temperature
- application of the epoxy powder
- continuity of coating
- coating thickness
- coating flexibility
- cathodic disbondment
- storage and handling of the epoxy powder
- handling and storage of the coated bars.

For fabrication and placing activities, there are four tenets that need to be followed. They are:

1. *Repair.* The material used to patch ends and damaged portions of the coated bars is to be compatible with the epoxy coating and capable of providing an acceptable level of protection from

277

corrosion. In practice, this means that the patch material must meet the same requirements of Appendix A of ASTM A775 as does the epoxy coating, the patch material is applied in accordance with manufacturer's instructions, and that the patch material is compatible with powder manufacturer's recommendations.

2. *Fabrication.* Contact areas of the fabrication equipment are to be covered with material such that coating damage is minimised. Practically, this means that drive rolls, mandrels and back-up barrels are covered with high-density plastic.

3. *Handling*: Coated bars are to be handled in a manner that minimises the likelihood of damage. For handling, this primarily means that contact areas for handling and storage equipment are padded, and coated bars are bundled in such a manner that the strapping will not damage or cut the coating. Also, coated bars are lifted in a manner that minimises bar-to-bar abrasion, and bars or bundles must not be dropped or dragged.

4. *Storage.* Coated bars are to be stored in a manner that minimises the likelihood of damage. This means that coated bars or bundles are to be stored above ground on wooden or other padded supports with timbers placed between bundles, coated and uncoated steel reinforcing bars are stored separately, long-term storage is avoided, and material delivery is scheduled to suit construction progress. If cumulative outdoor exposure, including previous uncovered storage time, is greater than 2 months, coated bars are to be covered with opaque polyethylene, or other suitable ultraviolet-light-protective material, and provisions are made to minimise condensation under the cover. Finally, coated bars and uncoated bars are to be stored separately.

References

ASTM A775/A775M (2009) *Standard specification for epoxy-coated steel reinforcing bars*, West Conshohocken, PA: ASTM International.

ASTM A882/A882M (2009) *Standard specification for filled epoxy coated seven-wire prestressing steel strand*, West Conshocken, PA: ASTM International.

ASTM A884/A884M (2009) *Standard specification for epoxy-coated steel wire and welded wire reinforcement*, West Conshohocken, PA: ASTM International.

ASTM A933/A933M (2009) *Standard specification for vinyl (PVC) coated steel wire and welded wire fabric for reinforcement*, West Conshohocken, PA: ASTM International.

ASTM A934/A934M (2009) *Standard specification for epoxy-coated pre-fabricated steel reinforcing bars*, West Conshohocken, PA: ASTM International.

ASTM A1055 (2009) *Standard specification for zinc and epoxy dual coated steel reinforcing bars*, West Conshohocken, PA: ASTM International.

ASTM D3953/D3963M (2009) *Standard specification for fabrication and jobsite handling of epoxy-coated steel reinforcing bars*, West Conshohocken, PA: ASTM International.

1. Clifton, J.R., Beeghly, H.F. and Mathley, R.G. (1974) *Nonmetallic coatings for concrete reinforcing bars*. Building Science Series 65. US Department of Commerce/National Bureau of Standards, Washington, DC.

2. Clear, K.C. (1976) *Time-to-corrosion of reinforcing steel in concrete slabs. Performance after 830 daily salt applications*. Federal Highway Administration, Washington, DC, vol. 3.

3. Russell, H. (2004) *Concrete bridge deck performance – a synthesis of highway practice*. NCHRP Synthesis 333. Transportation Research Board, Washington, DC.

4. Concrete Reinforcing Steel Institute (2004) *Thirty-years of success in corrosion protection*. Concrete Reinforcing Steel Institute, Schaumburg, IL.

5. Wikipedia. Epoxy: http://en.wikipedia.org/wiki/Epoxy.

6. Wikipedia. Cross-linking: http://en.wikipedia.org/wiki/Cross-linking.

7. Hare, C. (1994) *Protective coatings – fundamentals of chemistry and composition*, SSPC 94-17. Technology Publishing Company, Pittsburg, PA.

8. Munger, C.G. (1984) *Corrosion prevention by protective coatings*. National Association of Corrosion Engineers, Houston, TX.

9. Al-Amoudi, O.S.B., Maslehuddin, M. and Ibrahim, M. (2004) Long-term performance of fusion-bonded epoxy-coated steel bars in chloride-contaminated concrete. *ACI Materials Journal*, 101, 303–309.

10. Kondratova, I.L., Bremner, T.W. and Sakir, E. (1996) *Effect of type of surface treatment and epoxy coating on corrosion activity*. Special Publication 163. American Concrete Institute, Farmington Hills, MI.

11. McDonald, D.B., Sherman, M.R. and Pfeiffer, D.W. (1995) *The performance of bendable and nonbendable organic coatings for reinforcing bars in solution and cathodic debonding tests*, FHWA-RD-94-103. Federal Highways Administration, Washington, DC.

12. McDonald, D.B., Sherman, M.R. and Pfeiffer, D.W. (1996) *The performance of bendable and nonbendable organic coatings for reinforcing bars in solution and cathodic debonding tests: phase II. Screening tests*, FHWA-RD-96-021. Federal Highways Administration, Washington, DC.

13. McDonald, D.B., Pfeiffer, D.W. and Sherman, M.R. (1998) *Corrosion evaluation of epoxy-coated, metallic-clad and solid metallic reinforcing bars in concrete*. Report No. FHWA-RD-98-153. Federal Highways Administration, Washington, DC.

14. Darwin Transportation Research Board (2006) Use of epoxy-coated rebar in concrete bridge components: pros and cons. Darwin Transportation Research Board, Darwin.

15. Torres-Acosta, A.A. and Sagüés, A.A. (2004) Concrete cracking by localised steel corrosion-geometric effects. *ACI Materials Journal*, Nov.–Dec., pp. 501–507.

16. Transport Research Board (2007) *Strategic highway research proposal – project number R19. Bridges for Service Life beyond 100 Years.*

17. NACE (2005) *Design considerations for corrosion control of reinforcing steel in concrete*, RP0187–2005, NACE International, Houston, TX.

18. Samples, L. and Ramirez, J.A. (1999) *Methods of corrosion protection and durability of concrete decks reinforced with epoxy-coated bars. Phase I*, FHWA/IN/JTRP-98/15. Purdue University, Indiana.

19. Sohanghpurwala, A.A. and Scannell, W.T. (1998) *Verification of effectiveness of epoxy coated rebars*. Project 94-5. Pennsylvania Department of Transportation, Harrisburg, PA.

20. Fanous, F., Wu, H. and Pape, J. (2000) *Impact of deck cracking on durability*. Center of Transportation Research and Education, Iowa State University, Ames, IA.

21. Lee, S.K. and Krauss, P. (2003) *Service life extension of northern bridge decks containing epoxy-coated reinforcing bars*. Report prepared for the Concrete Reinforcing Steel Institute. Wiss, Janney, Elstner Associates, Northbrook, IL.

22. Concrete Reinforcing Steel Institute (2009) *Voluntary certification program for fusion-bonded epoxy coating applicator plants*, 10th edition, Schaumburg, Illinois.

7.3.2 Galvanised steel reinforcement

Desmond Makepeace, Galvanisers Association, UK

Galvanised steel reinforcement has been in use for a considerable period of time, its first application being in the 1930s, although it has only been used extensively since the 1960–1970s.[1–4] To fully understand the benefits of galvanised reinforcement, it is important to understand a little about the process and the properties of the coating.

7.3.2.1 Galvanising: the process

Hot dip galvanising is an in-line process which is carried out under factory-controlled conditions, with work being immersed in a series of tanks, as shown in Fig. 7.56. It is, however, important that the steel is free from heavy oil/grease contamination, paint or welding slag, which would not be removed during the normal pre-treatment process.

Typically, steel is degreased and pickled in hydrochloric acid to remove rust, mill scale and light oils. This is followed by immersion in a flux solution, which forms a coating on the steel surface that will aid the reaction between the iron and the molten zinc by increasing the surface wetability. Finally, after drying, the steel is immersed in a bath containing molten zinc, which is typically at a temperature of about 450°C.[5–7]

During the immersion, the iron reacts with the molten zinc, and a zinc–iron alloy coating is formed which is metallurgically bonded to the base steel. The alloy layer continues to grow during immersion, gradually becoming richer in zinc, until upon withdrawal from the galvanising bath some free zinc is pulled out with the steelwork and solidifies to form an outer layer of pure zinc.

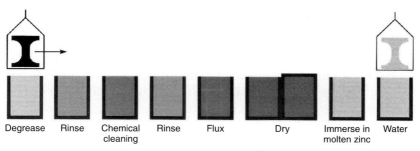

| Degrease | Rinse | Chemical cleaning | Rinse | Flux | Dry | Immerse in molten zinc | Water |

Fig. 7.56 Schematic lay-out of a galvanising plant[5]

7.3.2.2 Coating properties

The different layers within the galvanised coating offer a unique combination of mechanical properties.[5,6,8] The pure zinc outer layer is very soft, and cushions against impact, so protecting the coating from damage during handling and erection. Conversely, the alloy layer beneath is very hard and wear-resistant, typically being harder than the base steel itself, and will not be easily damaged by abrasion. As a result of these properties, a galvanised coating is robust and far more difficult to damage during normal handling and erection than other corrosion protection systems.

Once formed, the galvanised coating corrodes at a very slow predictable rate in an atmospheric environment, and so would not normally be affected by conditions of storage prior to installation. In addition, the coating has the unique property of providing sacrificial protection to any exposed areas of steel, so stopping these from rusting.

7.3.2.3 Protection of steel reinforcement by hot dip galvanising

Uncoated reinforcement forms a protective layer due to passivation of the steel surface at the high pH encountered in fresh concrete, and as a result a reinforced-concrete structure might be expected to last for a long period of time.[9] However, this is strongly dependent upon the performance of the reinforcement, which in turn can be affected by a variety of factors, including concrete composition, the water/cement ratio, the level of compaction, the amount of concrete coverage, the density and porosity of the concrete, and the service environment in which the structure will operate.[10]

It is fair to say that a structure cast from good-quality concrete which provides adequate coverage to the reinforcement will achieve a long life. However, in practice, variables such as concrete quality and accurate positioning of reinforcement to ensure adequate coverage cannot be guaranteed, and are strongly dependent upon the operatives conducting the work. Two issues encountered in concrete which may lead to corrosion of steel reinforcement are carbonation and chloride attack.[11–13]

Carbonation involves carbon dioxide dissolving into moisture in the concrete and forming an acidic solution. This then reacts with calcium hydroxide in the concrete, and causes a reduction in pH. Carbonation takes place over an extended period of time, and its affects will not usually be seen in the short term, as the carbonation front must reach the steel reinforcement for it to have an affect. However, as it penetrates further into the concrete and reaches the reinforcement, problems may

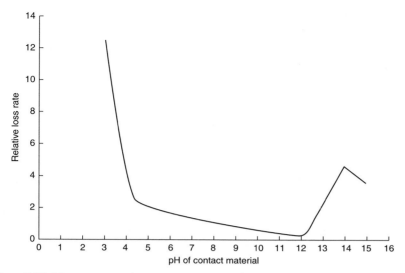

Fig. 7.57 Variation in the corrosion rate of zinc with pH of the service environment[17,22]

occur, as once the pH at the reinforcement concrete interface falls below about 11.5, uncoated reinforcement depassivates,[4,14,15] and corrosion may occur, resulting in delamination of the reinforcement and, ultimately, crack formation and spalling of the concrete.

While a galvanised coating will react with both acids and alkalis, it normally gives good performance in a pH range of 5.5–12.5.[4,16] It is, therefore, far more tolerant of alkalis, but when concrete is first cast it will typically have a pH of about 13, leading to a short period of attack during which about 10 μm of the coating may be lost.[17–19] However, as the concrete dries and cures, its pH falls, and once below about 12.5 the galvanised coating passivates against further attack due to the formation of a tenacious layer of calcium hydroxyzincates.[2,20,21]

Due to its tolerance of alkalis, the galvanised reinforcement withstands carbonation far more readily than does uncoated reinforcement.[17,19] While the coating will normally operate well within a pH range of 5.5–12.5, as shown in Fig. 7.57, work on galvanised reinforcement normally specifies a more conservative pH value of around 9.5, below which passivation of the galvanised reinforcement would cease.[3] Given the amount of calcium hydroxide present in concrete, a pH value of 9.5 cannot be easily reached, so the use of galvanised reinforcement virtually eliminates the risk of carbonation being an issue.

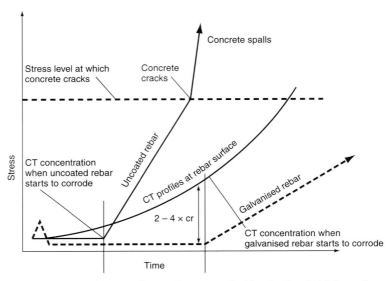

Fig. 7.58 Diagram showing the much-increased chloride threshold for galvanised steel reinforcement[2]

The chloride content of the concrete also affects the performance of the reinforcement with a number of potential sources, including calcium chloride, added to concrete as a hardening agent, use of estuary-sourced aggregates, sea spray and de-icing salts. The last two might potentially be encountered once the structure is in service, dependent upon the location and climatic conditions.

It is accepted that uncoated reinforcement has a relatively low chloride threshold limit, the American Concrete Institute giving a figure of 0.15% by mass or 0.4 kg/m^3, which if exceeded is likely to lead to corrosion of the reinforcement being initiated. Galvanised reinforcement is far more tolerant of chlorides,[1−4,6,9,14,15,17−19,22−24] and will significantly delay the onset of corrosion due to chloride attack, as shown in Fig. 7.58.

Typically, the threshold value quoted for galvanised reinforcement is between two and four times greater than that for uncoated reinforcement. However, in practice, galvanised reinforcement has been shown to offer excellent protection with little or no deterioration of the coating at values of six to eight times that for uncoated steel reinforcement. Indeed, a single reported case even indicates good performance of galvanised reinforcement where a chloride content 10 times greater than the accepted threshold for uncoated steel reinforce-ment was measured.[2]

Yeoman[17] has calculated the time to onset of corrosion due to chloride attack based upon conservative field data for galvanised reinforcement. This indicates an increase from 14 years for uncoated reinforcement to 44 years for galvanised reinforcement.

In addition to withstanding carbonation and higher chloride levels, the galvanised coating also offers the unique property of sacrificial protection. This means that should any small areas of the coating become damaged, the surrounding coating will sacrifice itself to prevent rusting of the exposed steel surface, and also prevent undercutting of the remaining sound coating.[4-6] This is of particular importance, as reinforcement may well be cut to length on site, so exposing steel at cut ends.

Even where corrosion of the galvanised coating takes place, the products produced have a smaller volume than rust[3,22] and diffuse within the concrete, so filling any porosity which might be present.[3,18,19] The net effect is that zinc corrosion products do not produce the same high tensile stresses as rust, and therefore do not result in disbondment of the reinforcement and cracking and spalling of the concrete.

7.3.2.4 Galvanised reinforcement bond strength

One of the other key advantages of galvanised reinforcement is the good bond strength achieved with the concrete.[1-2,6,16,20,22-27] Tests conduc-ted at the Building Research Establishment indicate that galvanised reinforcement has a bond strength of 3.3–3.6 MPa, which compares favourably to that for uncoated reinforcement of 1.3–4.8 MPa, which although giving some high results has a much wider spread. Additional tests have been conducted to the ACI standard at the University of California, and these also show the consistent nature of results for galvanised reinforcement, as shown in Fig. 7.59.

7.3.2.5 Advantages of galvanised steel reinforcement

Galvanising therefore offers a variety of advantages, as summarised below:

- Galvanising corrodes slowly in an atmospheric environment, and therefore prevents corrosion prior to being embedded in the concrete.
- The coating is robust and does not require any special handling.

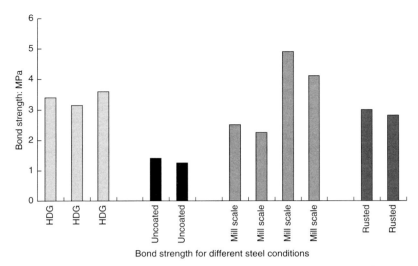

Fig. 7.59 Variation in bond strength for different reinforcement conditions as measured at the University of California[20]

- The galvanised reinforcement forms a strong bond with the concrete.
- Once embedded in the concrete it will withstand carbonation better than uncoated reinforcement, and has a higher chloride threshold.
- If passivation breaks down, the galvanised coating corrodes slowly, and the corrosion products formed migrate into the concrete away from the reinforcement filling any porosity.
- The corrosion products formed do not produce the same tensile stresses as rust.
- Due to the good performance of the galvanised coating, variation in concrete quality and level of coverage are less critical.

7.3.2.6 Cost of galvanised steel reinforcement

While hot dip galvanising offers many technical advantages over uncoated reinforcement, there is a cost implication. Recent data provided by a major steel reinforcement supplier and major galvanising groups indicated the costs for different forms of steel reinforcement[28–30] in the form of an index, as shown in Table 7.15. It is important to note that while these figures were accurate at the start of 2008, variations in

Table 7.15 Cost of different types of steel reinforcement

Reinforcement	Indexed cost
Uncoated	100
Galvanised (straight bar)	165
Galvanised (bent bar)	200
Fusion bonded epoxy coated	220
Stainless steel	510

the price of steel, stainless steel, zinc, etc., may lead to changes in the relative indexes.

Calculations have shown that galvanising would be expected to add 6–10% to the overall cost of the steel reinforcement. However, the cost of the structural frame and skin is typically only 25–30% of the total build cost, so the cost of galvanising all the reinforcement would only add 1.5–3.0% to the total build costs, a figure which may be further reduced by selective use of galvanised steel reinforcement.[4,6,22]

For a modest increase in initial cost, it is therefore possible to effectively eliminate any possibility of carbonation related issues and very significantly extend the maintenance-free life period of the structure where it is likely to be subject to chloride build-up in the concrete.

7.3.2.7 Case histories

The performance of galvanised steel reinforcement has been assessed in a number of structures, most notably bridges in North America, where its use is commonplace in concrete structures in marine environments or locations which are likely to be subject to high levels of de-icing salts during winter months.[2,31–33] The results of these assessments is summarised in Table 7.16.

The first five bridges are existing structures where the galvanised reinforcement is in pristine condition, the most notable feature being the high chloride levels measured in four of the cases, which exceeds the typically stated threshold limit of 2–4 times that for uncoated reinforcement.

In the case of the Algoa Bay Footbridge (now demolished), galvanised reinforcement was located on the seaward side of the bridge, again leading to a high build-up of chloride. The other point of interest though is that carbonation had progressed to a depth of up to 23 mm, which was beyond the first layer of reinforcement, which only had

Table 7.16 Summary of the performance of galvanised reinforcement in various bridges[2,32]

Bridge	Location	Erected	Last inspection	% chloride (x threshold)	Galvanised coating thickness: μm
Boca Chica	Florida Keys, USA	1972	1999	0.38–0.74 (2.5–4.9)	170
Tioga Bridge	Pennsylvania, USA	1974	2001	0.35 (2.3)	198
Curtis Road Bridge	Michigan, USA	1976	2002	0.14–0.96 (0.9–6.4)	155
Spring Street Bridge	Vermont, USA	1971	2002	0.43–1.14 (2.9–7.6)	191
Evanston Interchange	Wyoming, USA	1975	2002	0.85–1.55 (5.7–10.3)	236
Algoa Bay Footbridge	South Africa	1964	2005	0.15–1.26 (1.0–8.4)	240

15 mm coverage. The good condition of the galvanised reinforcement illustrated the non-effect of carbonation on its performance.

7.3.2.8 Global applications

Hot dip galvanised steel reinforcement is a global product which has been used in a diverse number of applications in a range of service environments on virtually all continents to varying degrees. Table 7.17 lists just a few select projects where galvanised steel reinforcement has been utilised in concrete[2,20,34–38] in environments varying from dry inland to coastal locations where the concrete is subject to immersion and/or sea spray.

7.3.2.9 Summary

Hot dip galvanised reinforcement has been used extensively in structures around the world. The increased cost of using galvanised reinforcement is minimal compared with the costs of potential future maintenance work, which might otherwise be required.

Table 7.17 A small number of projects where galvanised steel reinforcement has been used in concrete[2,35,37-40]

Location	Project
Africa	Kogel Bay Tidal Pool, South Africa
	Muizenberg Walkway, South Africa
Asia	Bahaii Lotus Temple, India
	Marine Research Centre, Okinawa, Japan
	Sewage Outfall Pipes, Singapore
	Chung Cheng Overpass, Taiwan
	Offshore Piers, Ominichi, Japan
Australasia	Sydney Opera House
	Parliament House, Canberra
	The High Court Building, Canberra
	New Zealand Parliament Buildings, Wellington
Europe	Barclays Bank Building, London
	National Theatre, London
	Eastbourne Congress Theatre
	Dunkirk Coal Quenching Tower
	Office of the German Chancellor, Berlin
	Theatre 'La Fenice', Venice
	Spijk Power Station
North America	Bank of Hawaii
	Crocker Building, San Francisco
	Levi Strauss Building, California
	Brooklyn Bridge (replacement bridge decks)
	Royal Bermuda Yacht Club
	Watford Bridge, Bermuda
	Veracruz Aquarium

Generally, its use is limited to high-profile projects where maintenance work would be unacceptable or structures which are liable to be subject to chloride build-up in the concrete due to their service environment (e.g. marine applications and use of de-icing salts). Given these limitations, it is most widely used in the USA and Australia, where there are many structures located in coastal locations.

References

1. Yeoman, S.R. (1991) *ILZRO project ZE341. Comparative studies of galvanized and epoxy coated steel reinforcement in concrete*. Research Report No. R103, University of New South Wales, Australia.

2. International Zinc Association (2006) *Hot dip galvanized reinforcing steel – a concrete investment*. International Zinc Association, Brussels.
3. Yeoman, S.R. (1993) *Considerations of the characteristics and use of coated steel reinforcement in concrete*, NISTIR 5211, University of New South Wales, Australia.
4. Yeoman, S.R. (2005) Corrosion protection for steel in concrete; the case for galvanizing. In: *6th Asian Pacific General Galvanizing Conference*. Cairns.
5. Galvanizers Association (2001) *Engineers and architects' guide: hot dip galvanizing*. Galvanizers Association, Sutton Coldfield.
6. Galvanizers Association Australia (1999) *After fabrication hot dip galvanizing*. Galvanizers Association Australia, Melbourne.
7. Galvanizers Association. *The galvanizing process*. Datasheet 1.2. Galvanizers Association, Sutton Coldfield.
8. Galvanizers Association. *Galvafact – toughness & abrasion resistance*. Galvanizers Association, Sutton Coldfield.
9. Yeoman, S.R. *Galvanized steel reinforcement in concrete: an overview*. University of New South Wales, Sydney.
10. Davies, H. (1990) Performance of concrete reinforcement products. *Transactions of the Institute of Metal Finishing*, 60(3), 103.
11. Testing & Consulting Services. *Structures datasheet 5 – carbonation of concrete*. www.testconsult.co.uk.
12. Concrete Experts International (2006) Carbonation of concrete: www.concrete-experts.com.
13. Testing & Consulting Services. *Structures datasheet 2 – chloride ion content*. www.testconsult.co.uk.
14. Yeoman, S.R. (2006) Corrosion of steel reinforcement in concrete. In: *CFC Rebar Workshop*. Singapore.
15. Yeoman, S.R. (1994) A conceptual model for the corrosion of galvanized steel reinforcement in concrete. In: *International Conference on Corrosion and Corrosion Protection of Steel in Concrete*. Sheffield.
16. Porter, F. (1991) *Zinc handbook: properties, processing and use in design*. Marcel Dekker, New York.
17. Yeoman, S.R. (2006) Processing and characteristics of galvanized steel for use in concrete. In: *CFC Rebar Workshop*. Singapore.
18. Nordic Galvanizers (2007) *Galvanizing handbook*. Nordic Galvanizers, Stockholm.
19. Fratesi, R. (2002) Galvanized steel reinforcement. In: *COST 521 Workshop*. Luxembourg.
20. European General Galvanizers Association. *Galvanized reinforcement*. European General Galvanizers Association, Caterham.
21. White, R., Goodwin, F.E. and Rourke, D. (2007) In: *ILZRO, 7th Asian Pacific General Galvanizing Conference*. Beijing.
22. White, R. *IZA Southern Africa. General galvanizing – a cure for rebar corrosion*. www.hdgasa.org.za.

23. Yeoman, S.R. (1998) In: *Corrosion 98 Conference*. San Diego, paper 653, University of New South Wales, Australia.
24. American Galvanizers Association (1996) *Galvanizing for corrosion protection: a specifiers guide to reinforcing steel*. American Galvanizers Association, Centennial, CO.
25. American Galvanizers Association (2006) *Hot dip galvanizing for corrosion protection: a specifiers guide*. American Galvanizers Association, Centennial, CO.
26. Kayyali, O.A. and Yeoman, S.R. (1995) Bond and slip of coated reinforcement in concrete. *Construction and Building Materials*, 9(4), 219–226.
27. Swamy, R.N. (1994) *Corrosion and corrosion protection of steel in concrete*. Sheffield Academic Press, Sheffield, vols 1–2.
28. Hyten (2008) Personal communication.
29. Beech, T. and Wedge, B.E. (2008) Personal communication.
30. Catley, S. (2008) Personal communication.
31. Galvanized Rebar Resource Center. Case histories: www.galvanizedrebar.com.
32. Galvanizers Association of Southern Africa (2005) Case history 04/2005. Hot dip galvanized steel reinforcement in concrete. *Hot Dip Galvanizing Today*, 2(3), 51.
33. Smith, T. *A case study of hot dip galvanized reinforcement*. www.hdgasa.org.za.
34. Mackie, K. www.hdgasa.org.za.
35. Bishop, R. (2006) Practical applications of HDG Rebar in marine exposure conditions. In: *CFC Rebar Workshop*. Cape Town.
36. Zinc Development Association. Galvanizing in action – reinforced concrete. Zinc Development Association, London.
37. White, R. (2006) More durable concrete structures using hot dip galvanized reinforcing steel. In: *CFC Rebar Workshop*. Cape Town.
38. Huckshold, M. (2008) Personal communication.
39. Yeoman, S.R. (2006) The use of galvanized reinforcement in building construction and infrastructure. In: *CFC Rebar Workshop*. Singapore.
40. American Galvanizers Association. *Galvanizing projects*. American Galvanizers Association, Centennial, CO.

7.3.3   Stainless steel reinforcement

David Cochrane, Consultant to the Nickel Institute, UK

7.3.3.1 Introduction
In order to turn steel into a 'stainless steel', a minimum of 10.5% chromium is included in the mix at the molten metal stage. When solidified, the chromium reacts with oxygen in the air to form a chromium oxide film at the surface that is transparent and tightly adhering. The passive film, as it is referred to, is highly protective and resistant to normal oxidation (rust).

Additions of nickel, molybdenum and nitrogen further toughen the passive film, to provide a range of stainless steels that can meet most corrosive working environments.

Discovered in Sheffield in 1913, the high corrosion resistance of stainless steel has become its most universally recognised property. It inspired the architects and engineers in the early 1920s to use it to save one of London's most famous buildings, St Paul's Cathedral, from possible demolition. Over 120 tons of stainless steel reinforcement was used to prevent further sinking of the Dome. Since its discovery, the number of grades of stainless steel has grown to over 100, as major industries have recognised the significant benefits from using this alloy. The construction industry, however, has been slow to recognise the true potential of a strong high-corrosion-resisting material – and at a significant financial and environmental cost. For example, the corrosion of carbon steel reinforcement has been the principal cause of premature failure in many thousands of highway bridges, necessitating costly repair and traffic delays, and creating disruption that could have been avoided by using stainless steel.

Recognising the extent and cost of the problem, highway authorities have sought solutions to the corrosion problems in bridges over the last two decades and investigated many different materials and design solutions. Nickel-containing stainless steel has been proven in tests[1,2] to have high resistance to the chloride ion, which is the principal cause of corrosion of carbon steel rebar. It has also been tested for any adverse galvanic reaction when in contact with carbon steel rebar.[3,4] The conclusion of feasibility studies in North America[5,6] and Europe is to use stainless steel reinforcement in the elements of the structure that are at high risk of corrosion, and carbon steel where there is little or no risk. This solution has proven, in practice, to provide increased durability cost-effectively, and has been adopted by highway authorities around the world. In Scandinavia, it is now mandatory to use stainless steel reinforcement in bridge parapets – a recognised corrosion-prone element.

7.3.3.2 British standards and other national acceptance documentation

The principal standard for stainless steel reinforcement is BS 6744: 2001,[7] *Stainless Steel for the Reinforcement and Use in Concrete.*
Other official standards and admissions are:

- USA: ASTM A955/A955M-03b
- Denmark: national standards and admissions DS 13080 and DS 13082
- Germany and Italy: admissions
- France: XP A 35-014
- Finland: SFS 1259.

Guidance documentation on the use of stainless steel reinforcement
In the UK, guidance on the use of stainless steel reinforcement was prepared by Arup R&D for the Highways Agency (HA), and is incorporated in the HA's *Design Manual for Roads and Bridges*, BA 84/02.[8]

These guidelines recognise the contents of BS 6744:2001 and that stainless steel can improve the durability of the structure, reduce maintenance, and minimise the costly effects of traffic delay and disruption. BA 84/02 recommends the use of stainless steel according to the location of the bridge, and indicates where stainless steel should be used, and the appropriate grade of stainless steel.

The guidelines also recognise that stainless steel reinforcing bar is issued with a CARES (a standards compliance certification body) certificate. The agency also accepts that stainless steel reinforcement can be used cost-effectively by siting the stainless steel in the high-risk elements of the structure and using carbon steel reinforcement in protected low-risk areas. The stainless steel content can vary, therefore, from a small percentage to 100%, depending upon the bridge location.

Due to the high corrosion resistance of stainless steel, the HA recognises that the rules developed to improve the durability of carbon-steel-reinforced structures can be relaxed when using stainless steel reinforcement. This relaxation, however, should not be taken to

Table 7.18 Relaxation rules for using stainless steel reinforcement in BA 84/2

Design condition	Relaxation
Cover	Cover for durability can be relaxed to 30 mm where stainless steel is used irrespective of the concrete quality or exposure condition
Design crack width	Allowable crack width increased to 0.3 mm
Silane treatment	Not required on elements with stainless steel

Data from the HA manual BA 84/2: reproduced by kind permission of the Controller of Her Majesty's Stationery Office

imply that the concrete quality and workmanship can be relaxed when using stainless steel, and it is important for these qualities to be maintained to realise the full durability benefits of the improved structure. The relaxation rules for using stainless steel, as given in BA 84/2, are given in Table 7.18.

Classification of structures and elements

Chapter 3 of the BA 84/2 provides guidance on the locations of structures and elements where the use of stainless steel reinforcement is recommended. The HA identifies the areas of potential corrosion risk and the potential disruption of carrying out future maintenance. It also identifies where stainless steel may not be appropriate.

A summary of Chapter 3 is given in Table 7.19.

Selection of steel grade

With regard to BS 6744:2001 and the HA BA 84/02 guidelines, the advised grade of material is shown in Table 7.20 for a range of exposure conditions.

7.3.3.3 The corrosion resistance of stainless steel in concrete

Research into the use of stainless steel in concrete has been carried out in the USA, Canada, Denmark, Italy and the UK over the past 30 years. Arguably, the most important research programmes are:

- the exposure tests carried out by the UK Building Research Establishment (BRE) that commenced in the early 1970s[1,2]

Table 7.19 Classification of structures and elements

Classification	Recommendation
Total substitution of stainless steel for major components	Major components of new structures where future repair and maintenance work would be very disruptive to traffic and costly. Examples: decks of bridges carrying heavily trafficked roads over busy railway lines and with limited possessions for repair; exposed piers and columns in centre reserves but not deeply buried elements; deck slabs where access for maintenance is difficult due to traffic levels. Whole life costs to include traffic delay costs
Elements of new structures exposed to seawater or are in the seawater splash zones	Total substitution with stainless steel for all structural elements above low-water spring tide level to 5 m above high-water spring tide. Consideration to soffits and edge beams also where they are subject to spray
Elements of new structures adjacent to the carriageway exposed to chlorides from road de-icing salts	For structures on motorways and trunk roads, elements are likely to include: bearing shelves and other elements below road joints, the faces of abutment or pier supports adjacent to carriageway, parapet edge beams
Where stainless steel is not appropriate	For highway bridge and structure components unlikely to be regularly exposed to high concentrations of chloride, components are remote from the highways, will not need traffic management for maintenance

Data from the HA manual BA 84/2: reproduced by kind permission of the Controller of Her Majesty's Stationery Office

- the 5 year research programme carried by McDonald of Wiss Janney Elstner (WJE) on behalf of the US Federal Highways Administration in the mid-1990s, into materials suitable for the full design life of bridge structures[5]
- the research and test programmes carried out by Pedeferri *et al.* at Milan University on stainless steel reinforcement in concrete, and galvanic corrosion issues when used with carbon steel reinforcement.[7]

A summary of this research is given below.

BRE (UK) exposure tests
Five hundred concrete prisms and beams were laid down in 1972 containing austenitic and ferritic stainless steels, high-yield carbon steel, galvanised steel and Corten.

Table 7.20 Selection of stainless steel in BA 84/02, Chapter 4

Material grade	Exposure condition
1.4301	Stainless steel embedded in concrete with normal exposure to chlorides in soffits, edge beams, diaphragm walls, joints and substructures
1.4301	As above, but where design for durability requirements are relaxed in accordance with Table 4.6 of BA 84/02
1.4436	As above, but where additional relaxation of design for durability is required for specific reasons on a given structure or component, i.e. where waterproofing integrity cannot be guaranteed over the whole life of the structure
1.4429 1.4436	Direct exposure to chlorides and chloride-bearing waters, e.g. dowel bars, holding down bolts and other components protruding from the concrete
1.4462 1.4429	Specific structural requirements for the use of higher-strength reinforcement and suitable for all exposure conditions

Data from the HA manual BA 84/2: reproduced by kind permission of the Controller of Her Majesty's Stationery Office

The concrete was cast with varying levels of calcium chloride (0, 0.32, 0.96, 1.9 and 3.2% by weight of cement), and the test pieces sited at coastal and industrial sites in the UK.

The condition of the reinforcement materials was reported after 10 years,[1] and showed that only the austenitic stainless steel remained unaffected in the prisms with the highest level of calcium chloride (3.2%). Not all the prisms containing the austenitic bars were broken open as a result, and they remained at the exposure sites until 1995. When broken open after 23 years of exposure, with 3.2% calcium chloride in the mix, the stainless steel bars still remained unaffected.[2]

The WJE programme for the US Federal Highways Administration
The Federal Highways Administration commissioned WJE to investigate suitable materials to meet the design life of highway bridges following a survey that revealed over 200 000 bridges were suffering from reinforcement corrosion. The cost of premature failure was estimated in billions of dollars.

WJE investigated 57 different materials during the 5 year programme.

Table 7.21 Strength and size range of stainless steel reinforcing bar

Strength grade	Nominal size: mm
500 N/mm^2	3, 4, 5, 6, 7, 8, 10, 12, 14, 16, 20, 25, 32, 40, 50
650 N/mm^2	3, 4, 5,6, 7, 8, 10, 12, 14, 16, 20, 25, 32, 40, 50

The conclusion was that austenitic stainless steel 1.4401 was the most suitable material for reinforcement to meet the design life of 120 years.

Research by Pedeferri at Milan University
The tests by Pedeferri *et al.* have shown that 1.4301 (grade 304) and 1.4436 (grade 316) stainless steels retain passivity and corrosion resistance at pH values typical in concrete and at chloride levels considerably higher than encountered in civil engineering structures.

Pedeferri has shown that carbon steel can tolerate a chloride level of 0.4% only at the highest level of pH, 13. At a pH of 11.2, the chloride level it can tolerate is 0.

Stainless steel grade 1.4301, however, can tolerate chloride levels of 6, and 3% at the same levels of pH, while stainless steel grade 1.4436 can tolerate levels of 7 and 4.5%, respectively.

7.3.3.4 Size range, material grades, and strengths
Stainless steel is produced to two strength levels and in a range of diameters, as shown in Table 7.21. It is also produced to a similar profile as carbon steel, and meets the same bond strength.

Stainless steel reinforcing bar is produced in the material grades shown in Table 7.22 with the recommended service conditions. (As given in Table B.1 in Annex B of BS 6744:2001.)

7.3.3.5 Selective substitution
Improved durability of a structure can be achieved by placing stainless steel reinforcement in the elements of the structure that are at greatest risk to corrosion, and by using carbon steel reinforcement where it is unlikely to be reached by chloride ions, hence the term 'selective substitution'. Highways authorities recognise this method of improving durability.

Table 7.22 Material grades of stainless steel reinforcing bar and service conditions

Grade	A	B	C	D
1.4301	1	1	5	3
1.4436	2	2	1	1
1.4429	2	2	1	1
1.4462	2	2	1	1
1.4529	4	4	4	4
1.4501	4	4	4	4

Note: It is anticipated that grades 1.4162 and 1.4362 will be added to this table at the next amendment.

Key to Table 7.22:
1   Appropriate choice for corrosion resistance and cost.
2   Over specification of corrosion resistance for the application.
3   May be suitable in some instances: specialist advice should be obtained.
4   Grades suitable for specialist applications which should only be specified after consultation with corrosion specialists.
5   Unsuitable for the application.

Key to service conditions:
A   For structures or components with either a long design life, or which are inaccessible for future maintenance.
B   For structures or components exposed to chloride contamination with no relaxation in durability design.
C   Reinforcement bridge joints, or penetrating the concrete surface and also subject to chloride contamination (e.g. dowel bars or holding down bolts).
D   Structures subject to chloride contamination where reductions in normal durability requirements are proposed (e.g. reduced cover, concrete quality or omission of waterproofing treatment).

Maximum durability, however, will be achieved with an all-stainless reinforced structure.

Principal elements
Table 7.19 indicates typical elements where it can apply.

Galvanic reaction issues
The selective substitution of stainless steel means that the different steels will be in direct contact in the structure. Civil engineers' concerns that a separating barrier be used to prevent galvanic corrosion can be discounted, as this phenomenon has been specifically studied and

Fig. 7.60 Broadmeadow Bridge, Republic of Ireland (Courtesy of Anaco Stainless (Edward James))

tested,[3-5] for adverse effects in Italy, Denmark, Sweden, the USA and Canada, and the results have unanimously concluded that there is no detrimental reaction to the materials being in direct contact. On the contrary, they concur that in repair, it is beneficial to use stainless steel rather than replacing with carbon steel.

7.3.3.6 Cost-effectiveness

Case study: Broadmeadow Bridge, Republic of Ireland
By adopting the principal of selective substitution, the Broadmeadow Bridge in the Republic of Ireland is an example of the cost-effectiveness of the method (Fig. 7.60). Spanning 313 m across mud flats, 180 tonnes of 1.4436 (grade 316) stainless steel reinforcement was used for the starter bars for the pre-cast parapets, the supporting columns and the parapet edge beams. This accounted for an additional 9% to the total cost of the reinforcing steel used in the project. However, the overall effect was an increase in the total project of only 2.9%, but the use of stainless steel means that no corrosion will occur in these crucial elements for the design life of 120 years.

Initial cost tool

A report produced by Arup for the British Stainless Steel Association is available on a CD-Rom.[9] This contains an initial cost tool for the use of stainless steel reinforcement in different elements of a structure for two different bridge designs.

7.3.3.7 *Summary of the benefits of using stainless steel*

- Resistant to high levels of chloride ions.
- Reduces maintenance, downtime and traffic delays.
- Requires no added surface protection against corrosion.
- Has the same rib configuration and bond strength as conventional carbon steel reinforcement.
- Produced to higher strengths than conventional carbon steel.
- Can be bent to tight radii.
- Can be used in conjunction with carbon steel to enhance the durability of a structure.
- Eliminates the need for concrete coatings such as silane.
- Requires no monitoring.
- Reduces concrete cover to 30 mm.
- Permits an increase in the design crack width.
- Environmentally friendly.
- Recyclable.

Fig. 7.61 Guildhall Yard East, London (Reproduced with permission of C.J. Abbott, Fixing Centre Ltd.)

7.3.3.8 Case study: Guildhall Yard East, London

An extension to the Guildhall was completed in 1997 at a cost of £50 million. Known as Guildhall Yard East, and consisting of four office levels and three basements, it is in keeping with the longevity required for the historical Guildhall (Fig. 7.61). More than 140 tonnes of stainless steel reinforcement grade 1.4301, ranging from 6 to 25 mm in diameter, was used in the construction. The upper basement incorporates the remains of a Roman amphitheatre discovered in 1987 during construction.

7.3.3.9 Acknowledgement

Permission to use extracts from BS 6744:2001 is granted by BSI.

Further reading

Concrete Society (1998) *Guidance on the use of stainless steel reinforcement.* Technical Report No. 51. Concrete Society, Camberley.

International Stainless Steel Forum: www.worldstainless.org/ISSF/Files/Rebar/ Flash.html.

References

1. Treadaway, K.W.J., Cox, R.N. and Brown, B.L. (1989) Durability of corrosion resisting steels in concrete. *Proceedings of the Institution of Civil Engineers*, 1(86), 305–331.
2. Cox, R.N. and Oldfield, J.W. (1996) The long term durability of austenitic stainless steel in concrete. In: *Proceedings of the 4th International Symposium on Corrosion of Reinforcement Construction, Robinson College.* Cambridge, 662–669.
3. Bertolini, L., Gastaldi, M., Pedeferri, M.P. and Pedeferri, P. (1998) Effects of galvanic coupling between carbon steel and stainless steel reinforcement in concrete. In: *Proceedings of the International Conference on Corrosion and Rehabilitation of Reinforced Concrete Structures.* Orlando, CD-Rom.
4. National Research Council Canada (NRCC) (2004) *Investigation of the effects of galvanic coupling between stainless steel and carbon steel reinforcements in concrete.* Report for Alberta Infrastructure, Quebec Ministry of Transport. Nickel Institute, Valbruna Canada Inc., Ontario.
5. McDonald, D.B., Pfeifer, D.W. and Sherman, M.R. (1998) *Corrosion evaluation of epoxy-coated, metallic-clad and solid metallic reinforcing bars in concrete.* US Department of Transportation, Washington, DC.
6. Knudsen, A., Jensen, F.M., Klinghoffer, O. and Skovsgaard, T. (1998) Cost effective enhancement of durability of concrete structures by intelligent

use of stainless steel reinforcement. In: *International Conference on Corrosion and Rehabilitation of Reinforced Concrete Structures*. Orlando.

7. BSI (2001) *Stainless steel bars for the reinforcement of and use in concrete – requirements and test methods*, BS6744. British Standards Institution, London.

8. Highways Agency (2002) *Design manual for roads and bridges. Use of stainless steel reinforcement in highway structures*, BA84/02. Highways Agency, London, vol. 1, section 3, part 15.

9. British Stainless Steel Association (BSSA) (2003) *Stainless steel reinforcement for concrete*. British Stainless Steel Association, Sheffield, CD-Rom.

7.3.4 Glass-fibre-reinforced polymer (GFRP) rebar

Douglas Gremel, Hughes Brothers, Inc., USA

7.3.4.1 Applications and uses of GFRP rebar

One of the primary failure mechanisms in reinforced concrete is corrosion of the steel reinforcing rebar. When steel corrodes, it expands, causing the adjacent concrete to spall, resulting in failure and limiting the useful life of the structure. The typical solutions to this problem are to find ways to protect the steel by adding concrete admixtures, increased clear cover, using coated reinforcing or non-ferrous steel and/or 'high-performance concrete'. Ultimately, the various protection methods will fail, and the steel will corrode, leading to the deterioration of the bridge.

A chloride-rich environment, such as a bridge deck that is subjected to de-icing salt or a marine structure exposed to salt water, accelerates the rate at which steel will corrode. Inevitably the concrete will crack, leading to avenues for chlorides to penetrate to the steel. Additionally, a chemical electrolytic cell or battery effect is often created within the structure, which accelerates corrosion of the reinforcing steel. GFRP rebar is impervious to attack by chlorides, and will not corrode and spall the concrete. GFRP eliminates the typical failure mechanisms and extends the service life of the structure. These same benefits are extended to reinforced-concrete elements subjected to other low-pH solutions such as waste-water treatment, chemical plants and containment vessels or simply concrete elements that are desired to be thin with inadequate cover for traditional ferrous reinforcing.

Other reinforced-concrete applications take advantage of the electro-magnetic neutrality of the GFRP bars should concrete members be adjacent to high voltages, magnetic fields, stray currents or radio or high-frequency signals. Typical applications include magnetic resonance imaging units in hospitals, high-voltage substations, transformer pads, cable ducting, inductance loops in highways and tollbooths, and many other uses where electromagnetic transparency is beneficial.

In addition to being non-corrosive and non-conductive, GFRP rebar is also thermally non-conductive, and a number of uses are made in concrete elements that take advantage of the lack of thermal transmissivity through the bar of varying temperatures such as thermal shunts for

Fig. 7.62 Floodway bridge near Winnipeg – two eight-span structures

cast balcony units and as connecting elements in insulated concrete sandwich wall construction.

Unlike steel, which is an isotropic material, equally strong in any given direction, GFRP bars are highly anisotropic, meaning they are strong primarily along the main axis of the bar. This weaker strength in the transverse direction of the bar makes them especially difficult to test and means they can be abraded away. Some important applications take full advantage of this anisotropic nature of the GFRP bars by incorporating the potential sacrificial nature of the bar into temporary use applications. Examples include zones for tunnel-boring machines to penetrate and launch in deep-foundation diaphragm walls to begin and end tunnel-boring runs; and sequential excavation tunnelling methods, whereby GFRP bars are used to temporarily stabilise sections of ground or earth and then consumed by tunnelling and mining equipment. This same concept facilitates the use of GFRP soil nails in passive removable situations where encroachment on adjacent land is not inhibited by the presence of GFRP bars that might have once been used to retain the earth for the construction of a high-rise building, for example.

GFRP bars have been used as concrete reinforcing for more than a decade, and invaluable experiences have been gained from the many projects. For example, to date over 60 bridge decks have been built with GFRP reinforcing in the deck, just in the USA and Canada (Fig. 7.62). All bridges that have been built thus far are performing as intended, and continue to perform successfully.

7.3.4.2 *Composite materials and properties of GFRP bars*

Fibre-reinforced polymer (FRP) bars are made from continuous fibres, typically carbon, glass or aramid, contained in a polymer resin, usually epoxy, vinyl ester or polyester. The fibre acts as the structural element, and the resin is a binder of the fibres. The resin serves an important role in transferring stresses from fibre to fibre as well as protecting and encapsulating the fibre from the environment. FRP bars are somewhat analogous to reinforced concrete itself, whereby the concrete acts as the resin and the reinforcing behaves as the fibre.

The physical and mechanical properties are mostly dictated by the producer's choice of fibre. Glass fibres are the most common commodity and the least expensive type of composite fibre. Large multinational companies, such as Owens Corning, Pittsburgh Plate and Glass, Vetrotex CertainTeed, St Gobain and others, in plants that have overhead and capital investment rivalling or exceeding most mini-steel mills, produce them. Multinational global companies such as Ashland Chemical, Dow Chemical, Reichhold Chemical, DSM and several others also produce the resins. The role of the rebar fabricator is to combine the fibres and resins into a pultruded composite that is the final product. Typically, the manufacturing process used to fabricate FRP bars is the pultrusion process, whereby filaments are impregnated with resin and drawn through a heated die from which they emerge as a final or semi-final product. Thereafter, various manufacturing techniques are used to enhance the FRP bar to introduce some type of bond mechanism to the concrete, which distinguishes FRP rebar from other pultruded shapes. The process is called pultrusion since the final formed product is pulled upon, which is the mechanism that draws the fibres through the resin and mechanically produces the product.

When producing the FRP bar, the fabricator has many fibre resin choices to make, and there are many variables to the pultrusion process. Fortunately, once defined and controlled, the pultrusion process is a relatively stable and continuous one.

The fibre choice is dictated by considerations of the end use. The emphasis of the commercial FRP bars is currently in the area of glass FRP bars for new construction. Due to the economics of the GFRP bars, they can compete closely in cost to coated steel bars (epoxy or galvanised), and are typically much less expensive than stainless steel bars. Other possibilities include the use of carbon fibres (CFRP bars). CFRP bars are generally used in special applications such as the strengthening of existing structures or in rare instances as prestressing or post-tensioning tendons, assuming a suitable anchorage for tensioning the tendons is

Table 7.23 Size designation of FRP round bars

Bar size designation		Nominal diameter		Nominal area	
in.-lb	Metric	in.	mm	in.^{2}	mm^2
2	6	0.250	6.4	0.05	32
3	10	0.375	9.5	0.11	71
4	13	0.500	12.7	0.20	129
5	16	0.625	15.9	0.31	199
6	19	0.750	19.1	0.44	284
7	22	0.875	22.2	0.60	387
8	25	1.000	25.4	0.79	510
9	29	1.128	28.7	1.00	645
10	32	1.270	32.3	1.27	819

used. Aramid fibres have been used successfully as tendons as well. But, for use as reinforcing bar in new concrete construction, due to economic considerations, GFRP is the predominate fibre type, with the carbon and aramid fibres being more in the realm of exotic uses.

The calculated diameter of a FRP bar is equivalent to that of a smooth round bar having the same area as the FRP bar measured by ASTM D 7205/D 7205M (Table 7.23).[1] Unlike steel bars, FRP bars exhibit a size effect, as loads cannot be transferred with 100% efficiency from fibre to fibre through the resin matrix. Minimum 'guaranteed tensile strengths' for FRP bars are published by the ACI 440 materials standards,[2] and are shown in Table 7.24.

Table 7.24 Minimum guaranteed tensile strengths for FRP bars

Bar size designation		Minimum guaranteed tensile strength			
in.-lb	Metric	GFRP		CFRP	
		ksi	MPa	ksi	MPa
2	6	110	760	210	1450
3	10	110	760	190	1310
4	13	100	690	170	1170
5	16	95	655	160	1100
6	19	90	620	160	1100
7	22	85	586	N/A[a]	N/A
8	25	80	550	N/A	N/A
9	29	75	517	N/A	N/A
10	32	70	480	N/A	N/A

[a] Indicates that CFRP bars of these sizes are currently not available.

The American Concrete Institute (ACI) further prescribes the tensile modulus of elasticity shall be determined based on the bar nominal diameter. The nominal tensile modulus of elasticity of GFRP bars shall be at least 5700 ksi (39.3 GPa) regardless of bar size or geometry. The tensile modulus of elasticity of CFRP bars shall be at least 18 000 ksi (124 GPa) regardless of bar size or geometry. The tensile modulus of elasticity is derived from specimens tested in accordance with ASTM D 7205/D 7205M at a prescribed frequency and number of specimens.

7.3.4.3 Testing

To ensure that GFRP bars supplied are furnished with properties used by the design engineer, various test standards are available. Testing of GFRP bars is generally performed based on the test methods outlined in ACI 440.3R-04,[3] *Guide Test Methods for Fiber-reinforced Polymers for Reinforcing or Strengthening Concrete Structures*. ACI 440.3R was developed to facilitate the 440.1R-06 design guide and to transition to ASTM standards. Many individual test methods have already been published as ASTM standards, such as ASTM D7205/D, *Standard Test Method for Tensile Properties of Fiber Reinforced Polymer Matrix Composite Bars*. The standardised tests allow for easy determination of relevant engineering properties, and give the engineer and bridge owner confidence that the bars will meet the properties that were assumed for design calculations. These documents also enable independent confirmation of properties and quality assurance testing and enable multiple FRP rebar suppliers to bid on individual projects.

Test methods are important with these new materials as they enable verification of individual bar properties and give confidence to the engineer the design will be properly built. With standardised tests the engineer can easily determine if a prospective supplier meets the near-term design properties and if the bar will perform in the long term as anticipated. The standardised testing procedures also eliminate proprietary FRP 'systems', which allows a designer to accept multiple bids from suppliers. The guidelines are a great tool to 'pre-qualify' manufactures so that multiple bids can be received with confidence that all the bar properties meet the specifications required. In Canada, the ISIS network has taken the lead in this area by publishing *Specifications for Product Certification of FRP's as Internal Reinforcing in Concrete Structures*.[4]

Very recently, the ACI has published material specifications describing in mandatory language provisions governing testing, evaluation and acceptance of FRP bars and describing the permitted constituent materials, limits thereof and minimum performance requirements for FRP bars to be used as concrete reinforcing. A companion specification[5] covers construction aspects of the use of FRP bars such as bar placement, preparation, repair of FRP bars, field cutting and concrete placement. These documents enable the owner and engineer to pre-qualify potential FRP bar suppliers for approved bidding and aid site engineers and contractors in successfully monitoring and installing FRP bars in the field.

7.3.4.4 Design issues

Based on the experience gained from past projects, and a voluminous amount of academic research, several authoritative consensus design guidelines have been published or are in the draft stages. One of the more important documents, in its third iteration, is the ACI Committee 440 document 440.1R-06,[6] *Guide for the Design and Construction of Concrete Reinforced with FRP Bars*. In Canada, two important design documents have been published, CSA S6-06,[7] *Canadian Highway Bridge Design Code*, which now includes provisions for the use of FRP bars, and CSA S806-02,[8] *Design and Construction of Building Components with Fibre-Reinforced Polymers*. Worldwide, several other guidelines are in various stages of publication such as that by FIB Task Group 9 in Europe, the GB in China and the JSCE *Recommendation for Design and Construction of Concrete Structures using Continuous Fiber Reinforcing Materials*. For the American bridge design community, drafts of AASHTO LRFD, *Bridge Design Guide Specifications for GFRP Reinforced Concrete Decks and Deck Systems*[9] have been circulated for review to Committee T6. These documents offer the bridge designer authoritative design guidance using traditional reinforced-concrete design methodology with variations described to account for the different physical mechanical properties of FRP bars. The basic principles of FRP reinforced-concrete design are very straightforward and easily adapted by the structural engineer.

Design of GFRP reinforced members is very similar to design using conventional steel reinforcing. The mechanics principles are the same, but differences in the material properties require slightly different computations. The biggest change is that GFRP is linear elastic up to failure and does not yield (Fig. 7.63).

Typically, in place of the ductile steel reinforcing being mandated to be the weak link in the reinforced concrete, with FRP bars the designer must choose a failure mode of either rupture of the FRP bar or

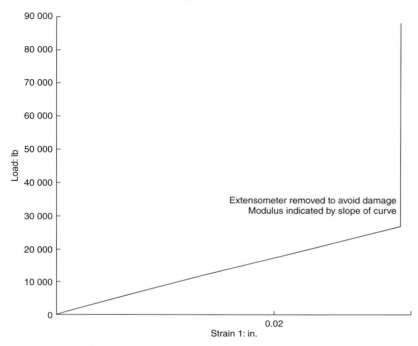

Fig. 7.63 *Typical stress–strain curve for GFRP bars*

compression failure of the concrete. Safety factors differ between the two failure modes, with a transition zone, to ensure there is no shift in failure modes should actual concrete strengths exceed design concrete strengths. In the case of compression failure of the concrete, a ϕ factor of 0.65 is inherent to the design. If the chosen mode of failure is bar rupture, a ϕ factor of 0.55 is used (Fig. 7.64). In either case, due to the low modulus of the FRP bars, tremendous service load deflections and large crack widths will be present prior to failure. In fact, design for deflections and crack control most often control the design.

Careful effort was made in the various design guidelines to keep the design procedure as similar to steel-reinforced concrete design as possible. The ACI 440.1R-06 guide, for instance, addresses ultimate limit states for shear and flexure as well as all of the serviceability issues. Utilisation of FRP bars is very conservative, due to the relative novelty of FRP reinforcing. For example, the ACI 440.1R-06 guide limits the sustained stresses on a GFRP bar to just 20% of the guaranteed short-term properties.

The 2006 *Canadian Highway Bridge Design Code* (CSA S6-06) has a chapter devoted to GFRP-reinforced concrete design. It addresses

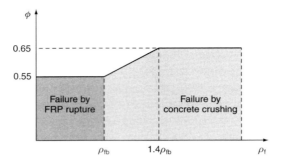

Fig. 7.64 The strength reduction factor as a function of the reinforcement ratio

similar topics as ACI 440.1R-06, but also provides for the use of an empirical design method. The empirical method assumes that the deck can be modelled as a simple truss. The compression forces of the truss are carried through arching action, by the concrete and either internal reinforcing or an external steel strap carries the tensile force. When an external steel strap is used, it is known as the 'steel free deck design'. The Red River Floodway project, mentioned earlier, was designed using the steel free deck design method.

In general, the design philosophy and load and resistance factors for FRP-reinforced concrete beams are the same (ACI 318 compared with ACI 440.1R-06) and most standards organisations conform to limit states design procedures. Strain compatibility and equilibrium are used to determine internal forces in an FRP-reinforced cross-section. Some notable differences between steel and FRP design include:

- *Bar properties.* FRP bars are anisotropic whereas steel bars are isotropic. FRP bars exhibit high tensile strength and low tensile modulus.
- *Resistance factors.* FRP bars are not ductile, and failure modes are compression failure of concrete or bar rupture, with varying factors of safety.
- *Shear capacity.* FRP bars have lower shear capacity than steel, but the ACI 440 and ACI 318 guidelines are being unified with common equations.
- *Moments of inertia.* As with shear capacity, a general trend is towards unifying equations for FRP and steel reinforcing, but currently modification factors are used.
- *Minimum reinforcing requirements.* Minimim requirements ensure that the ultimate moment exceeds the cracking moment in FRP reinforced members.

- *Environmental reduction factors.* These are dictated by varying the fibre-type and environmental exposure conditions.
- *Design for serviceability.* Due to the lower tensile modulus of FRP bars, in many instances deflections will control design. Crack widths are also often a controlling design parameter, and are limited mainly for aesthetic reasons.
- *Sustained loads and limits on service load stresses.* Due to the phenomenon of creep rupture, long-term sustained loads are limited on FRP bars to a threshold much lower than for steel bars.

7.3.4.5 Installation and construction

Generally, installation and construction using GFRP rebar is very straightforward, with very few differences from coated steel rebar. Due to the low modulus of the materials, bars need to be supported with chairs at a spacing of two-thirds that of steel rebar. It is recommended that 50% of bar intersections be tied using plastic-coated tie wire. GFRP is less dense than concrete, and in theory could float in the concrete, and so it is generally recommended that the mats be tied down every once in a while. Experience has shown that during concrete placement there is enough foot traffic on the bars to keep them from floating up (Fig. 7.65). Floating of bars can occur on large

Fig. 7.65 Deck pour – Miles Road Bridge, Cuyahoga County, Ohio

vibrating beds in a precast plant, but this phenomenon has yet to be observed during the construction of a bridge deck. Of chief concern should be the potential for 'abrasion' of GFRP bars on site. ACI construction specifications describe remediation in such instances.

7.3.4.6 Successful projects

As mentioned earlier, there have been many successful projects where GFRP bars have been used. The largest single use of FRP bars in a structure being the Floodway Bridge over the Red River near Winnipeg. It consists of two bridges with eight spans each and a total length of 348 m for each span. The deck was built using the steel free deck concept, and used GFRP for both the top and bottom mats as well as the barrier walls. The project required over 150 tonnes of GFRP rebar (GFRP is a quarter the density of steel, and would be equivalent to approximately 500 0000 kg of steel). Only two areas of the deck had steel, a stainless steel connector for the barrier walls and a steel strap located underneath the deck. A full life cycle analysis was done on the bridge, and GFRP reinforcing was chosen based solely on life cycle costs.[10] The Floodway bridges were some of the first GFRP reinforced bridge decks to be constructed without receiving some form of grant money: this is important to note, as there were no outside factors influencing the decision to use GFRP, and it was chosen simply based on cost versus benefit analyses.

It would be impossible to mention all of the various FRP-reinforced concrete projects in a single chapter. Since GFRP bars are considered for use mainly as a means of improving the longevity of the reinforced concrete structure, the ISIS research network published some key metrics on the performance of bridge structures in 2005. Their series of publications *Durability of GFRP Reinforced Concrete from Field Demonstration Structures*[11] involved extracting GFRP bars from bridges throughout Canada that had been in service for between 5 and 8 years. Detailed analysis showed that there was no degradation of the internal GFRP bars in these series of bridge structures, leading a group of leading civil engineers from around the world to declare the 'Winnipeg Principles'[12] advocating the use of FRP bars in bridge decks.

7.3.4.7 Conclusions

The extended service life anticipated by the use of GFRP rebar benefits all parties involved, from the owner to the end user. When life cycle costs are examined, it is less expensive to use GFRP reinforcing over traditional steel or epoxy-coated steel. The initial cost premium to use GFRP bars is of the order of just a couple of percentage points. Designing structures with GFRP internal reinforcing is no longer a novel and new concept. There have been many successful projects, and every year more and more bridges are built with GFRP reinforcing. Designing with GFRP does not require any special education, as it is very similar to conventional reinforced concrete. This is a huge benefit for the engineering community, as it allows for the same design avenues currently used. The procurement stages are the same, and no changes need to be made constructing the elements using GFRP. With the durability of GFRP, the possibility to build structures exposed to highly corrosive environments achieving a service life of 75–100 years is possible.[13]

References

1. ASTM (2006) *Standard test method for tensile properties of fiber reinforced polymer matrix composite bars*, D7205/D7205M. American Society for Testing and Materials, West Conshohocken, PA.

2. ACI (2008) *Specifications for carbon and glass fiber reinforced polymer (FRP) bar materials for concrete reinforcing*, 440.y. American Concrete Institute, Farmington Hills, MI.

3. ACI (2004) *Guide test methods for fiber-reinforced polymers (FRPs) for reinforcing or strengthening concrete structures*, 440.3R-04. American Concrete Institute, Farmington Hills, MI.

4. ISIS Canada (2006) *Specifications for product certification of FRPs as internal reinforcement in concrete structures*. ISIS Canada, University of Manitoba, Winnipeg.

5. ACI (2008) *Specifications for construction with fiber-reinforced polymer reinforcing bars*, 440.x. American Concrete Institute, Farmington Hills, MI.

6. ACI (2006) *Guide for the design and construction of structural concrete reinforced with FRP bars*, 440.1R-06. American Concrete Institute, Farmington Hills, MI.

7. CSA (2006) *Canadian highway bridge design code*, CAN/CSA-S6-06. Canadian Standards Association, Mississauga.

8. CSA (2002) *Design and construction of building components with fibre-reinforced polymers*, S806-02. Canadian Standards Association, Toronto.

9. AASHTO Task Group Committee T-6. *LRFD bridge design guide specifications for GFRP reinforced concrete decks and deck systems*. Working draft. American Association of State Highway and Transportation Officials, Washington, DC.

10. Eden, R. Personal communication.
11. Onofrei, M. (2005) *Durability of GFRP reinforced concrete from field demonstration structures.* ISIS technical report. University of Manitoba, Winnipeg.
12. ISIS Canada (2005) ISIS Canada research network, newsletter, July 2005. In: *International Workshop on Innovative Bridge Deck Technologies.* Winnipeg.
13. Dejke, V. (2001) *Durability of FRP reinforcement in concrete.* Chalmers University of Technology, Goteborg.

8

Construction processes for improved durability

Ken Day, *Independent Concrete Technologist, Australia*[4]

For concrete to be durable, the designer's intentions must be translated into the actual structure. The concrete must be produced to give a consistent final in-situ product, allowing for any variations in the constituent materials, the production conditions and site practices. It is not sufficient for test specimens to give consistent results, although this is certainly a necessary first step. The concrete must be capable of being placed and fully compacted without segregation, must not develop cracks or other faults either before or after hardening, and must be initially maintained in such an environment as will allow it to develop its intended properties.

It must not be forgotten that most concrete is reinforced with corrodible steel. There is a huge difference between the durability of plain concrete and that of reinforced concrete. Many ancient structures survive after millennia for no other reason than that they were *not* reinforced. It is clear that durability is significantly dependent on the properties of the 'covercrete' – that concrete between the reinforcing steel and the elements to which the structure is exposed.

Of course durability is also significantly dependent on the service conditions to which the concrete is exposed. These can include a requirement to resist salt spray, chemical attack, severe surface abrasion, and fire exposure. Most such influences are surface attacks, and resistance to them may be as dependent on the treatment of the concrete during placing and curing as on its basic nature. Particular examples are formwork details and the finishing of floor surfaces. Another situation is where the surface concrete is deliberately removed to expose the coarse aggregate.

This chapter will examine aspects of the production, placing and curing of concrete which can lead to a lack of durability. In doing so, it will also be necessary to refer to aspects of the specification of concrete which require attention in order to focus the attention of the contractor

Concrete durability
978-0-7277-3517-1

and concrete producer on the elimination of such faults. Other chapters have dealt with deterioration mechanisms and testing and designing for durability; this chapter considers what can go wrong with well-designed concrete and how to avoid this happening.

8.1 Factors affecting durability

Other chapters examine a more comprehensive list of factors, but here we need to examine what factors can affect the relative durability of individual mixes and particular parts of the structure of nominally identical concrete. This will include excessive water content in the initial mixing or by subsequent addition, segregation during placing and compaction, bleeding, and surface conditions and treatment.

In terms of the concrete as delivered, it is often assumed that the difference between different batches as delivered is adequately represented by their difference in compressive strength. While strength may not be the ultimate criterion, it is strongly related to the water/cement (w/c) ratio, which probably is the major factor. Assuming a reasonably consistent cementitious content, this essentially means a difference in water content. However, regarding a difference in strength between successive deliveries of the same mix as indicating their likely difference in durability does not mean that a difference in strength between different mixes necessarily has any significance for their relative durability. For example, the substitution of an equal mass of fly ash for a portion of the cement in a mix is very likely to both reduce its strength and increase its durability.

8.1.1 Selection and control of aggregates

Usually the most significant feature of aggregates causing batch-to-batch variation in water content at a given slump is the grading and silt content of the fine aggregate. Other than this, variation of slump is usually due to failing to adjust the added water to take account of variation in fine aggregate moisture content.

8.1.1.1 Fine aggregates

It is not simple, or, rather, it is rarely convenient, to very frequently check sand grading at the point at which it enters the batching process. Often, it is found that grading is checked, if at all, during extraction and prior to stockpiling. If silt content is a problem with the local supplies, it is a little easier to check this on each delivery. This is because a sand with higher silt content is often visually different. It is also very quick and simple to shake up a sample of the sand with a 3% solution of caustic soda (NaOH) and allow it to stand. The percentage of clay (which is what increases the water requirement) is visible as a surface layer within a few minutes. Any percentage above 6% will cause an increase in the water requirement. The colour of the supernatant liquid after standing overnight will reveal the presence of organic impurities, if darker than a pale amber colour. It is important to note that this colour test establishes whether an organic impurity is present, but does not establish whether such an impurity will cause a problem if present. The potential problem is of retardation of set and strength development. If the colour test is failed, comparative mortar cubes at the same w/c ratio, using the sand as is and after washing, should be made. If there is no substantial delay of set, and 1 day strengths are similar, the sand is suitable for use (with the possible exception of fair-faced work, where the colour of the concrete may be affected). If early strength is not essential for the use of the concrete in question, further cubes at a later age may (or may not!) show that there is no significant effect on eventual strength. Where the possible problem is with the amount of silt affecting the water requirement or bond, rather than any organic impurity, mortar cubes can again be used to investigate, but this time using equal workability rather than equal w/c ratio in the comparative mixes.

8.1.1.2 Coarse aggregates

The effects of variation in a sand, if deleterious, tend to be relatively immediately reflected in strength tests or setting time. So, such variations are very much in the province of day-to-day control systems. Coarse aggregate variations can also have an immediate effect through variations in dust content, particle shape, surface coatings reducing bond, or the inclusion of weak particles, but the most significant property for durability is volume stability. Some coarse aggregates are subject to moisture movement, causing potentially disruptive swelling and shrinkage over a period of time. Others may be subject to the alkali–aggregate reaction, causing a disruptive expansion over an extended period. Control of such features is a matter of thorough investigation of new sources of aggregate by geologists rather than day-to-day control by concrete technologists.

8.1.2   Selection and control of cementitious materials

There is a tendency for modern cements to be ground finer and to be aimed at maximum early age strength, or at latest at 28 day strength. This arises from strength being specified at such an age, and the financial value of the cement to the purchaser (the concrete producer) being directly assessed as the amount of cement required to produce that strength, there being no financial benefit in subsequent strength gain. In former times, the quantity of cement was specified rather than the strength of the concrete, and cement was more coarsely ground, gained substantial strength after 28 days, and was more resistant to cracking. It has been argued (see below) that specification of a cement content is undesirable since it results in there being no incentive for a producer to learn or apply either mix design or quality control. Some solutions to this dilemma are to specify strength at a later age or to require a specified percentage of fly ash or slag to be incorporated. In private correspondence with the writer, Professor Swami of Sheffield University has written: 'No concrete should be made without incorporating mineral admixtures or other pozzolanic cement replacement materials. Indeed, if one uses Portland cement alone in the cementitious system, then it should be justified.'

Obviously, accurate batching of cementitious materials is an essential, but quality testing of the cement by the concrete producer is not frequently done, and, in any case, test results on already received cement are often too late for mix adjustments to be made. Another complicating factor is that cement flow in a silo is often not a simple matter of first in at the top, first out at the bottom. Cement from the top can often funnel down the centre of the silo, leaving substantial amounts of the previous delivery still around the sides.

Test certificates are often required to be provided by the cement manufacturer to the concrete producer, but there is tendency for such information to be provided too late and for it to relate to average production rather than specific deliveries. Another tendency is for concrete producers to point out that a strength turndown in the cement requires an increase in cement content, and therefore the cement is of lower financial value, meriting a cost deduction. This obviously does not encourage the cement manufacturer to be diligent in promptly advising the concrete producer of such downturns. It is important that such matters be agreed between the parties in order to minimise concrete variability caused by delayed reaction to cement quality variations. The world over, cement manufacturers tend to receive unjustified complaints from inexpert producers. Experience is that if a concrete producer operates a really top-class quality control programme, the cement producer will come to rely on them as an adjunct to their own quality control programme, and a fruitful relationship will be established.

8.1.3 Water

The regulation of water content in concrete is a significant ongoing problem, responsible for much of the variability experienced. While it is easy to control the amount of water added to concrete during batching, it is more difficult to allow for the moisture in the aggregates. It is important that concrete has the required workability, so water addition is often regulated by assessing its workability. However, the water content necessary to achieve a given workability is affected by many factors, especially temperature and sand grading and silt content. Where the effect of temperature has not been allowed for, timeline graphs of strength and temperature usually show as a mirror image of each other (Fig. 8.1), making the inverse relationship very clear. Where compensation by varying the cement content is not convenient, variation of the water-reducing admixture dosage can be equally effective.

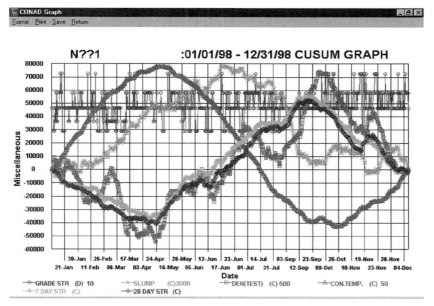

Fig. 8.1 Multigrade, multivariable cusum plot showing the effect of temperature (client's results before control commenced)

8.1.4   Batching

These days, any significant batching plant is equipped with hardware and software to record actual batch quantities of every mix. The information generated is usually too extensive for primary inspection to be useful, but software is available to automatically detect trends in variation from intentions and to incorporate batch quantity information in the analysis of test data.

8.1.5   Mixing

Central mixing has always been regarded as likely to result in better control, but, these days, there are moves to equip mixer trucks with sensors and software so as to almost use them as rheometers.

8.1.6   Cover

The most frequently encountered cause of deterioration in concrete structures is corrosion of the reinforcing steel. The assumption is that this will be protected by the surface layer of concrete, known as the cover.

The protection provided depends on both the thickness and the properties of this concrete. It is not desirable to provide an excessive thickness of cover because, apart from the extra cost and weight, this concrete is essentially un-reinforced, and can develop substantial cracking through shrinkage or loading stresses. Therefore, although adequate cover thickness is essential, the quality of the cover concrete is also critical. The factors affecting this quality are the potential quality of the fresh concrete as supplied, the workmanship in placing the concrete, the nature of the formwork, and the subsequent curing of the concrete.

8.1.7   Surface permeability tests

Tests can be undertaken on the surface of in-situ or precast concrete.

The original Figg tests originated in the UK but have subsequently been neatly combined into a single instrument by James Instruments in USA. A hole is drilled into the concrete (which may be in-situ concrete or a test specimen) and a plastic plug inserted, to create a cell below the surface of the concrete. A hypodermic needle is inserted through the plug to provide access. The first test involves applying a suction to the cell, so as to draw in air through the surrounding concrete. The (very small) volume of air is measured by the movement of mercury in a tube through which the suction is applied. The second involves filling the cell with water and using movement in the same tube (but in the opposite direction) to measure the rate at which water is absorbed into the surrounding concrete.

The Wexham variant identifies two problems sometimes encountered with the above test. One is that air permeability is substantially affected by moisture content. The other is that air may be entering via defects in the concrete or a leaking plug rather than via permeable concrete. These two potential problems are solved firstly by using a slightly larger diameter hole and including an instrument to measure humidity in the hole. Secondly, pressure rather than suction is employed, so that any leaks can be detected by bubbles in a soapy water film on the surface.

An additional advantage of these kinds of in-situ tests is that they can be used to measure the adequacy of curing (which has a large effect on permeability). Potentially, a contractor could be required to continue or resume water curing until an acceptable permeability is achieved.

It is simple to measure actual cover to steel using an electromagnetic covermeter. The extensive use of such an instrument often reveals wide variations and a proportion of very inadequate values. Specifications tend to require a minimum cover, but not to provide effective means of enforcement. Theoretically, demolition can be required, but is rarely enforced. Depending on the extent of the shortfall, the quality of concrete being supplied, the degree of conservatism in the specified value, the risks to which the particular part of the structure is to be exposed; and the required visual quality of the surface, a variety of rectifying measures can be considered. These can include a requirement for extended curing, a surface treatment ranging from painting (whether with paint (bituminous, epoxy or other) or a colourless polymer absorbed into the surface) to applying gunite or a sheet material. It is essential that the cost of any remedial work should be sufficient to ensure maximum effort by the contractor to avoid future occurrences.

Of course, some measures that may be effective in compensating for reduced cover from the steel corrosion viewpoint may be completely ineffective from the fire protection viewpoint, for which some kind of an applied surface layer would be essential if the risk appeared substantial.

It would appear relatively simple and inexpensive to make a point of measuring cover (using a covermeter) immediately on stripping formwork and to specify a substantial cash penalty for infringements. It would seem likely that this might be more economical and more effective than employing an inspector to check reinforcement for adequate provision of cover prior to casting.

8.1.8 Placing and vibrating concrete

Most concrete is placed using poker vibrators, but even before the concrete is vibrated, placing techniques can themselves cause segregation or leave the concrete in an unsuitable condition. Discharging from skips or pumps at an excessive height above the bottom of the formwork or surface of already placed concrete, leaving the concrete in substantial heaps, or discharging into reinforcement which may essentially sieve the concrete, are examples of poor placing techniques.

Insertion of a poker vibrator at the top of a heap is likely to cause a lack of compaction at the foot of the heap, as liquefied concrete flows out over concrete not affected by the vibration, entrapping air voids. The correct practice is to vibrate at the foot, expelling air as it flows outwards.

Causing concrete to flow along a formwork presents a segregation risk, depending on whether the concrete has been designed to withstand this. Excessive vibration can cause coarse aggregate to sink in the mortar and a layer of mortar to rise to the surface. This can cause low surface wear resistance in floors. The greater shrinkage of such a surface layer can cause surface cracking or curling of the slab. These effects are substantially exacerbated if the concrete has a higher water content than it was designed to withstand.

The writer's mix design system, not presented here, is based on a parameter he calls MSF (mix suitability factor), which combines the specific surface of the combined aggregates with the effect of the cementitious content.[1] An initial version of this system automatically linked this factor to the desired slump. This proved too inflexible for continued use, but made the point that the higher the required slump, the greater the fines content of the mix needs to be to resist segregation. Where a particular batch has a higher than intended water content, it may well not have a sufficient fines (MSF) to handle the higher slump, and so may have a segregation tendency. The MSF does not depend only on the quantity of sand but also on its fineness. If the sand goes coarser, it has the same effect as reducing the quantity and risks segregation. If it goes finer, the water requirement for a given slump increases, so either the slump is reduced or the concrete is weaker. There are limits to the extent to which increasing the sand percentage can compensate for increasing coarseness. Especially where the required strength is low, resulting in a low cement content, a lack of fines in the sand can result in severe bleeding. Increasing the sand percentage in such circumstances effectively reduces the cement content in the mortar, and only intensifies the problem. As noted, the point at which the problem may be experienced depends on the cement content; also, the inclusion of fly ash will reduce the problem (because it is finer than cement and also of lower specific gravity, so providing a greater volume and a greater surface area per unit weight). Depending on these factors, the problem may be experienced where between 50 and 60% of sand is needed to provide the desired MSF.

Vibration, especially form vibration, attracts water to the vertical face of the formwork, and may carry air voids with it. If that face is impermeable, steel or plastic coated, the voids may adhere to the surface, creating the well-known 'bug holes'. The water may flow up the form surface or horizontally towards a point of bleedwater leakage (such as an even slightly ill-fitting joint in a formwork sheet), creating the black areas known as 'hydration staining'. This effect arises because cement grains are actually not grey but black, coated with a very fine grey powder. The upward flow of the bleed water can wash the grey powder away, leaving a black surface. The same effect, but caused mechanically rather than by water flow, is responsible for the shiny black surface of burnished concrete floors.

If the formwork is of new wood, not completely saturated with mould oil, it will absorb at least some of the excess water, avoiding bug holes and hydration staining and leaving a slightly stronger surface layer. However, because this absorbency is likely to be variable, the concrete surface, when exposed, is likely to show variations in colour or shade because the lower the w/c ratio, the darker the colour of a cement paste. I have often witnessed the discomfort of an architect when asked whether he wants bug holes or variable colour in his fair-faced concrete (hydration staining does not occur in low-bleeding concrete, so an impermeable form can give a uniform colour). The use of controlled permeability formwork, described below, is a further exploitation of the absorbency effect.

8.1.9 Bleeding, settlement and cracking

Even in concrete which does not have an excessive water content, a certain amount of bleeding is normal. Essentially, the solids in concrete, including the cementitious material, are heavier than water, and so tend to settle in it when the placed concrete comes to rest. The displaced water rises in the paste and tends to form voids under anything in its upwards passage. This includes coarse aggregate, steel reinforcement, and plumbing and electrical fittings. These voids are a substantial factor in the permeability of the concrete in a horizontal direction. They also partly explain the reduced bond of top steel; however, this is further exacerbated by the general settlement of the concrete away from the lower surface of the bars as the bleeding takes place.

Two kinds of 'pre-setting cracks' can occur in concrete. The term is not very helpful, because the two are due to entirely opposite causes. One type, usefully described as settlement cracking, is caused by concrete 'breaking its back' over obstructions to bleeding settlement. The obstructions often include top reinforcement, plumbing and electrical fixtures, and changes in depth of the concrete due to beams, column capitals or abrupt changes in thickness in general. Diagnosis of such cracking is simple if the configuration of the bottom of the formwork and the reinforcement and other cast-in items is known. Closing such cracks by revibration is simple if done at the correct time. However, such cracks may appear to be closed by non-vibratory surface trowelling but, since this only closes them at the surface, they will often subsequently open under drying shrinkage.

The other type of pre-setting crack is more usefully described as an evaporation crack. This occurs when water evaporates from the surface layer of the concrete, reducing its volume and so initiating tension. Incompetent paviours who cannot be bothered to take precautions such as wind barriers, mist sprays or chemical evaporation retarders have been known to complain that the concrete should be made to bleed more!

8.1.10 Controlled permeability formwork (CPF)

Seeing the beneficial effects of slightly absorbent formwork led to the development of techniques aimed at deliberately de-watering the outer layer of the concrete. This can go all the way from merely inserting sheets of substantially absorbent material in otherwise normal formwork, to using a vacuum pump to suck water out of the concrete through a specially designed form incorporating a double skin, with the inner skin perforated and covered by a filtering cloth. The latter technique can have the dual benefit of allowing immediate stripping (after half an hour's vacuum processing) and producing a very low w/c ratio, perfectly compacted, concrete exactly where it matters most. It has also been used for large vertically precast pipes (with inner and outer processing) and by placing processing pads on the surface of floors to obtain high wear resistance. The writer was involved in development work on this technique in the 1950s but, to the best of his knowledge, it has not attained any widespread use.

Between the two extremes of absorbent sheeting and vacuum extraction has arisen the technique of CPF. This entails the use of a form liner sufficiently permeable to allow excess mix air and water to pass through it, but fine enough to retain the cement and fine particles in the mix. The writer observed the possibility of this process by seeing the substantial amount of water still issuing from the equipment on one occasion when the vacuum pump failed to function during his pipe experiments. However, the kind of equipment used for vacuum processing was far too bulky and expensive to consider using it as CPF. With CPF attached to the internal face of formwork, the concreting process is undertaken in the normal way. A clear distinction should be made between CPF and other open-textured materials such as geotextiles, sacking and absorbent liners, which may remove water from the concrete surface, but in an uncontrolled and variable manner.

The adoption of CPF as a workable and affordable technique awaited the development of suitable liners. These needed to be highly permeable to water, yet resistant to infiltration by cement, so as to be easily washable and re-usable. One type now available comprises a filter membrane of woven polypropylene fibres, spun and thermally bonded and with a latticed plastic backing to provide both stiffness and a drainage channel. Due to the polypropylene filter, the liner easily debonds from the concrete whilst remaining attached to the formwork, so that release agents are not required. The liner is robust, and comes in the form of rolls of varying length to sites, ready to fix to the formwork, which must incorporate drainage provision.

CPF improves the quality of concrete exactly where it matters most:

- it produces a more uniform and attractive appearance
- it reduces the w/c ratio of the 'covercrete', giving a higher strength and a less permeable cover to steel
- if left in place, it provides moist curing for a time, holding more water than a sheet of Hessian.

For further details on CPF, see McKenna.[2] McKenna has contributed to the above, having studied the technique in a PhD investigation and subsequently used it on bridges in Scotland.

8.1.11 Curing

Moist curing of concrete in general is important because:

- Strength and impermeability development are hydration processes, and stop when the concrete is dry. In most cases, durability is much more dependent on impermeability than on strength.
- The development of impermeability is even more strongly dependent on the w/c ratio than is strength. At w/c ratios below 0.45, it may take place in a few days, but at more than 0.50 w/c it would take weeks of curing, and above 0.60 it would require months.
- It is important that concrete dries as uniformly as possible, because drying of surface layers causes differential stresses, potentially leading to cracking and to curling in floor slabs.
- Shrinkage is mostly drying shrinkage, commencing as concrete dries. However, high-strength concretes with a low w/c ratio can be subject to autogenous shrinkage in which all free water is consumed by the chemical reaction at a much earlier stage than the concrete would otherwise dry. Since such concrete rapidly becomes substantially impermeable, it is difficult to introduce further water, and it is essential that at least none of the original water is allowed to escape. Attempts to combat this situation include CPF formwork left in place and the use of a small proportion of saturated lightweight aggregate, providing small internal reservoirs of water. A similar function has been claimed for brucite (magnesium hydroxide), formed from magnesium oxide (magnesia) and water. This development is particularly interesting because many specifications place a strict low limit on the content of magnesia in cement, on the basis that it can produce delayed expansion, disrupting in-place concrete. However, it appears that, just as fly ash prevents disruption by the alkali–aggregate reaction by consuming all the available alkali prior to setting, a large excess of sufficiently reactive magnesia is beneficial.

8.1.12 Floor slabs

Floor slabs are a particular case where surface treatment is even more important than the basic concrete. Formerly, it was essential for floor slab concrete to be of very low slump and low fines content to ensure that it could be consolidated without bringing a layer of mortar to the surface. Now, perhaps, it is even possible to use self-compacting concrete, but this would have to be of a very high strength grade since the surface would not contain much coarse aggregate. This may mean the difference between using 40–50 MPa concrete and 80–100 MPa concrete.

After striking off the surface to the required level, it is essential that the surface is not worked until it is sufficiently stiff to permit foot traffic without leaving any mark. This is especially the case if the concrete bleeds. The surface must not dry before this stage or evaporation cracking will occur (if necessary, an evaporation retardant or a fine mist spray should be used). Equally, any remaining bleed water must be removed by squeegee or other means before the surface is worked. Working the surface before bleeding is complete will seal the surface against the passage of bleed water, and result in a layer of such water just below the surface, causing eventual spalling. Working the surface in the presence of water will stir such water into the surface, creating a high w/c ratio paste of low strength and abrasion resistance.

Another effect of using concrete which bleeds substantially is that, in bleeding, the concrete settles and can 'break its back' over embedded items such as reinforcement and plumbing and electrical conduits. With high workability concrete there is no alternative to having the reinforcement suspended on chairs, but with stiffer concrete, a two-layer construction can be used, laying reinforcing mesh on the first layer and permitting clear access for concrete trucks. Another recent development is to incorporate glass or steel fibres in the concrete mix in lieu of mesh reinforcement, or to use post-tensioning cables in ducts, when the concrete can then only be placed by pump.

Having avoided evaporation and settlement cracking in slabs, the remaining hazards are shrinkage cracking and curling. Shrinkage can be either drying shrinkage or thermal contraction, and cracking can only occur if there is restraint to the shrinkage. Restraint may arise through failing to isolate such items as column bases, floor pits or raised areas. Curling occurs when shrinkage of the surface layers is resisted by the bottom layers (of slab on ground), either through bleeding or segregation, having created different shrinkage properties in concrete from the same truck, or through the surface layers being permitted to dry while the lower layer remains saturated, or to cool while they remain warm.

It follows from the above that freshly finished floor slabs need to receive a spray of curing compound and also a plastic sheeting (or more substantial insulation in very cold climates). Curing compounds are not 100% effective, and their real function is to allow concrete to lose moisture slowly enough to avoid a substantial moisture difference over the depth of the slab. The plastic sheeting achieves a degree of thermal insulation, if only by avoiding evaporative cooling, but also by some air entrapment. Abrupt termination of curing under a plastic sheet is asking for surface cracking, if no curing compound is used.

Delamination of a surface layer (spalling) has been reported on floor slabs using air entrainment. The writer has not had personal experience of this, but considered that it could be due to the known fact that air entrainment slows the rate of bleeding.

8.1.13 *Influence of specifications on achieved control*

In Australia, assurance of durability in specification is sought primarily by requiring a minimum strength grade be used according to the anticipated exposure conditions, e.g. a minimum of 32 MPa is required for any external reinforced concrete. Individual specifiers may then add to this by requiring the use of a particular cement (sulfate-resisting or low heat) or a minimum percentage of fly ash or slag or particular admixtures (including air entrainment). Some specifiers like to include a permeability test, such as the volume of permeable voids assessed by an absorption test, and/or a shrinkage test. However, while recognising that strength may not be the most important property, it is recognised as the best means of assessing variability, and particularly of detecting any change in the quality of the concrete (and therefore the cement) being supplied at the earliest possible moment.

In the USA, a movement called P2P (prescription to performance) is underway, and the RMC Foundation has commissioned a 150-page report by Bickley *et al.*[3] However, delay is resulting from disputes about the possibility of directly specifying a durability test. The RCP (rapid chloride permeability) test is favoured, but both the timescale and the substantial variability of such tests make it impractical as an initial acceptance/rejection criterion. This writer has proposed that RCP or any other durability test could be the subject of a subsequent bonus/penalty clause, leaving strength as the primary acceptance criterion.

In Europe, a standard (EN 206) for concrete specification, performance and production, has been agreed between participating countries, but its adoption permits each country to modify its provisions to some extent. In the UK, EN 206 has been issued as BS EN 206-1:2000 along with a complementary standard, BS 8500-2:2006, and concrete is further controlled by QSRMC, the Quality Scheme for Ready Mixed Concrete.

QSRMC was established to provide product conformity certification for ready mixed concrete supplied throughout the UK. It is a specialist body which brings together the producers of ready mixed concrete and a broad range of their customers, to set the standards for certification and for the assessment of the producers. The scheme is governed by an independent board which reflects the partnership between the producers and the specifiers and purchasers of concrete.

The UK/European scheme is very comprehensive in its provisions for the selection and testing of materials, mix designs, trial mixes and assessment of supplied concrete for conformity over a period. The one thing that it fails to do is to initiate mix revision promptly when a change in quality occurs. An example (some years ago and possibly now rectified) was that it took several weeks to adjust to the intended mean strength when, as should be obvious, water requirement increases and strength reduces in summer, and the reverse happens again in winter. As Bryant Mather was so keen to point out, slump is not actually initially lower at a higher temperature, but the loss of slump in the first few minutes (essentially while the concrete is still in the mixer truck) is much more rapid, so that the water content at a given slump during placing is actually higher at higher temperatures. It is important to realise this, because concrete may be rejected on account of high slump on a cold morning when concrete of higher water content but lower slump is accepted on a hot afternoon.

Low variability is obviously desirable in attaining durability. As noted, to date it has not proved practicable to assess relative variability through a direct test on durability, due to both the timescale of any durability test and the substantial variability of such tests. Two features strongly related to durability are permeability and strength. Of these two, strength is the property most conveniently used to assess variability. This is assessed as a standard deviation (SD); however, to the writer's knowledge, no code of practice or individual specifier has imposed a limit on SD. A financial pressure to operate at reduced variability potentially arises from the specification of a minimum strength, requiring a mean or target strength a number of SDs above the mean.

Two factors affect whether there is any such pressure and, if so, the extent of the pressure. The first factor is the use of prescription specifications, such as has been common for many years in USA. Such specifications provide no incentive whatsoever to achieve low variability. The second factor is the percentage defective on which the minimum strength is assessed. If this is 10%, as in USA, the required margin is 1.28 times the SD; if 5%, as is currently common in most of the world, the margin is 1.65SD; if the figure was 1%, as used to be the case many years ago in the UK and elsewhere, the margin would be 2.33SD; and if a figure of 0.1% were used, it would be 3.09SD. So, a 5% basis provides $1.65/1.28 = 1.29$ times the incentive to reduce variability compared with a 10% basis.

There is no essential relationship between the choice of the percentage defective and the factor of safety provided in the structure. Without changing current factors of safety, the required minimum strength could be specified in the form $X + A \times SD - B$, and B could be increased to compensate for any increase in A. A could then be selected on the basis of the relative value placed on mean and SD. There is no particular merit in using values of 1.28, 1.65, 2.33 or 3.09. A choice could be made between 1.5, 2.0, 2.5 or 3, depending on the incentive to be provided to achieve low variability.

8.1.14 *Immediacy of control action*
Strength test data are usually assumed to have a normal distribution about a consistent mean, but in fact there tend to be mean strength change points in a string of data. It is when such a change point occurs that substandard concrete is at risk of being produced. The time taken to recognise and compensate for such a change (if negative) is an important, perhaps the most important, feature of the control system in use. It has long been realised that simply reacting to failed test results is inadequate. The concrete tested is only representative of very substantial quantities of concrete which have not been tested and can be distinctly better or worse than that concrete – particularly if it is high variability concrete. So what has to be detected is a change in mean strength and variability of all the concrete being produced. Even for compressive strength, it is relatively ineffective to attempt to detect and reject defective concrete being supplied to a particular project by testing deliveries of such concrete. The only really effective way of ensuring defective concrete does not reach a particular project is to ensure that no defective concrete is being produced at the supplying plant.

Although effective systems can be devised for numerical cusum analysis, it is usually better to use graphical systems, particularly now that any desired graphing can be provided effortlessly and almost instantaneously by computer. Various systems of increasing effectiveness have been devised over the course of many years. Firstly, there were Shewhart charts, which were standard time sequence plots of individual

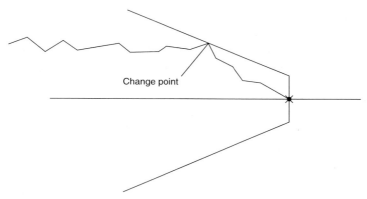

Fig. 8.2 Use of a V mask on a cusum chart. (Reproduced from Day[1], Concrete mix design quality control and specification © 2006 Taylor and Francis)

results, but provided with statistically calculated 'warning' and 'action' limits. Then it was found that more reliable action could be based on plotting running means of three or four consecutive results, since this reduced the influence of variability as a masking factor. A further improvement was the introduction of 'cusum' analysis. A cusum is the cumulative sum of differences between successive results and a selected target value (Fig. 8.2). A change of slope on a cusum graph is about three times as effective in detecting change as is a Shewhart chart.

Cusum analysis originated in the UK chemical industry, and was first applied to concrete by Testing Services Ltd, a subsidiary of RMC. The technique was adopted by QSRMC. A technique known as a V mask (Fig. 8.2) was developed, whereby a precise mathematical test is applied to determining the probability that a downturn could have occurred by chance, with no requirement for knowledge or intelligence on the part of the operator. The problem with this approach is that it tends to require 12–15 results to determine a marginal downturn. If these results are of 28 day tests, it may take several weeks, or even months, for them to be accumulated on any particular grade of concrete. The QSRMC regulations even refer to mixes being adjusted on the basis of an analysis of not less than 15 results accumulated over not more than 12 months!

The view of this writer is that control action, where necessary, needs to be taken within a few days of the need for it arising. Accordingly, it was necessary to devise techniques to predict 28 day results from early age tests, and to combine results from several grades of concrete onto a single cusum graph. QSRMC envisages prediction being by means of a graph of 28 day strength against 7 day strength for the grade in question.

Such a graph, or the common assumption that 7 day strength is some percentage of 28 day strength, assumes that a low 7 day result would also have a lower strength gain from 7 to 28 days. The experience of this writer is that this is often not the case, and that it is better to add the average 7 to 28 day gain to the 7 day result.

The EN 206/BS 8500/QSRMC technique for getting results from several grades onto the same graph is described as to have 'families' of grades with fairly similar strengths, constituents and properties. One such grade is selected as the 'control grade', and a transformation is applied to the strength results from each other member of the family so that they can be analysed as though they were results from the control grade. Other properties such as density and workability are not graphed, and so do not require transformation. This system requires the origination of the transformations and checking from time to time that members of the family are truly conforming.

This writer has evolved a technique described as multigrade, multi-variable, cusum analysis. This involves the computer automatically maintaining a separate running mean value for every measured property of every grade of concrete, including strength, density, workability and temperature, among many others. The appropriate mean value is then subtracted from each subsequently obtained test result, and a cumulative sum is then maintained of these differences for each measured property, regardless of which grade it originates from. The advantages of this technique are:

(a) Having purchased the computer program, it requires no setting up or subsequent checking.

(b) All results from potentially many hundreds of grades can appear on the same cusum graph.

(c) Cusum graphs of many other properties can appear, along with the strength results.

(d) Recognition that a change point in strength has occurred no longer needs to be based on the strength result alone. If other cusum graphs exhibit a change point coinciding with that in strength, then the change is verified without requiring statistical verification from the strength results alone. An additional benefit is that several of the other cusum graphs run ahead of even the 7 day strength cusum (density, slump, temperature), so that any change could be confirmed with fewer results. Combined with point (b) above, this meant that control action could be initiated in days rather than the weeks or months required under QSRMC/EN 206.

(*e*) A further major advantage is that, by observing which other cusums reflect the change point in strength, the cause of the change is often immediately obvious.

It is clear that specified control systems need to have two distinct purposes. One of these is to provide for accurately assessing the quality of the concrete supplied over a period and providing an incentive (whether acceptance/rejection or otherwise) for this to be above a required level. The other is to include a technique for detecting any change in that quality at the earliest possible moment.

It is again pointed out that, although durability may, in many circumstances, be more important than strength, strength is, in nearly all circumstances, the most effective means of detecting quality changes and assessing variability. Having used strength as an initial means of controlling concrete production and supply, it is then possible to apply longer-term tests for quality and use them as a basis for cash penalties or bonuses.

Any attempt on the part of the purchaser to use a very limited number of early age tests to predict longer-term performance and to use this to demand control action or reject concrete is likely to lead to legal or other disputes and expensive in-situ testing. If a clear basis is provided for a long-term assessment of quality based on a substantial volume of evidence and leading to a cash penalty or bonus, then any sensible concrete producer will set up an effective quality control system, and take action on its own initiative long before there is sufficient evidence for the purchaser to require it to take action.

8.2   Significance of the foregoing

Other chapters have examined the factors affecting durability and the means available to test for it. The purpose of the current chapter is to examine how this knowledge can best be brought to bear on the task of ensuring that concrete of the required durability will actually be continuously produced. It is, of course, desirable that materials from an untried source should be tested before being incorporated, and trial mixes may be desirable when starting to use a new material. However, it is clear that there is no real substitute for so specifying concrete as to bring about a situation where it is financially worthwhile for producers to develop expertise, to acquire a high standard of plant and equipment (including their own laboratory), and to devote maximum effort to consistently producing concrete of the required quality. Of course, even such producers may not be aware of all aspects of producing high-durability concrete, so researchers and specifiers do still have a role to play. It is particularly desirable that producers develop suitable standard mixes, so that concrete backed by a comprehensive range of tests over a long term can be available. For this reason, it is undesirable to require a new mix to be developed unless there is substantial certainty that it represents an improvement. The EN 206/ BS 8500/QSRMC regulations are clearly aimed at bringing about such a situation, and it is unfortunate that they do not also incorporate early reaction to unintentional change.

8.2.1 Control systems

It has been pointed out that effective control must be by the concrete producer, since it is not feasible for an external organisation to impose really effective and timely control on an unwilling, incompetent or ill-equipped producer without taking complete control. Specification techniques to encourage producers to invest the necessary time and money to attain good control (and to ensure that competitors saving money by not responding to such pressure do not prosper thereby) have been discussed. It remains to describe the techniques that the producers should use. Illustrative examples of these are taken from the writer's ConAd system and from his textbook,[1] but it is emphasised that it is the principles involved rather than the exact details of the system that are important.

It is worthwhile to begin by considering the two terms 'quality control' (QC) and 'quality assurance' (QA). There is a tendency to consider QA to be a newer development, superseding QC, but this can be misleading. QA can be thought of as actions taken in advance of the actual production of the concrete to positively ensure that the concrete produced will have the required properties. It can also be thought of as *documented* QC. This writer's concepts could, to some extent, be thought of as *reactive* QC, concentrating on very prompt reaction to marginal changes in quality. If changes in the concrete produced can be detected very quickly, and if the cause of the change can be seen, then any change can be quickly compensated for, and variability reduced. Of course, if change could be avoided in all matters, that would give the lowest variability of all, but that has not proved physically possible. The writer is currently investigating an alternative concept he calls 'just-in-time mix design', in which the exact contents of a truck of concrete are to be determined only a few minutes before the truck is batched, taking into account all information that can be made available on a timely basis. Such information would certainly include ordered slump, travel time and concrete temperature, and efforts would be made to update analysed data on current input material and concrete test results within minutes of their availability. 'Analysed data' are emphasised, because adjusting the mix on the basis of the single last test result may well increase rather than reduce variability, since all test data are inevitably subject to variability.

The basis of any effective system must necessarily include recording every piece of relevant data. There is a huge amount of such data, and the very volume is counter-productive unless an efficient storage, retrieval and analysis system is employed. Fortunately, computer speed and capacity is no longer a limiting factor. In addition to technical data, there is also a large amount of administrative data, to which little reference will be made here. It should be realised that, in some cases, data will be being received from dozens of batching plants spread over a wide geographical area, and that there is substantial advantage to be gained by the combined central analysis of all such data. Indeed, the writer has suggested that advantage could be gained by the centralised analysis of international data, so that a multinational organisation could learn from the experiences of all its branches.

One class of data will be test data on incoming materials. While some such tests may constitute acceptance tests and lead to the occasional rejection of unacceptable material (the QA approach), the more usual use of such data will be to detect change as an explanation of variations in concrete performance and to modify mix designs to accommodate such change (the reactive QC approach).

Then there will be the batch quantity data, with the batch quantity of every ingredient of every batch of concrete recorded and connected to data on the concrete by the delivery ticket number. The batch data will be analysed to reveal and correct any trends in batching error, some clients advising that this facility alone has been sufficient to recover the purchase price of a total system. Such data are also of substantial value in production and commercial analysis. Although this aspect need not concern us here, it can effectively constitute a time and motion study on the entire operation, and also highlight which customers and which grades of concrete are most important, and how much concrete is being carried by each truck in a given period (useful for maintenance, and also driver payment, etc).

Putting together the aggregate grading and batch quantity data enables the combined grading and w/c ratio of each truck to be on record, along with a calculated strength and density for comparison with any actual test on the concrete. Of course, it will be objected that the water content is not really known, since the aggregate moisture may not be known, and water may have been subsequently added. However, some producers do arrange to record such items, and the water content can be calculated from a knowledge of the aggregate gradings, slump and temperature of the concrete. Especially where variability is higher than desirable, it can be of substantial value to graph calculated water and calculated strength and density alongside measured slump, strength and density. While such graphs may not be in exact agreement, matching or not matching variations from average can be very interesting.

Obviously, concrete test data will be fully recorded, and it is important that the testing operator should also be recorded. Variability of the actual concrete is one thing, but it should be realised that the final result also includes the within-sample variability (which can be referred to as the testing error). The latter can be estimated from the average difference between pairs (or triples) of cylinders or cubes from the same sample of concrete tested at the same age. The best attainable average difference, in this writer's experience, is 0.5 MPa, and up to an average of 1.0 MPa is a reasonable limit for good testing. Some testing operators (and some testing laboratories) have been known to exceed 1.5 MPa average difference, and such operators should be retrained and laboratories threatened with de-registration, or avoided. These average sample ranges can be converted into within-sample SDs by dividing by 1.13 in the case of pairs or 1.69 in the case of ranges of three.

Graphing testing error is effective in detecting any relaxation of testing standards. Of course, such error is, in most organisations, the result of actions by several different operators. However, if an increase is noted and applies to all field testers, it is likely that the problem is in the laboratory, whereas if it applies to specimens from only one field tester, or if it correlates with an increased difference in density between specimens from the same sample, the problem is likely to be in the field. The writer recalls, decades ago, setting up a large-scale wall graph in the laboratory on which the average pair difference of each testing officer for the past week was recorded. No threats were made, or bonuses offered, but the average pair differences declined to almost half of their previous values over a short period of time. It appeared that no-one liked to be the one with the highest pair difference!

As previously noted, the most effective detector of change in a series of data is a cusum graph (Fig. 8.3). With a suitable control system, it takes only a few minutes to display a multigrade, multivariable, cusum

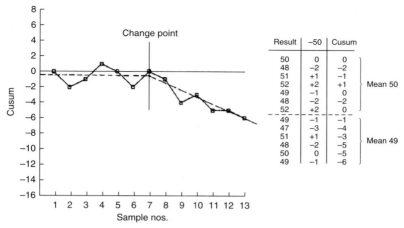

Fig. 8.3 *Simple cusum control chart. (Reproduced from Day[1], Concrete mix design quality control and specification © 2006 Taylor and Francis)*

graph, covering all specimens tested to date, at the conclusion of each day's testing (see Fig. 8.1). If such a graph shows no downturn, no further action need be taken. If a downturn is seen, the attention of the person in charge is drawn to it. Experience shows that a typical operator, knowing nothing about concrete, draws the attention of the senior person about twice as often as necessary. This is fortunate, because an unnecessary call wastes only a few minutes, whereas if calls were only made on 99% of occasions meriting them, the senior person would need to check every day. It has also been the writer's experience that an initially non-technical person often acquires considerable expertise in a relatively short time – this type of control is truly an excellent learning experience.

Where (as is typical, although perhaps undesirable) hundreds of grades of concrete are in use, there is a possibility that one or more grades of concrete represented by very few results will suffer a downturn potentially obscured by a mass of other satisfactory results. One aspect of this is that the selection of which loads to test over any short period of time must include one or more tests on grades including every material and every plant in use. Another aspect is that the analysis system, in addition to the graphical analysis, must output a statistical analysis table showing every grade of concrete in use arranged in order of departure from intended target strength (Fig. 8.4).

The above two items are sufficient to detect the onset of any problem, and we can now consider how to investigate a revealed potential

337

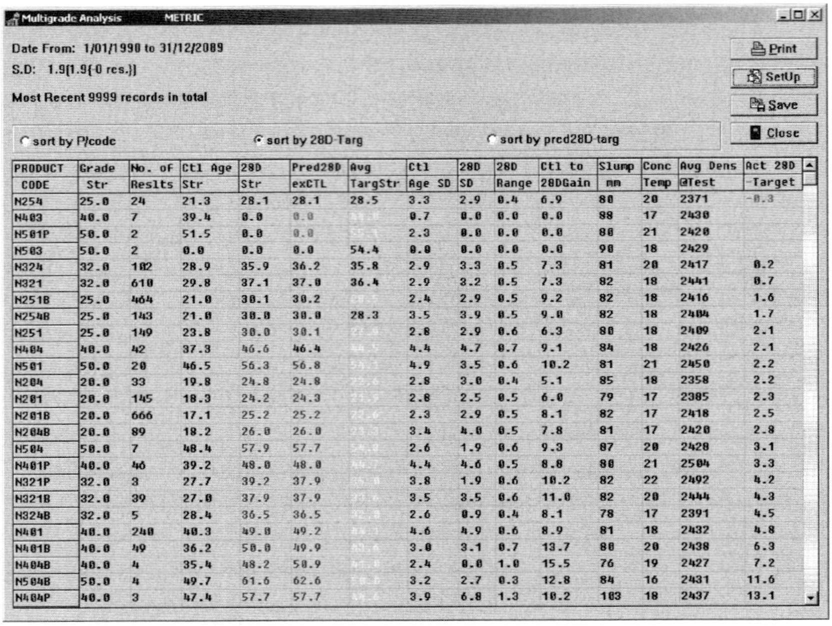

PRODUCT CODE	Grade Str	No. of Reslts	Ctl Age Str	28D Str	Pred28D exCTL	Avg TargStr	Ctl Age SD	28D SD	28D Ctl to 28DGain	Slump mm	Conc Temp	Avg Dens @Test	Act 28D -Target
N254	25.0	24	21.3	28.1	28.1	28.5	3.3	2.9	0.4	6.9	80	20	2371	-0.3
N403	40.0	7	39.4	0.0	0.0		0.7	0.0	0.0	0.0	88	17	2430	
N501P	50.0	2	51.5	0.0	0.0		2.3	0.0	0.0	0.0	88	21	2420	
N503	50.0	2	0.0	0.0	0.0	54.4	0.0	0.0	0.0	0.0	90	18	2429	
N324	32.0	102	28.9	35.9	36.2	35.8	2.9	3.3	0.5	7.3	81	20	2417	0.2
N321	32.0	610	29.8	37.1	37.0	36.4	2.9	3.2	0.5	7.3	82	18	2441	0.7
N251B	25.0	464	21.0	30.1	30.2		2.4	2.9	0.5	9.2	82	18	2416	1.6
N254B	25.0	143	21.0	30.0	30.0	28.3	3.5	3.9	0.5	9.0	82	18	2404	1.7
N251	25.0	149	23.8	30.0	30.1		2.8	2.9	0.6	6.3	80	18	2409	2.1
N404	40.0	42	37.3	46.6	46.4		4.4	4.7	0.7	9.1	84	18	2426	2.1
N501	50.0	20	46.5	56.3	56.8		4.9	3.5	0.6	10.2	81	21	2450	2.2
N204	20.0	33	19.8	24.8	24.8		2.8	3.0	0.4	5.1	85	18	2358	2.2
N201	20.0	145	18.3	24.2	24.3		2.8	2.5	0.5	6.0	79	17	2305	2.3
N201B	20.0	666	17.1	25.2	25.2		2.3	2.9	0.5	8.1	82	17	2418	2.5
N204B	20.0	89	19.2	26.0	26.0		3.4	4.0	0.5	7.8	81	17	2420	2.8
N504	50.0	7	48.4	57.9	57.7		2.6	1.9	0.6	9.3	87	20	2428	3.1
N401P	40.0	46	39.2	48.0	48.0		4.4	4.6	0.5	8.8	80	21	2504	3.3
N321P	32.0	3	27.7	39.2	37.9		3.8	1.9	0.6	10.2	82	22	2492	4.2
N321B	32.0	39	27.8	37.9	37.9		3.5	3.5	0.6	11.0	82	20	2444	4.3
N324B	32.0	5	28.4	36.5	36.5		2.6	0.9	0.4	8.1	78	17	2391	4.5
N401	40.0	240	40.3	49.0	49.2		4.6	4.9	0.6	8.9	81	18	2432	4.8
N401B	40.0	49	36.2	50.0	49.9		3.0	3.1	0.7	13.7	80	20	2438	6.3
N404B	40.0	4	35.4	48.2	50.9		2.4	0.0	1.0	15.5	76	19	2427	7.2
N504B	50.0	4	49.7	61.6	62.6		3.2	2.7	0.3	12.8	84	16	2431	11.6
N404P	40.0	3	47.4	57.7	57.7		3.9	6.8	1.3	10.2	103	18	2437	13.1

Fig. 8.4 Extract from a table listing every grade of concrete in order of departure from the target

problem. It is possible that the initial multigrade graph will already be clearly indicating the cause. If a strength downturn is due to an excess of water, then the density graph will also show a downturn, since water is lighter than all other constituents. Furthermore, if (as is strongly recommended but often not provided for in codes of practice) specimens are weighed and measured on receipt at the laboratory rather than at test, the density graph will run several days ahead of the early strength graph. This enables it to be seen whether the strength downturn is a genuine change or only a statistical aberration. If the density graph confirms that the change is due to excess water, then the slump or temperature graphs may provide an answer. Failing this, the fine aggregate grading must be checked along with the admixture quality and dosage.

While cusum graphs are incomparable in their ability to detect change and correlation, they do so by de-emphasising the effect of individual variations. If a low strength is the result of some individual error, then it will best be revealed on a direct plot.

A particularly interesting combination of graphs is one including cusums of 7 and 28 day strength along with direct plots of actual and predicted 28 day strength minus either specified strength or current average strength (the latter is the only kind of direct plot useful on a multigrade basis, and predicted 28 day strength is essentially the early age strength plus the anticipated gain). If low or marginal individual results are seen when the cusums show a downturn, that is to be expected; however, any low or marginal results during a period with a level or rising cusum are individual sample problems, and should be investigated. Possibilities are batching error, the truck delayed on site or otherwise subject to added water, inept sampling of the concrete, poorly cast specimens, delayed collection of specimens, failure to protect on site, delayed entry to laboratory curing or inept testing in general.

The basic point is that any such explanations point to a need to discipline staff rather than adjust future mix quantities. Of course, we are looking here at data including 28 day strength, so these techniques are not part of the early detection routine for a well-run organisation but rather ways of approaching problem-solving in an organisation currently experiencing problems.

While multigrade graphs are very powerful in rapid change detection, it has already been noted that they have a tendency to gloss over some particular problems. Examples of this are one plant malfunctioning among many which are not, and grades of concrete or particular plants using materials not in wide use by other grades. So, while the first approach should be to simply include all results in one analysis, it is also desirable to think carefully about setting up groups which include or exclude particular materials. This might be seen as retrogression to the EN 206 'families' approach, but the differences are that all results have first been combined to detect a problem more rapidly, and that, most of the time, only the one automatic analysis is necessary to establish that further action is not necessary at that time. When a downturn is detected, the automatic statistical table may show which grades are affected, but the system operator can rapidly redraw the cusums for selected groups, to see which are affected and which not. If a problem is due to a particular material or a particular plant, this is likely to be quickly revealed.

8.2.2   Prediction of 28 day strength

Another factor in early detection of problems is the prediction of 28 day strength from earlier tests. The world over, the most frequent assumption is that 7 day and earlier results can be taken as some percentage of 28 day strength. It is certainly recognised that this percentage will be different for different mixes, being generally higher for high-strength mixes and lower for mixes containing pozzolanic material. The QSRMC system advocates reading the percentage from a graph constructed from previous results in the same grade. What such assumptions fail to realise is that the lowest early results in a grade quite often show a larger than average gain to 28 days. This is because early results are usually of a single specimen, and are thus more often subject to error. An incorrect result is almost always lower than a correct result. It is difficult to accidentally obtain a higher than merited result, but all too easy to obtain a lower one. This means that concern over a single low early result is often greater than it need be. The best predictions are obtained by adding the continuously updated current average gain to the early result.

The foregoing technique works well with early ages down to 3 days, especially on major projects in hot countries where the 3 day maturity is usually higher and more uniform, and testing is rigidly at 3×24 hours. At less than 3 days, and especially where the early test age is less than 24 hours, a different approach is necessary, and must involve measuring the actual maturity or equivalent age of the early test specimen. There are two alternative methods in use for this purpose.

One of these uses the maturity concept of degree hours, and relies on the prior establishment of a graph of strength against maturity for the particular grade of concrete in use. Measuring the temperature history of an early specimen and the age at test provides a point for comparison with the graph, and enables an estimate of how much the eventual strength will be above or below the strength of the standard mix used to construct the graph. This method is actually mainly used for the determination of in-situ strength rather than prediction of strength from an early age. Here, the assumption is that the standard mix will follow the standard strength/maturity curve and the in-situ strength can be estimated by measuring the strength and temperature history of the particular concrete.

The alternative method uses the Arrhenius equivalent age concept, equivalent age being the age at which a standard cured specimen would have attained the same strength as the specimen (or in-situ concrete) in question. The concept involves a rather complicated exponential formula:

$$EA = \sum \{t \exp[-Q(1/T_a - 1/T_s)]\}$$

where

$EA =$ equivalent age (hours)

Q = activation energy divided by the gas constant
T_a = temperature (K = °C + 273) for the time interval t
t = time (hours) spent at temperature T_a
T_s = reference temperature (K)

The reference temperature (T_s) is the standard curing temperature at which test specimens are kept. In many parts of the world, it is 20°C (293 K); in Australia it is 23°C in temperate zones and 27°C in tropical zones; and it may be that 30°C would be appropriate in some tropical countries (if this is the average temperature of unheated curing tanks). The Q value can range from below 4000 to over 5000, depending on the characteristics of the particular cement. It is often taken as 4200.

ASTM C 1074 sets out an experimental routine for determining the value of the constant Q in this equation, and it is quite important that an accurate value is used. However, the fact that the relationship is assumed to be exponential means that a graph of strength against the logarithm of equivalent age should be a straight line, the slope of which can be determined by measuring the strength and equivalent age at any two points, and the line can then be projected to predict the strength at any other equivalent age. Actually, the situation is not as simple as that: the straight line often turns out to have a change of slope somewhere along it, and the method does not work very well at directly predicting 28 day strength. However, very accurate assessment is possible if the point of any change of slope is first established (by taking several specimens from a single mix) and the method is used to predict the 7 day strength (or earlier if there is an earlier change of slope), and the average gain from this age, the 'control age', to 28 days is then added. The onerous task of determining the correct Q value is avoided if any reasonable value is assumed, and the slope of the graph joining the early results to the control age is automatically averaged. An incorrect Q value will simply result in an incorrect complementary slope to give correct predictions.

The early age maturity determination is of interest because it can be used to enable a faster reaction to change (where the additional cost is considered justified), but is of especial interest from the durability viewpoint in avoiding early damage to the concrete. Prior to the development of this method (by the writer in the 1980s), specifications used to call for curing to be continued for a particular number of days, e.g. to avoid frost damage, and early strengths, e.g. for prestressing readiness, were assessed by site-cured test specimens – which the new method showed too often to have a widely different maturity to a slab of concrete they were sitting on, or a beam alongside which they were left. It is now possible to cast a thermocouple into a test specimen and test it at 24 hours or less to establish the quality of the concrete and establish the equivalent age of any in-situ concrete, using other cast-in thermocouples, at which that concrete will reach any particular strength.[1]

8.3 Conclusion

Value judgements can be made about the conditions to be resisted by a concrete, the design life required, and the extent to which extra expense on providing higher quality is justified. However, the durability of the whole is only as good as that of its lowest-quality part, and the importance of this chapter is that it is now possible to produce low-variability concrete.

References

1. Day, K. (2006) *Concrete mix design, quality control and specification*. Taylor and Francis, London.
2. McKenna, P. (2007) Effect of controlled permeability formwork on concrete structures in marine environments. In: *FIB Symposium*. Dubrovnik.
3. Bickley, J., Hooton, R.D. and Hover, K.C. (2006) *Preparation of a performance based specification for cast-in-place concrete*. RMC Research Foundation, Silver Spring, MD.
4. www.kenday.id.au

9

Design aspects that can reduce risks from deterioration mechanisms

Bryan Marsh, Arup, UK

9.1   Introduction

The design form of a structure can be fundamental to its durability performance, and many examples can be seen from history such as gargoyles channelling water away from the vulnerable building fabric. Nevertheless, much of the available detailed guidance concerned with the achievement of durability of concrete tends to concentrate on materials aspects. Whilst undeniably of major importance, this is often not the complete story, and the design of the element itself can play a vital role. Indeed, there are circumstances where appropriate design can relieve the demand for technologically challenging materials solutions and reduce dependence on high standards of skilled workmanship. The lack of attention paid to the durability aspect of design is held responsible for the premature loss of serviceability of many highway structures.[1]

This chapter is intended to introduce unfamiliar readers to the concept of how design aspects can be used to minimise the risk of deterioration of concrete structures in various exposure environments. It does not attempt simply to give standard solutions but rather to provide an extra tool for the designer's 'toolkit'. Moreover, nothing in this chapter is new or original; the need for consideration of durability at the design stage has long been pointed out within concrete design standards.[2]

In most 'normal' applications, adequate durability can be achieved by the simple combination of good-quality concrete, sufficient cover to reinforcement and good workmanship. A more considered approach, however, may be needed for any of a number of reasons, including:

- particularly aggressive exposure conditions
- unavailability or prohibitive cost of suitable materials

- very long intended working life
- unavailability of skilled or trained labour
- unusually high consequences of failure.

It is difficult to define the position of any dividing line between design and materials aspects of durability if, indeed, there is one. In many cases, the optimum solution is likely to involve a combination of both design and materials measures. It is useful, therefore, for designers to have some knowledge of the deterioration mechanisms of reinforced concrete and the properties and limitations of the range of available materials. Likewise, it is useful for materials advisors to be aware of where a simple or effective solution to a potential durability problem may be achieved through appropriate design measures. For the purposes of this chapter, design measures are considered to be anything beyond the selection of the composition of the concrete (e.g. cement type, free water/cement ratio, admixtures) and depth of cover to reinforcement, although the latter has obvious consequences for the structural design.

To set the scene, one can do little better than to quote, somewhat extensively, relevant parts of the European design standard, Eurocode 0.[3] This standard sets down a very clear basic requirement that:

> A structure shall be designed and executed in such a way that it will, during its intended life, with appropriate degrees of reliability and in an economic way:
> - sustain all actions and influences likely to occur during execution and use, and
> - remain fit for the use for which it is required

It adds that a structure

> shall be designed to have adequate:
> - structural resistance
> - serviceability, and
> - durability

and that these basic requirements

> should be met:
> - by the choice of suitable materials,
> - by appropriate design and detailing, and
> - by specifying control procedures for design, production, execution, and use relevant to the particular project.

344

There should be no doubt after reading this that durability is as much a part of the responsibility of the designer as is structural loading, and not simply an 'add-on' to be dealt with by the concrete technologist much further down the line when the structural form and details have already been decided.

9.2   The decision process

Durability must be considered right at the onset of design, even at the concept stage. It is necessary at this stage to identify the performance requirements and any particular durability threats that affect the proposed structure or element, and also to identify any parts that are vital to the function of the structure – the durability critical areas. Once these have been identified, a durability strategy and plan can be formulated. Again quoting Eurocode 0:[3]

> In order to achieve an adequately durable structure, the following should be taken into account:
>
> - the intended or foreseeable use of the structure
> - the required design criteria
> - the expected environmental conditions
> - the composition, properties and performance of the materials and products
> - the properties of the soil
> - the choice of structural system
> - the shape of members and the structural detailing
> - the quality of workmanship, and level of control
> - the particular protective measures
> - the intended maintenance during the design working life

9.2.1   *Performance requirements*

Before informed decisions can be made about durability strategies, it is essential to know the required performance for the structure or element under consideration. Most deterioration mechanisms are time-dependent, so the intended working life should be defined, at least in broad terms. Clearly, a public monument will generally have a much longer required service life than the hard-standing for temporary site buildings associated with the construction of that monument. Eurocode 0[3] recommends that the intended or design working life for a structure should be stated, and defines a number of categories (Table 9.1).

Table 9.1 Design life categories from Eurocode 0

Category	Indicative design working life	Examples
1	10 years	Temporary structures
2	10–25 years	Replaceable structural parts (e.g. bearings)
3	15–30 years	Agricultural and similar structures
4	50 years	Building and other common structures
5	100 years	Monumental building structures, bridges and other civil engineering structures

Specific design measures are most likely to be required for the higher categories, although, under exceptional circumstances of extreme exposure conditions or very high consequences of failure, they may be required even for the lowest category.

9.2.2 Acceptable or possible maintenance

Although there is little that can be done directly to maintain concrete itself, other than perhaps reapplication of any protective coating that may have been applied, overall maintenance of a structure can have a large influence on durability. Particular examples include ensuring the continued efficacy of joint sealants and drainage channels. Thus, an essential element to consider within the durability design is the acceptability of regular maintenance to the owner or operator of the structure and, perhaps more importantly, the practicality and cost of such maintenance. A motorway bridge, for example, that requires frequent carriageway closures to facilitate maintenance of joints could hardly be said to be a good or efficient design. Moreover, the nature of some structures may make maintenance effectively impossible, such as hazardous waste containment structures or deep foundations.

9.2.3 Threats to durability

Once the intended working life is known, it is necessary to determine the threats to the achievement of the required durability. These will generally be in the exposure environment and may include the following:

- chlorides originating from de-icing salts or seawater

- sulfates or other aggressive chemicals in the ground
- freezing of water-saturated surfaces
- aggressive chemicals from industrial processes
- deleterious chemical reactions within the concrete itself.

Most, if not all, these threats are dependent upon the presence of water, so this, although not inherently deleterious in itself, should perhaps also appear in the list. Likewise, temperature might appear because increased ambient temperature will accelerate most chemical reactions as well as permeation rates and the rate of corrosion of steel.

9.2.4   Durability critical areas

Certain elements or parts of a structure will be more crucial to its function than others, and the consequences of a durability failure will be greater. Where failure of a certain element or part of a structure would seriously inhibit the ability of the structure to perform its required function, this part can be classed as a durability critical area. Such areas clearly warrant special consideration with respect to durability.

9.2.5   Durability strategy or plan

The means by which the required durability can be provided may vary depending on the particular requirements (e.g. intended working life, acceptability or ease of maintenance) and the durability threats provided by the exposure environment. Generally, the durability strategy will comprise either deterioration-proofing or deterioration management. Deterioration-proofing is achieved by the selection of measures that will effectively remove the risk of deterioration over the required working life by removal of the degradation mechanism or its deleterious effects. Deterioration management does not remove the degradation mechanism or its deleterious effects but controls the rate of deterioration such that its effect does not become critical over the required working life. The normal materials measure of the combination of suitable concrete quality and adequate cover to reinforcement in external environments is an example of deterioration management. As few solutions are entirely foolproof, it may be argued that deterioration-proofing is often a special case of deterioration management, but where the expected effective life of the measure is usually well in excess of the required working life. Some examples are given below.

9.2.5.1 Deterioration-proofing

- Remove the degradation factor by changing the exposure environment (e.g. over-cladding, extension of the building envelope to bring the element into the indoor environment, waterproofing).
- Make the structure immune to the degradation factor (e.g. non-corroding reinforcement, cathodic prevention, unreinforced concrete).

9.2.5.2 Deterioration management

- Use corrosion-resistant reinforcement to prolong the propagation time from corrosion initiation to manifestation of damage.
- Use controlled permeability formwork to create a dense, low-penetrability layer at the surface to slow the ingress of any aggressive media.
- Use coatings or impregnations to slow the ingress of any aggressive media.
- Use additional cover or enhanced concrete quality to slow the rate of progress of aggressive media towards the reinforcement.
- Control water movement to reduce the mobility of water-borne aggressive media or divert them away from the concrete surface.

9.3   Deterioration mechanisms and appropriate durability strategies

The main mechanisms of deterioration are covered in depth in other parts of this book and elsewhere. Nevertheless, an understanding of the basic mechanisms by which concrete can deteriorate in different exposure conditions is essential for the selection of appropriate design measures. For example, it is important to understand where the main threat to durability is corrosion of reinforcement or where attack of the concrete itself is more likely. Eurocode 0[3] states that the 'environmental conditions shall be identified at the design stage so that their significance can be assessed in relation to durability and adequate provisions can be made for protection of the materials used in the structure'. The following sections review each of the major deterioration mechanisms affecting concrete and its reinforcement, and suggest appropriate durability strategies. The actual choice of strategy will depend on many factors, not least of which will be cost. Assessment of cost-effectiveness, however, should be based on the expected benefit over the whole required working life of the structure rather than just the initial up-front cost. In performing a cost–benefit analysis, consideration should be given to the possible costs of disruption caused by a durability failure in service. The reasons behind decisions to provide specific measures beyond what might be considered normal should be clearly recorded and communicated along the construction chain to ensure that vital measures are not removed at a later stage under the guise of otherwise well-intentioned 'value engineering'. An example, best not identified, involved the specification of stainless steel reinforcement in a vulnerable detail which it was felt might be difficult to ensure achievement of the required cover during construction. Post-construction checks revealed the pessimistically predicted lack of cover, but unfortunately the specification of stainless steel reinforcement had been lost somewhere along the way.

9.3.1   Corrosion of reinforcement

Initiation of corrosion can occur when the protective action of the concrete cover is compromised, usually either by reduced alkalinity (carbonation) or by a build-up of chlorides above the threshold level for corrosion. Nevertheless, corrosion will only then occur if there is a sufficient supply of both oxygen and moisture. Design measures to prevent reinforcement corrosion should therefore address at least one of these factors:

- maintenance of alkalinity
- prevention of excessive chloride build-up, including raising the chloride threshold
- reduction in moisture content
- reduction in oxygen availability
- reduction in the susceptibility of the reinforcement to corrosion.

Or, alternatively, deal with corrosion once it starts by utilisation of a cathodic protection system installed at the time of construction (cathodic prevention).

9.3.1.1   Carbonation-induced corrosion

Carbonation-induced corrosion is readily dealt with for normal design lives (50–100 years) by provision of an adequate depth of good-quality concrete in accordance with design code recommendations. Specific design measures are only really necessary where there is a requirement for an exceptionally long design life or exceptionally high CO_2 levels.

Measures to reduce oxygen availability are likely to be largely impractical, but moisture content can be reduced by enclosing any otherwise externally exposed surfaces. Where this is impractical or undesirable, the susceptibility of the reinforcement to corrosion can be reduced by the use of galvanised or stainless steel reinforcement. Anti-carbonation paints are available which assist in the maintenance of alkalinity but have a relatively short life before re-application is necessary, so are best suited to remedial applications rather than new construction.

9.3.1.2 *Chloride-induced corrosion*
Chloride-induced corrosion can be a greater problem than carbonation-induced corrosion, and can be more difficult to deal with by materials measures alone, particularly where a long service life is required or where exposure conditions are severe such as in the Middle East. Materials measures such as cement type and free water/cement ratio combined with an adequate cover to reinforcement and sometimes a hydrophobic surface impregnation can be adequate to deal with many conditions. Where such conditions are likely to prove inadequate or where, for example, it is not desirable to provide deep cover it may be necessary to employ design measures. These measures may involve changing the exposure environment, for example, by the provision of permanent formwork, cladding or waterproofing. Or they may proof against deterioration by, for example, the use of stainless steel reinforcement whose chloride threshold level may be greater than the level of chloride that can be expected to ingress to the depth of the outer bars within the intended working life of the structure.

9.3.1.3 *Chemical attack (external)*
Certain chemicals found in natural groundwater, industrial processes or contaminated land can attack concrete. The severity of any attack will depend on many factors including the aggressiveness of the chemical, its concentration, the groundwater mobility, the temperature and, of course, the chemical resistance of the concrete. Most naturally occurring ground conditions can be resisted adequately by careful selection of cement type and free water/cement ratio. More severe conditions or very long design lives, for example, may require specific design measures. These may involve protecting the concrete surface from contact with the aggressive environment or providing extra concrete cover to act as a sacrificial layer and thus permit some

degree of attack to occur but without threatening the structural integrity. Reduction of groundwater mobility through the provision of drainage can greatly reduce the rate of attack and bring it down to a manageable level.

9.3.1.4 Chemical attack (internal)

Internal chemical attack, including alkali–silica reaction and delayed ettringite formation (DEF) can usually be avoided by materials measures such as selection of cement type and minimisation of cement content. Nevertheless, this type of attack is dependent upon the ready availability of water, so design measures that remove the concrete from contact with water can be effective in avoiding problems. An example would be the provision of effective waterproofing to below ground concrete. DEF is also dependent on a high temperature developing within the concrete during hydration or accelerated curing, so design measures which reduce the temperature to an acceptable level ($<70°C$) will be effective. An example of such a measure would be the incorporation of cooling pipes within a massive pour, which may also be employed to avoid cracking due to early age thermal contraction effects.

9.3.1.5 Salt recrystallisation

Salt recrystallisation can be a major problem in marine structures in hot climates or in land-based structures in hot dry climates where the ground contains high concentrations of salts such as chlorides or sulfates, as is the case in much of the Middle East. Damage can occur when salt-laden water is absorbed into the concrete below ground level but is left behind in the pore structure when the water evaporates to the atmosphere above ground after having risen up through the concrete by wick action. Concrete is inherently porous, so this form of potential deterioration can only be prevented by blocking water access to the pore structure. A possible materials solution is the incorporation of hydrophobic and pore-blocking admixtures in the concrete mix. A design solution is to prevent ingress of the salt-laden water into the concrete surface by effective tanking of the concrete below ground. Special details may be required to cope with piles that cannot practically be tanked. This may comprise coating the top of the pile with a material such as a polymer-modified cementitious layer which will be impermeable to the ground water but which will still allow transfer of forces between pile cap and pile.

351

9.3.1.6   *Freezing and thawing*

So-called freezing and thawing damage, or frost damage, can occur where water-saturated concrete is allowed to freeze. Expansion of the water when changing to ice causes tensile stresses to develop when there is no empty space into which it can expand. If these stresses exceed the strength of the concrete, then damage will occur. Materials solutions involve either the use of a certain minimum strength of concrete or, more effectively, the provision of empty space in the form of an entrained air void system. The necessary conditions for damage are saturation of the concrete surface and, of course, freezing temperatures. Design measures should therefore address either of these factors, although in most cases the materials solutions should be adequate. The surface can be prevented from becoming saturated by some form of waterproofing or surface impregnation, or even by the simple provision of shelter. Alternatively, freezing can be prevented, again by the provision of shelter or by the provision of integral heating, such as has been done with some bridge decks to avoid the use of de-icing chemicals.

9.3.1.7   *Cracking and crack widths*

The influence of cracking and crack width on the corrosion of reinforcement has been the subject of much debate. Codes of practice have long provided limits for acceptable crack widths related to the perceived severity of the exposure environment, although the most recently developed code, Eurocode 2,[4] has bucked this trend, to some extent, to reflect the findings of much research and experience that show there is no general link between the reinforcement corrosion rate and crack width. Control of cracking within defined limits may, however, be desirable for reasons such as appearance and water-tightness and for durability in environments of high chemical aggressiveness or severe wave action, particularly when containing sand particles.

9.4   Specific design and additional protective measures

Although the term 'additional protective measure' (APM) came to prominence through BRE Special Digest 1[5] for the design of concrete in aggressive ground, it is a logical term for any measure deliberately applied to reduce the dependence for durability on the concrete alone.

Measures introduced into BRE Special Digest 1 in order to provide durability for buried concrete under conditions of high risk of the thaumasite form of sulfate attack (TSA) comprise:

- *Address drainage* – the risk of serious chemical attack of concrete surfaces is much increased if there is a ready supply of mobile groundwater to constantly replenish the aggressive ions as they become consumed by reaction with the concrete. Good drainage design can be employed to interrupt groundwater flow towards the structure.
- *Protective coating* – as described elsewhere in this chapter, a waterproof coating will effectively change the exposure environment of the actual concrete surface from aggressive to benign by preventing the aggressive groundwater from actually making contact with the concrete surface. Some waterproofing measures, however, are not always totally effective and their protective properties may decline with time. For these reasons, BRE Special Digest 1 still requires a good-quality concrete to be used in conjunction with the waterproofing, to act as a second line of defence if necessary.
- *Sacrificial cover* – TSA has been found to transform the outer layer of concrete into a weak incohesive material reminiscent of toothpaste. Where the surface of the concrete is not important structurally, the cover to reinforcement can be increased by a thickness of concrete intended to be sacrificial should such degradation occur. Provided the thickness of the sacrificial layer is chosen correctly, the deleterious reaction can be tolerated over the intended working life of the structure without fear of distress. Based on early observations of structures severely affected by TSA, BRE Special Digest 1 suggests a sacrificial layer thickness of 50 mm. This measure is not suitable for application to piles that rely on skin friction, as this would be seriously diminished by loss or weakening of the concrete surface.
- *Permeable formwork* – controlled permeability formwork uses a fine mesh type of lining, similar to a geotextile, usually fixed to the inside of conventional formwork. The permeable liner allows excess water and air to pass through and escape but retains the cement, aggregates and sufficient water for hydration. This can result in a significant reduction in the effective free water/cement ratio over a few millimetres depth at the surface of the concrete. The actual depth of the effect will depend on the concrete composition and the efficiency of the formwork liner. The lower free water/cement ratio at the surface increases the impenetrability of

the surface, and hence reduces the extent of any chemical attack. The improved concrete surface should also provide enhanced resistance to chloride ingress.

- *Increased concrete quality* – resistance to chemical attack increases with increasing quality of concrete, as dictated by the free water/cement ratio and cement type. BRE Special Digest 1 lists four qualities of concrete, from design chemical class DC-1 (which has no specific limitation on composition) to DC-4 (which permits only a limited range of cement types and generally at very low water/cement ratios dependent on the actual cement type). One means of providing enhanced protection is thus to use a concrete of a higher design chemical class, although this is strictly more of a materials measure than a design one.

9.4.1   Unreinforced concrete

In some circumstances, it may be possible to design the structure without any reinforcement, hence removing any risk of damage from corrosion. The Highways Agency in the UK encourages this approach where it can be achieved in elements of highway structures such as abutments, wing walls and retaining walls. Clearly, this is an option in only a very limited number of structures, but it is one that is likely to be particularly effective.

9.4.2   Provision of adequate falls and drainage routes

As shown previously, most of the deterioration mechanisms affecting concrete and its reinforcement are dependent upon the presence of water. Failure to manage water to move it away from vulnerable parts of structures has been the cause of much premature deterioration. This is particularly the case where the water contains chlorides or aggressive chemicals.

9.4.3   Waterproofing (including integral) and tanking

Prevention of contact between concrete and water can be an effective means of deterioration-proofing an element or structure, provided the waterproofing remains effective over the whole required working life.

Following the example of the UK Highways Agency, BS 8500-1[6] allows the exposure class for concrete protected by waterproofing to be reduced from an onerous classification such as XD3 where subject, for example, to direct splashing with de-icing salt-laden water to XC3/4, which is the normal exposure class for externally exposed superstructure elements remote from any sources of chloride. XC3/4 may be argued not to be the technically correct exposure classification for concrete protected by waterproofing, but it provides the quality of concrete felt to be appropriate. Of course, if the waterproofing is fully effective over the required working life of the structure or element, the quality of concrete need not be particularly high, as there is no significant degradation factor directly affecting it. Nevertheless, such an approach would generally be held to be unwise, as it allows little by way of a safety factor should the waterproofing prove not to be 100% effective over the whole area or for the full required working life.

9.4.4 Corrosion-resistant reinforcement

9.4.4.1 Stainless steel reinforcement

The composition chemistry of stainless steel can give it a much greater chloride threshold value at which corrosion is initiated than normal carbon steel reinforcement. This means it can tolerate a much greater degree of chloride ingress before damage results. Nevertheless, this also means that stainless steel is not completely immune to corrosion should the chloride reach very high levels, and the design must therefore ensure that this threshold value is not reached during the intended working life. Moreover, the chloride threshold value will be significantly reduced should the surrounding concrete fall to a near-neutral pH through carbonation. For these reasons, it is recommended that the minimum cover should not be reduced below 40 mm[7] and, clearly, the concrete should be of low penetrability to ensure that chloride ingress and carbonation are restricted.

A barrier to the use of stainless steel reinforcement has traditionally been its greater initial cost than carbon steel. Nevertheless, the combination of whole-life costing and the selective use of stainless steel can prove its worth, particularly in very aggressive chloride environments and where long life is required with a high degree of reliability. Stainless steel bars usually only need be used in the most vulnerable locations. Generally, there is little risk of problems due to bi-metallic corrosion between the two types of steel.

The term 'stainless steel' covers a wide range of materials with varying compositions, properties and suitability for use as reinforcement. The most suitable types for use in concrete are austenitic (grade 1.4436) and duplex (grade 1.4462) stainless steels as covered by European standard EN 10088-1.[8] Martensitic and ferritic stainless steels are generally not suitable.

9.4.4.2 Epoxy-coated reinforcement

Fusion-bonded epoxy-coated reinforcement improves the corrosion resistance of normal carbon steel reinforcement by the provision of an epoxy resin barrier between the concrete and the reinforcement. Its success therefore relies on the impermeability of the epoxy coating to protect the steel surface from chlorides, oxygen and moisture. The quality of the coating is thus an important factor in the effectiveness of epoxy-coated reinforcement, and it has gained a poor reputation after a number of incidences of failure, particularly in bridge decks in the USA. The bond between the reinforcement and concrete may be lower than conventional carbon steel, and should be considered in the structural design.

9.4.4.3 Galvanised reinforcement

Despite its promise, the performance of galvanised steel reinforcement is such that it is generally regarded as unsuitable for use in concrete in chloride environments. It can, however, be useful where there is a risk of carbonation-induced corrosion that cannot readily be dealt with by the normal method of providing an adequate thickness of good-quality cover. An example of such a situation could be where a particularly slender element is required to have a long working life.

9.4.4.4 Non-ferrous reinforcement

Reinforcement made from fibre-reinforced polymer is immune from corrosion in both carbonation and chloride environments. Its properties as structural reinforcement are, however, quite different from conventional carbon steel reinforcement, which means it cannot be used as a simple substitution. Possibly because of this, its use has largely been restricted to specialist applications where advantage has been made of some of its other properties, such as being non-magnetic and non-conducting.

9.4.5 Cathodic protection/prevention

Cathodic protection has a long track record of protecting structures suffering from reinforcement corrosion from further deterioration through application of a small electrical current to counteract the electrochemical corrosion process. This process requires that the reinforcement is electrically continuous, which often requires intrusive works to make connections between bars. It also requires repair of any concrete already damaged by cracking and spalling and the installation of an anode system by drilling in or by the provision of an overlay. Cathodic prevention is a special form of cathodic protection whereby all the necessary features, such as reinforcement electrical continuity and a suitable anode system, are installed during construction, thus avoiding much of the expense of a retro-fitted cathodic protection system.

9.4.6 Coatings and impregnation

The application of a hydrophobic, pore-lining impregnation to an exposed concrete surface can significantly reduce the rate of ingress of water-borne chlorides impinging on that surface, for example by splashing. Indeed, such treatment has been required for all highway structures in the UK adjacent to carriageways.[9] The impregnation was originally specified as a silane, but, because of health and safety concerns, alternative materials such as siloxanes may be preferable. By lining the walls of the pores for just a small depth, water molecules are repelled and the normal uptake of water through capillary absorption is greatly reduced. This has the effect of also greatly reducing the rate of ingress of chlorides, as this is dependent on the presence of water to act as a carrier. This type of surface impregnation is only really effective against splashing or spray, and is not suitable for use on parts of structures in direct contact with water, such as in the tidal or submerged zones of marine structures.

9.4.7 Integral bridges

The UK Highways Agency design standard BD 57/01[10] requires that, unless inappropriate for the particular circumstances, bridges up to 60 m in length and with skews up to 30° should be designed as integral bridges. An integral bridge is one where the deck is connected directly to the abutments without any movement joints (Fig. 9.1). Experience has shown that failed or imperfect joints have been a major cause of deterioration of bridges in the UK because they have allowed the passage of salt-laden water onto the unprotected concrete surfaces of piers and abutments.

Fig. 9.1 A concrete bridge with integral abutments and piers. (Courtesy of Birse)

9.5 Examples of the application of design measures for durability

9.5.1 Stonecutters Bridge

The Stonecutters Bridge in Hong Kong was required to have a design life of 120 years in an aggressive marine environment of high heat and high humidity (Fig. 9.2). Informed use of predictive modelling showed that certain corner reinforcement bars were much more vulnerable to corrosion than the rest of the outer layer of reinforcement. This was largely because of the two-dimensional ingress of chlorides possible at

Fig. 9.2 Stonecutters Bridge, Hong Kong, incorporating selective use of stainless steel reinforcement. (Courtesy of Arup)

corner locations. The concrete materials specification was already very high, and it was felt undesirable to further increase the cover to these bars. The consequence of corrosion of these bars was an unacceptable risk, so it was decided to use stainless steel reinforcement in these durability critical areas. In the other areas, the normal carbon steel reinforcement was felt to be adequately protected by the combination of cover depth and concrete quality.

9.5.2 Sub-arctic CGS

The design brief for a concrete gravity platform in the Russian sub-arctic in an area of very high annual ice movement (approximately 5000 km/ year) called for a 30 year working life with no opportunity for maintenance or repair to the concrete surfaces. A study of the available published information on ice abrasion of concrete revealed that there was no reliable way to predict the likely loss of cover to reinforcement over that period. Moreover, even when using concrete designed to provide the greatest resistance to ice abrasion, it was not possible to be absolutely confident that even, say, 100 mm of sacrificial cover would be adequate over 30 years with no opportunity for maintenance or repair.

The design solution put forward was to provide a sacrificial steel jacket together with a generous cover depth of highly abrasion-resistant concrete.

9.5.3 Underground structures in the Middle East

Structures in the Middle East with continuity in the concrete from the groundwater level to above ground are at risk from salt recrystallisation damage. Where possible, concrete below ground is fully tanked to prevent water uptake. The use of integral permeability-reducing admixtures might also be considered as an appropriate alternative to tanking where the aggressivity of the ground permits. Where tanking is not possible, such as piles, the continuity of the water path can be broken by use of a waterproof seal (e.g. a polymer-modified cementitious layer) at the top of the pile where it enters the pile cap.[11]

9.5.4 A remedially applied design measure

A design solution was post-applied to a ground-bearing slab in the construction of a school where an insufficiently sulfate-resistant concrete had either been specified or supplied in error. Drainage trenches were installed beyond the boundary of the slab to interrupt the flow of groundwater and effectively change the exposure environment to one of lower aggressivity through removal of mobile groundwater.

9.5.5 An example of where design has contributed to deterioration

Attack of concrete in the foundations of bridges on the M5 motorway in the UK by TSA led to a major revision of guidance on concrete in aggressive ground, culminating in the publication of BRE Special Digest 1.[5] TSA had seldom been identified as a deterioration mechanism in concrete structures prior to this time, and no specific avoidance measures were given within the UK guidance at the time. The severity of the attack was surprising, but would probably have been much less had the design of the drainage been better. Instead of taking aggressive groundwater away from the concrete bridge foundations, the drainage actually created a sump where water collected and contributed to an otherwise largely unforeseeable deterioration mechanism. It is possible that the deterioration may not have occurred, or have been much less severe, had the drainage design been effective in moving water away from the concrete surfaces.

9.6　Summary

There is rarely a single means of achieving durability in a concrete structure. Consideration should therefore be given at the design stage to the protection that can be achieved both by the use of suitable design aspects and through appropriate materials measures. Design aspects may be effective in reducing the reliance on expensive or technically difficult materials measures or the need for high quality of workmanship where it may not readily be available. This chapter has attempted to provide an insight into the types of design measures that are available and how they should be targeted at the specific deterioration mechanisms identified in the exposure environment.

References

1. HA (2001) Design for durability. *Design manual for roads and bridges*, BA 57/01. Highways Agency, London, vol. 1, section 3, part 8.
2. BSI (1985) *Structural use of concrete. Code of practice for design and construction*, BS 8110-1. British Standards Institution, London.

3. CEN (2002) *Eurocode 0: basis of design*, EN 1990. European Committee for Standardisation, Brussels.
4. CEN (2004) *Eurocode 2: design of concrete structures. General rules and rules for buildings*, EN 1992-1-1. British Standards Institution, London.
5. BRE (2005) *Concrete in aggressive ground*. Special Digest 1. Building Research Establishment, Garston, Watford.
6. BSI (2006) *Concrete – production, performance and conformity criteria. Guidance for the specifier*, BS 8500-1. British Standards Institution, London.
7. Concrete Society (1998) *Guidance on the use of stainless steel reinforcement*. Technical Report 15. Concrete Society, Camberley.
8. CEN (2005) *Stainless steels. List of stainless steels*, BS EN 10088-1. British Standards Institution, London.
9. HA (2003) The impregnation of reinforced and prestressed concrete highway structures using hydrophobic pore-lining impregnants. *Design manual for roads and bridges*, BA 43/0. Highways Agency, London, vol. 2, section 4, part 23.
10. HA (2001) Design for durability. *Design manual for roads and bridges*, BD 57/01. Highways Agency, London, vol. 1, section 3, part 7.
11. Concrete Society (2008) *Guide to the design of concrete structures in the Arabian Peninsula*, CS 163. Concrete Society, Camberley.

10

Case studies: durability problems, repair strategies or proper consideration to durability design?

Marios Soutsos, University of Liverpool, UK

The owners of structures are beginning to recognise the high cost of repairs. Unfortunately, it is not only the engineering achievements that are making the headlines: it is also collapses or failures of structures to meet their design life. Durability problems arising during the design life of a structure may be considered to be a failure in the serviceability design of the structure. The owners of infrastructure are therefore expecting a longer design life for newly built structures, and design consultants have recognised that to achieve this longer design life, careful consideration must be given to the selection of mix proportions and constituent materials for the concrete.

This chapter presents case studies of durability problems that have arisen in major structures to raise awareness of the extent and cost of repairs that may be needed for the structures to remain operational. Such repair costs may be avoided by identifying at the design stage the durability property required for the structure. This may not necessarily be associated with steel reinforcement corrosion: for example, improved abrasion resistance is the property that is sought for durable factory floors. The last two case studies show that 100 years or more of service life are achievable with careful selection of concrete mix proportions and constituent materials.

10.1   Deterioration of a dry dock facility
Donald Wimpenny, Halcrow Pacific Pty, Australia

The dry dock facility was constructed in the Middle East between 1973 and 1979. The dock had three basins, with the widest able to take tankers up to 1 million dwt (Fig. 10.1).

Concrete durability
978-0-7277-3517-1

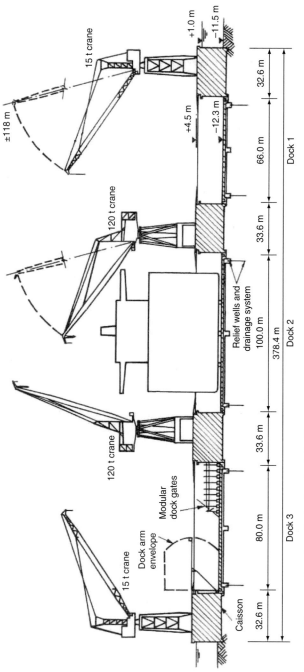

Fig. 10.1 Cross-section through the dry dock facility

Table 10.1 Requirements for the caisson concrete

Parameter	Value
Compressive strength	$30\,N/mm^2$
Minimum cement content	$360\,kg/m^3$
Actual cement content	$390\,kg/m^3$
Actual water/cement ratio	0.51
Target slump	75 mm
Nominal cover	60 mm

The best contemporaneous practices were employed in its design and construction, and considerable diligence was exercised during its construction. However, premature deterioration occurred to the facility within 10 years of it opening, requiring extensive remedial works.

Other dry docks in hot climates have also required premature repairs.[1]

10.1.1 Construction

The walls of the dock were formed from multi-celled slipformed caissons 17 m wide, 13.6 m in height and 31 m long. These cells were floated in position, and ultimately ballasted with marine-dredged sand. A grouted connection was made between the caissons and then crane beams, a service gallery and slab were cast on the caissons to provide the dockside arms. The key requirements for the concrete caissons are indicated in Table 10.1.

There was some uncertainty over the achievable strength and the water/cement ratio values, and both were subject to the results of trials. The water/cement ratio used was required to be within 10% of these trial mixes. Chemical admixtures were not widely used in the early 1970s, and reticence over their use is reflected in a specification requirement that admixtures were *only permitted at the express permission of the engineer.*

Figures 10.2 and 10.3 show the dock under construction and at completion respectively.

There were significant challenges during construction, including:

- workability loss at high ambient temperatures
- control of water content
- dusty gabbro aggregate
- handling of heavy reinforcement.

Fig. 10.2 View of the casting yard for the slipformed caissons (the truck mixer provides scale)

Fig. 10.3 View of the completed dock

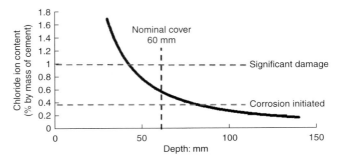

Fig. 10.4 Profile of the chloride level in the concrete with depth in 1980

The absence of admixtures considerably increased the difficulties of controlling the water demand of the concrete posed by the high temperatures and dusty aggregate. Great care was exercised to monitor and control the dust content of the aggregate, but even with these precautions the water/cement ratio occasionally increased to over 0.55.

Portland cement at the time of construction was in short supply and of variable quality. Secondary cementitious materials such as slag or fly ash were not readily available within the region, and their benefits in terms of enhanced resistance to chloride ingress were not widely recognised.

10.1.2   Deterioration

Chloride profiles measured in the concrete in 1980 are shown in Fig. 10.4. The profiles indicated that chloride levels of over 0.4% by mass of concrete had penetrated beyond the reinforcement, indicating a risk of corrosion having been initiated. Chloride levels of 1.0%, representing a risk of significant corrosion damage,[2] had penetrated to beyond 40 mm depth compared with a nominal depth of cover of 60 mm. The latter represents a minimum cover value of 45 mm for a typical reinforcement fixing tolerance for in-situ concrete.

By 1991, serious corrosion problems had manifest at the facility, including:

- chloride ions >1% at 50–75 mm depth
- cracking up to 0.2 mm wide
- hollowness to 20% of the caisson surface
- reinforcement pitting at cracks
- up to 50% loss of section of the reinforcement over short lengths.

Removal of the sand ballast from inside the caissons revealed deterioration to internal surfaces in the form of hollowness under tapping, and

Fig. 10.5 Deterioration inside the caisson showing areas of hollowness (H)

spalling of the concrete cover with corrosion to the reinforcement (Figs 10.5 and 10.6).

The apparent chloride diffusion coefficient of the concrete was estimated as approximately $8.9 \times 10^{-12}\,\mathrm{m^2/s}$ by solving Fick's law of diffusion for the chloride profile measured in 1980, using an in-house program called SALTS (Table 10.2). This apparent chloride diffusion

Fig. 10.6 Exposure of reinforcement showing exposure in areas of hollowness (left) and corrosion and cracking (right)

367

Table 10.2 Chloride diffusion characteristics for concrete

Characteristic	Value
SALTS best fit curve	
Assumed background level	0.15% by mass of cement
Apparent surface chloride level	2.78% by mass of cement
Apparent chloride diffusion coefficient in 1980 at the service temperature (assumed 35°C)	8.9×10^{-12} m^2/s
Predicted chloride diffusion coefficient at 20°C	
28 days	1.7×10^{-11} m^2/s
4 years (1980)	3.65×10^{-12} m^2/s
20 years	1.5×10^{-12} m^2/s

coefficient is equivalent to a value of 3.65×10^{-12} m^2/s. The latter is consistent with published values for Portland cement concrete.[2]

The apparent diffusion coefficient was used as an in-house corrosion prediction model (CORRPRED) to estimate the time to corrosion-related cracking (Fig. 10.7). The model is based on published guidance,[3] and is able to make adjustments for a large number of factors, including:

- temperature and relative humidity
- effect of age on the chloride diffusion coefficient
- type of reinforcement (stainless, carbon steel, prestressing)

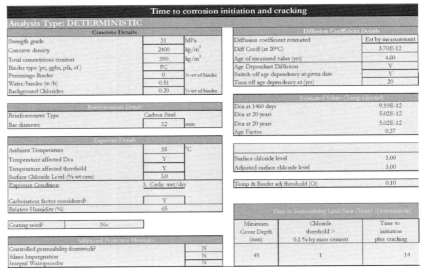

Fig. 10.7 Output from CORRPRED analysis showing 14 years to cracking

- additional durability enhancement measures (silane, corrosion inhibitors, controlled permeability formwork).

The modelling indicates a time to cracking at 35°C of less than 15 years. This predicted time to corrosion-related damage is consistent with the level of deterioration observed in 1991. The equivalent predicted time for temperate conditions is over 35 years.

10.1.3 Discussion

The limiting proportions and compressive strength of the concrete specified for the caisson are in accordance with the contemporaneous requirements of CP110,[4] which required a minimum cement content of $360 \, kg/m^3$ for concrete *exposed to alternating wetting and drying or freezing while wet or to seawater.*

However, as can be observed from the predictions above, the Middle East environment is much more severe than that in temperate climates, not least due to the higher ambient temperature. Furthermore, a dry dock is potentially a more severe exposure condition than normal marine environment, as the concrete is subject to long periods of drying followed by inundation with seawater. This leads to rapid chloride ingress by capillary absorption.

There was little recognition in the 1970s of the importance of cement type and the water/cement ratio in achieving durability. By 1981, the recommendations for concrete both in the UK and the Middle East had become much more stringent in respect of cover and the water/cement ratio, but by this stage the dry dock had already been constructed. Typical recommendations for the Middle East are summarised in Table 10.3, and include a requirement to use ASTM Type II, a moderate sulfate resisting Portland cement, in reinforced concrete rather than sulfate resisting Portland cement (SRPC). This

Table 10.3 Recommendations for concrete in Middle East exposure conditions[5]

Parameter	Value
Minimum cement content	$370 \, kg/m^3$
Cement type	ASTM C150 Type II
Maximum water/cement ratio	0.42
Nominal cover	100 mm

Table 10.4 Recommendations for concrete in a hot marine climate for 50 years service life[6]

Parameter	Value
Minimum cementitious content	425 kg/m^3
Cement type	25% Portland cement, 5% silica fume, 70% slag
Maximum water/cementitious ratio	0.34–0.38
Nominal cover	75–100 mm

appears to be an acknowledgement of the detrimental effect of low tricalcium aluminate levels in SRPC on chloride binding capacity.

More recent recommendations are shown in Table 10.4, which indicate the progressive lowering of water-cement ratios, using high range water reducing admixtures, and the incorporation one or more secondary cementitious materials.

10.1.4 Conclusions

A dry dock environment in a hot climate represents an extreme environment, and use of best contemporaneous practice from temperate marine climates will not ensure durability. Chlorides can build up rapidly by diffusion and capillary absorption, leading to chloride-induced corrosion damage within 15 years. The corresponding time in the UK would be over 35 years. The achievement of durability requires a low water/cementitious ratio, using high-range water-reducing admixtures, and incorporation of one or more secondary cementitious materials, such as slag or silica fume. Durability models can be used to get an accurate indication of the onset of corrosion damage.

References

1. de Almeida, I.R., Cordeiro, T.J.R.B and Costa, J.P.V.M. (1996) The Setenave dry dock rehabilitation. *Concrete International*, 18(3), 30–33.
2. Bamforth, P.B. (1994) Admitting that chlorides are admitted. *Concrete*, 28, 18–21.
3. Bamforth, P.B. (2004) *Enhancing reinforced concrete durability: guidance on selecting measures for minimising the risk of corrosion of reinforcement in concrete*. Technical Report No. 61. Concrete Society, Camberley.
4. BSI (1972) *The structural use of concrete, part 1: design, materials and workmanship*, CP110. British Standards Institution, London.

5. Kay, E.A., Fookes, P.G. and Pollock, D.J. (1981) Middle East concrete 3, deterioration related to chloride ingress. *Concrete*, 15(11), 22–28.
6. Slater, D. and Sharp, B.N. (1998) The design of coastal structures. In: R.T. Allen (ed.), *Concrete in coastal structures*. Thomas Telford, London, ch. 6.

10.2 Delayed ettringite formation in a Malaysian highway structure: investigation and management strategy

Donald Wimpenny, Halcrow Pacific Pty, Australia

The highway structure in Kuala Lumpur, comprising a 1.5 km viaduct carrying a six-lane carriageway, was opened in 2002.

The structure has split decks formed from externally prestressed precast concrete box girders resting on T-shaped piers, comprising a 18.7 m long crosshead tapering from 3.5 m to 2 m in depth, and a 3.6 m-wide octagonal column (Figs 10.8 and 10.9). The abutments are also T-shaped, similar to the piers, but are, for the most part, buried.[1]

Cracking to several piers and one abutment was reported 6 months after opening the viaduct. A detailed survey of the crack mapping was undertaken in 2004, by which time some of the most severely cracked crossheads had cracks several millimetres wide. Some of the cracking could be attributed to inadequate reinforcement. However, the variability, severity and presence of a lateral displacement across some cracks suggested other mechanisms, such as an expansive reaction, may be contributing to the cracking (Fig. 10.10). A thermal analysis using published guidance indicated values in excess of 80°C

Fig. 10.8 View from beneath the viaduct showing the form on construction

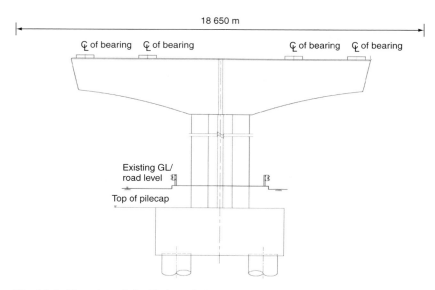

Fig. 10.9 Elevation of the T-shaped piers

could have been reached during the initial curing, leading to a risk of delayed ettringite formation (DEF).

DEF is an expansive reaction which occurs due to the delayed formation of ettringite in concrete which has been subject to high temperatures (typically over 65°C) during curing. These high temperatures can be generated during steam-curing or in thicker in-situ concrete members.[2] The delayed formation of the ettringite in

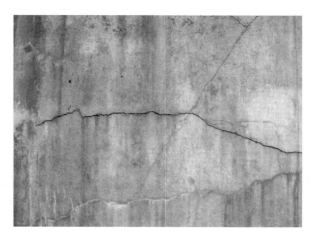

Fig. 10.10 Cracking to a crosshead

hardened concrete exposed to moisture leads to expansion and cracking. The pattern of cracking is superficially similar to that generated by the alkali–aggregate reaction, leading to a risk of misdiagnosis.

10.2.1   DEF management strategy
The proposed management strategy has three stages, as indicated in Table 10.5. Partial closure of the bridge meant there was considerable

Table 10.5 DEF management strategy for the Malaysian highway structure

Action	Purpose
Stage 1 – confirmation of diagnosis	
Crack movement and reinforcement strain monitored	Confirm progressive expansion and localised reinforcement yielding unrelated to live loading
Cores from crossheads for petrographic examination and scanning electron microscopy work	Confirm DEF is present from the distribution and nature of reaction products
Cores from crossheads for Duggan expansion testing	Confirm the potential for DEF based on expansion levels
Cores from crossheads for British Cement Association (BCA) expansion testing	Confirm the occurrence of DEF based on expansion levels and petrographic examination of the specimens after expansion testing
Stage 2 – numerical modelling	
Finite element analysis of the temperature within sections during construction	Confirm the likely zone of concrete at risk of DEF
Spreadsheet model developed to assess critical sections of a crosshead	Predict the likely structural impact of DEF and determine the strengthening requirements
Non-linear finite element analysis of stresses and damage based on the predicted expansion	Confirm the likely structural impact of DEF and the strengthening requirements in more detail
Stage 3 – repair	
Grout existing cracks with an epoxy resin-based grout	Seal the existing cracks and restore the tensile strength of the concrete
Apply waterproof coating to crossheads	Exclude water ingress from rain and ponding water
Strengthening scheme	Prestressing and plate bonding to address the present and possible future loss of strength

pressure to resolve the problem, and this meant several of the stages occurred in parallel.

DEF was unexpected, and a number of steps were taken to confirm the finding. Potentiometers and strain gauges were installed to measure crack movement and the response of the reinforcement for selected crossheads. These were remotely interrogated via a GSM modem. This indicated progressive expansion which did not coincide with traffic loading but showed a higher rate in the rainy season. Cores were taken for petrographic examination, which found ettringite in cracks and rims around aggregate particles characteristic of DEF. The reaction products were confirmed using SEM and microprobe analysis. Cores were subject to expansion testing using the BCA and Duggan test methods.[3,4] These confirmed damaging expansion levels of up to 0.2% at 20 days, compared with a safe limit of 0.05%. The primary cause of the expansion in the cores subjected to BCA expansion testing, petrographic techniques and electron microprobe analysis was indicated to be DEF.[5]

A two-dimensional finite element analysis was undertaken to predict the temperature in the crosshead during construction (Fig. 10.11). No concrete temperature records were available, except at placing (approximately 32°C), and assumptions were made on worst case and average ambient conditions, including solar gain. The worst case and average case analyses indicated temperature values in excess of 80 and 70°C, respectively, for over 100 hours at the core of the section. The temperature values from this analysis were used to predict DEF expansion at different locations in the crosshead, which were in turn used within the structural analysis.

The significant structural effects of ongoing DEF expansions were found to be:

- cracking around the perimeter of the section
- loss of ultimate moment capacity due to cracking of the soffit
- failure of laps in main tension and in shear links.

The issue of lap failure was a particular concern, as two of the four layers of main flexural reinforcement in the top of the crosshead were lapped at the centre of the crosshead with laps that were not full-strength laps. In order to assess the strength of DEF-affected crossheads and to identify suitable strengthening works, a spreadsheet model was developed for analysing critical sections of the crossheads from first principles. A three-dimensional non-linear finite element analysis was also undertaken. The spreadsheet gave similar results to conventional section analysis software, but in addition was able model the variation in DEF

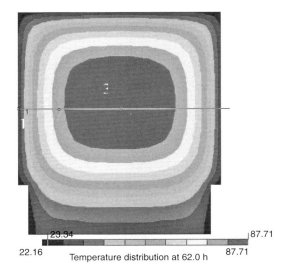

Temperature distribution at 62.0 h

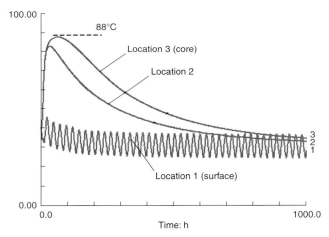

Fig. 10.11 Section through a crosshead showing the predicted temperature development

expansion across the cross-section, the effect of the stresses due to DEF expansion on the concrete and reinforcement, and the loss of stiffness and strength of the concrete section due to ongoing DEF. The spreadsheet model included options for modelling the tensile strength of the concrete, the stress–strain response of lapped bars and the effect of yielding of bars and strengthening by plate bonding and external prestressing.

The spreadsheet and non-linear finite element analysis models were used to predict the damage to the structure for different DEF expansion levels. There was particular concern about ongoing expansions leading to yielding of the reinforcement, failure of laps and, potentially, prejudicing the strengthening work at higher levels of expansion. At 0.1% DEF expansion, it was found that there was significant cracking around the top and sides of the section, but that the soffit remained intact and the effect on the ultimate moment capacity was negligible. However, as the modelled expansion increased, the moment capacity reduced due to cracking in the soffit and lap failure, necessitating strengthening work to reinstate the moment capacity at the root of the cantilever. The non-linear finite element analysis produced a pattern of cracks in the crossheads over the column very similar to that observed on site. Simulating the effect of increasing expansion lead to cracking developing further down the vertical face of the crosshead until it extended into the soffit.

The crossheads were subject to frequent wetting due to the longitudinal gap between the split decks, leaking drainage, and wind-driven rain. The rapid development of the reaction may have been promoted by the high ambient temperatures and humidity in Malaysia. The waterproofing of the structure was regarded as essential to control future expansion, and specifications were developed for a waterproof coating, which included requirements for:

- tolerance to damp substrate and high relative humidity during application
- ability to transmit vapour transmission not less than $3 \, g/m^2/day$ out of the concrete
- transparency to permit cracks to be viewed and monitored
- a minimum elongation to failure of 100% and crack bridging of up to 2 mm width.

The strengthening scheme developed for a typical crosshead is shown in Fig. 10.12. The longitudinal prestressing is required to strengthen the crosshead at the main lap regions primarily by providing a relieving moment, such that if the two top layers of bars fail at the laps, the external prestress in combination with the remaining reinforcement would be sufficient to develop the design moment. The transverse-bonded steel plate straps are required to strengthen the crosshead due to deficiencies in the transverse reinforcement. These straps also provide reinforcement to control ongoing cracking from the limited residual DEF expansion expected after waterproofing. The

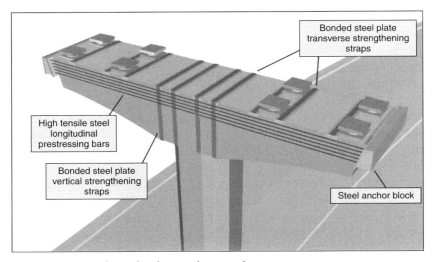

Fig. 10.12 Typical crosshead strengthening scheme

vertical-bonded steel plate straps are required to control ongoing cracking in the weakened crosshead–column connection region, where the existing reinforcement has yielded and lost its stiffness at wide cracks.

10.2.2 Conclusions

DEF can occur in both steam-cured and in-situ concrete, leading to serious cracking. There are few published cases relating to bridges, which may be due the deterioration having been diagnosed as resulting from the alkali–aggregate reaction. High curing temperatures during the initial curing, and exposure to moisture in service, are critical risk factors, and it has been found that the structural effects of DEF can be significantly different from the effects of the alkali–silica reaction.

The DEF investigation and management strategies have to be tailored to the particular structure, but include the following elements:

- removal of samples to confirm the cause of cracking
- monitoring of crack width movement and expansion testing of cores to assess future expansion and movement for selection of the repair products
- measures to exclude moisture in the form of coatings and sealing cracks
- strengthening or replacement of the affected elements.

378

A number of techniques were applied to assist in the investigation and management of DEF, including SEM to confirm the distribution of ettringite in the concrete, monitoring of crack movement, and modelling of temperature, stresses and deterioration.

References

1. Buckby, R.J., White, P.S., Mills, C.A. and Quillin, K. (2006) Severe cracking in in-situ concrete substructures due to delayed ettringite formation (DEF): diagnosis, assessment and remediation. *Hong Kong International Conference on Bridge Engineering.*
2. Quillin, K. (2001) *Delayed ettringite formation: in-situ concrete.* BRE Information paper IP11/01. CRC, London.
3. BCA (1992) *The diagnosis of alkalis-silica reaction, report of a working party,* 45.042. British Cement Association, Crowthorne.
4. Grabowski, E., Czarnecki, B., Gillot, J.E., Duggan, C.R. and Scott, J.F. (1992) Rapid test of concrete expansivity due to internal sulfate attack. *ACI Materials Journal,* 89(5), 469–480.
5. Eden, M.A., White, P.S. and Wimpenny, D.E. (2007) A laboratory investigation of concrete with suspected delayed ettringite formation – a case study from a bridge in Malaysia. *11th Euroseminar on Microscopy Applied to Building Materials, 5–9 June 2007.* Porto, Portugal.

10.3   Construction of a deep-water port
Donald Wimpenny, *Halcrow Pacific Pty, Australia*

Construction of the deep-water port, in North Africa, commenced in 2004. The port includes a 1.1 km long breakwater in approximately 30 m depth of water. The original design was for a rubble mound break-water, but an alternative, comprising reinforced-concrete caissons on a 10 m deep rubble mound, was proposed by the contractor. This alternative significantly reduces the volume of rock required, but requires careful consideration of durability given the environment, the difficulty of under-taking any maintenance work and the 100 year service life of the facility.

10.3.1 Design
The breakwater consists of 40 caissons, approximately 35 m in height and 28 m in diameter. The caissons are principally multi-celled 'quadra-lobes', with external perimeter walls 0.5 m in thickness and internal dividing walls 0.4 m in thickness. An upper attention chamber is also included to dissipate wave energy (Figs 10.13 and 10.14).

A probabilistic durability design was undertaken by the contractor, using the principles outlined in the pan-European Duracrete project.[1]

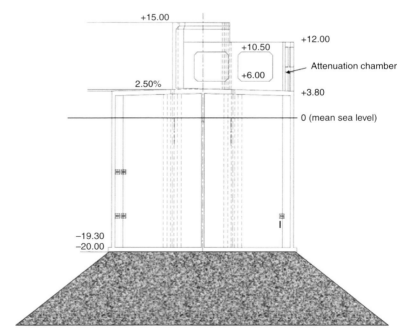

Fig. 10.13 Cross-sectional elevation of a typical caisson

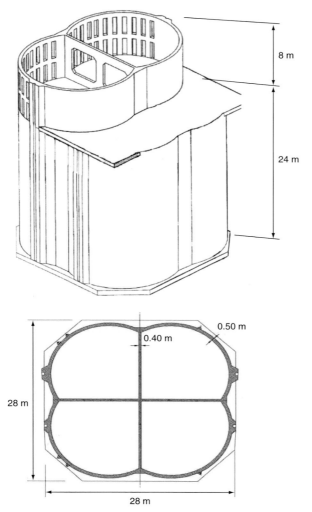

Fig. 10.14 Isometric sketch and cross-sectional plan of a 'quadralobe' caisson

This approach is similar to structural design in that a resistance and load are calculated with the probability of failure being defined as probability the resistance is less than the load. A frequency distribution is typically defined by the arithmetic mean and standard deviation for each of the variables, such as cover, to allow the failure probability to be estimated. A reliability index is calculated, which is the converse of the failure probability using the inverse normal function. For example, probabilities of failure of 1 in 1000 and 1 in 100 give corresponding reliability indices of 3 and 2.3. The serviceability limit state is reached when the reliability

381

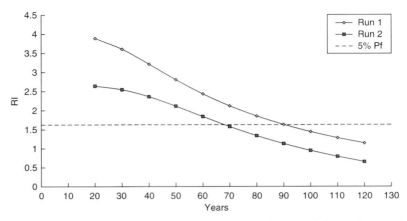

Fig. 10.15 Reliability index for different scenarios (horizontal line indicates 5% failure probability)

index drops below a pre-determined acceptance level. The acceptance level usually lies between 1 and 5, depending on the cost of construction and repair, the consequence of failure and the deterioration mechanism.[2] The proposed design was reviewed, and an independent check was made on the probabilistic design using an in-house model (CORRPRED) for corrosion-induced corrosion, with add-in proprietary software (@risk) for statistical modelling. Typical output from the model is shown in Fig. 10.15.[3] Some of the key input parameters are summarised in Table 10.6. The ageing factor, which defines the rate of change of the diffusion coefficient with time, has not been included, but should be selected with great care.

A number of potential shortcomings were identified in the proposed durability design. In particular:

Table 10.6 Key input parameters in the model

Parameter	Mean value and (standard deviation or range)	
	Run 1	Run 2
Temperature: °C	25	25
Cover: mm	70 (55–85)	65 (50–75)
Diffusion coefficient ($\times 10^{-12}$ m^2/s)	5.1 (1.1), 28 days	1.6 (0.5), 6 months
Threshold level: %	0.6 (0.15)	0.4 (0.15)
Surface chloride level: %	4 (1)	5 (1)

Table 10.7 Durability measures

Additional durability measures	Submerged (below −4 m)[a]	Tidal and splash (above −4 m)[a]
Minimum cover: mm	70	80
Cementitious blend	20% fly ash	20% fly ash and 5% silica fume
Maximum water/cement ratio	0.35	0.35
Maximum chloride migration coefficient: m^2/s	5×10^{-12}	1.5×10^{-12}
Cathodic protection	Installed	Provision for future installation, e.g. electrical continuity of reinforcement
Maximum loss of mass by wear, ASTM C1138 test	5%	5%

[a] Datum at 0 m is lowest astronomical tide.

- no allowance for construction defects
- no allowance for wear
- overly optimistic assumptions including unsubstantiated chloride diffusion characteristics of the concrete
- no provision for future cathodic protection.

An addendum was inserted to the durability plan to include abrasion testing using the ASTM C1138 method and chloride migration testing using the NTB 443 and NTB 492 methods. The NTB 443 method is a bulk diffusion test in which the chloride profile is generated by grinding the sample in depth increments of at least 1 mm. The NTB 492 method is an accelerated chloride migration test in which an electric current is passed though the sample to promote the movement of chloride ions. However, unlike the AASHTO T277 rapid chloride migration test, the current is moderated to prevent heating of the sample, and the chloride penetration is assessed directly by application of a silver nitrate indicator.[4] A chloride migration coefficient can be estimated from the depth of penetration. The durability measures for the caissons were enhanced by making provision for cathodic protection, imposing limits on chloride migration coefficient and requiring appropriate cementitious blends and minimum cover values, as summarised in Table 10.7.

10.3.2 Mix development

The contractor's proposed method of construction was slipforming the caisson and conventional casting the wave attenuation chamber using self-compacting concrete. Slipforming imposes a number of additional requirements on the concrete compared with conventional casting. In particular, the setting time of the mix must be carefully regulated to ensure the rate of concrete placing and slipform movement are such that the concrete has sufficient strength when it exits the bottom of the slipform to be self-supporting, but has not stiffened so much that there is a risk of high friction and hardened plates of concrete building up at the face of the slipform or of cold joints. In order to regulate workability and setting, a high-range water-reducing admixture and a separate retarding admixture were used.

Laboratory trials were undertaken comprising over 50 different mix designs and the following variables:

- 7 cement types
- 2 fly ash sources
- 2 silica fume sources
- washed and unwashed sand
- 3 different superplasticiser/retarder combinations.

The final mixes proposed for construction are summarised in Table 10.8. Chloride migration testing indicated that both mixes met the required limits using laboratory cast and cured specimens. Resistivity measurements were made on the concrete using the two-electrode method to

Table 10.8 Summary of caisson concrete

Parameter	Value	
28 day compressive strength: N/mm^2	75	
Cementitious content: kg/m^3	480	
Maximum water/cement ratio	0.35	
Target slump: mm	200	
Admixtures	High-range water reducer Retarder	
Cementitious type	20% PFA and 5% silica fume (tidal/splash)	20% PFA (submerged)
Approximate chloride migration coefficient (m^2/s) by the NTB 492 method	3×10^{-12} at 28 days 1.5×10^{-12} at 56 days 1×10^{-12} at 91 days	6×10^{-12} at 28 days 3×10^{-12} at 56 days 2.5×10^{-12} at 91 days
Resistivity (Ω m) by TEM	103 at 28 days 276 at 56 days	70 at 28 days 108 at 56 days
Wear by ASTM C1138	1.5–2.5%	NA

PFA, pulverised fuel ash; TEM, tunnelling electron microscopy.

 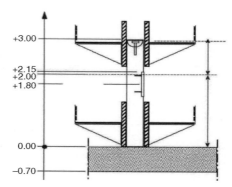

Fig. 10.16 Slipform field trial, showing a completed wall (left) and a section through a slipform (right)

give a rapid measure of concrete quality. Chloride diffusion coefficients from the NTB 443 bulk diffusion test were slightly less than the equivalent migration values estimated from the NTB 492 method (approximately 3.24×10^{-12} m^2/s compared with 4×10^{-12} m^2/s). In addition to the laboratory work, slipforming trials were undertaken using a slipform 3 m high and 5 m long (Fig. 10.16).

10.3.3   Construction

The construction sequence involved casting the base slab for the caissons on a bespoke casting bed (Fig. 10.17). The outer walls and any cross-walls were then formed by slipforming to approximately 10 m in height (Fig. 10.18). The partially completed caisson was then slid to a launching platform on the quayside on rail-mounted trolleys (Fig. 10.19). After launching, the slipforming continued from the quayside (Fig. 10.20). When the slipforming was complete, the caisson was floated into position and ballasted onto a prepared bed using sea-dredged sand. The connecting deck slab and the upper attenuation chambers were then conventionally cast to complete the breakwater (Fig. 10.21).

A number of problems were experienced during casting:

- poor compaction of the concrete, especially at the top of the slip-formed section
- fallout of concrete due to an excessive rate of rise of the slipform
- cracking to the walls due to wave action and early age thermal contraction.

Fig. 10.17 Preparing the casting bed for the base slab of the caisson

Fig. 10.18 Portion of the slipform being lowered into place for slipforming the walls

Fig. 10.19 Launching the caisson

Fig. 10.20 Completing the slipforming at the quayside

Fig. 10.21 Caissons and the deck slab in place

It was found that prior to casting the deck slab, the open-ended caissons were vulnerable to damage due to deflection of the walls during heavy seas, and the contractor modified the method of working to address this risk. Patch repairs were undertaken to the identified areas of poor compaction and fallout. Cracks over 0.2 mm in width were subject to resin injection. As indicated in Section 3.2, the durability modelling did not allow for construction defects. However, some mitigation was provided for these defects by three forms of cathodic protection, as shown in Table 10.9.

The cathodic protection measures complicated the construction process. In addition to dealing with the cabling and need to ensure electrical continuity of the reinforcement, the provision for cathodic protection in the upper tidal zone involved casting reservations or

Table 10.9 Summary of cathodic protection measures

Location	Cathodic protection measures
Spray (above 4.5 m)	Reservations/ducts for future installation
Tidal (−4 m to 4.5 m)	Ribbon anodes
Submerged (below −4 m)	Sacrificial anodes

Fig. 10.22 Cathodic protection system showing sensors, mixed metal oxide titanium ribbons and titanium conductor

ducts within the concrete. The intention being that these ducts would be used to install anodes at a later stage, should sensors indicate a risk of corrosion occurring. This strategy relies on being able to seal the ducts from seawater ingress until they are needed and then being able to detect corrosion sufficiently early to install the cathodic protection system. Figure 10.22 shows the cathodic protection system for the tidal zone installed prior to concreting.

10.3.4 Discussion

Developments in concrete technology have increased control over workability and the water/cement ratio and have expanded the range of cement types to enhance resistance to chlorides. The self-compacting concrete mixes used in the caissons have a low water/cement ratio, and include the use of 20% pulverised fuel ash and selective use of silica fume. These measures would be expected to increase durability and reduce the impact of workmanship.

Despite these measures, construction defects in the form of cracking and poor compaction occurred. These defects were not accounted for in the probabilistic durability modelling, and the repairs undertaken could not be regarded as restoring the durability to that of the parent concrete. Cathodic protection provides a means of mitigating the risk of premature deterioration associated with these features and other undisclosed defects. However, the ducts for the future installation of cathodic protection are a potential path for chloride ingress, and there is risk of promoting the reinforcement corrosion that the system was intended to protect against.

Durability testing has been undertaken to assess the characteristics of the concrete mix, but this does not necessarily represent the performance of the most important concrete, that is, the concrete in the actual works. The practical difficulty of slipforming large areas of exposed concrete should not be underestimated, especially for a marine facility with a long service life.

10.3.5   Conclusions

The use of concrete caissons to form a 1.1 km breakwater to a deep-water port presents significant challenges in order to achieve a 100 year service life.

A probabilistic durability design was proposed, based on a recognised approach. However, it was unrealistic in its assumptions, and, in particular, did not account for construction defects. Additional durability measures were required, including use of 20% fly ash, limiting the chloride migration coefficient and making provision for cathodic protection.

Difficulties encountered during construction, such as poor compaction and cracking, underlined the importance of making provisions for cathodic protection. However, this strategy presented its own challenges, including the need to provide corrosion sensors and sealed ducts within the concrete to allow for timely installation of anodes.

References

1. Duracrete (1999) *Probabilistic methods for durability design*. The European Union – Brite Euram III. CUR, Gouda.
2. Edvardson, C. and Mohr, L. (1999) Duracrete – a guideline for durability-based design of concrete structures. In: *FIP Symposium 1999, 13–15 October*, Prague.
3. Knights, J.C. (2005) Private communication.
4. Stanish, K.D., Hooton, R.D. and Thomas, M.D.A. (2000) *Prediction of chloride penetration in concrete*. Final Report to the Federal Highway Administration, DTFH-97-R-00022. Department of Civil Engineering, University of Toronto, 113–145.

10.4   Thaumasite sulfate attack in a UK highway structure: investigation and management strategy

Donald Wimpenny, Halcrow Pacific Pty, Australia

In February 1998, during bridge strengthening works, attack was discovered to the foundations of a 30-year-old overbridge to the M5 in Gloucestershire. The bridge has three slender columns at each pier, $750 \times 450\,mm$ in cross-section and $13\,m$ long, resting on a spread footing approximately $5.5\,m$ below ground level (Fig. 10.23).

The thaumasite form of sulfate attack (TSA) was suspected in view of the high quality of the concrete, and the Building Research Establishment confirmed the diagnosis. Prior to this discovery, the known incidence of TSA in the UK had been limited to a small number of cases in non-structural concrete exposed to cold wet conditions. However, the potential for such attack had been established in the laboratory, and the UK guide on sulfate attack, BRE Digest 363, published in 1996 briefly referred to the phenomenon.[1]

TSA is distinct from conventional sulfate attack in that the calcium silicate hydrates react with sulfates to form calcium sulfate carbonate silicate hydrate instead of the conventional reaction with calcium

Fig 10.23 Excavation of the slender bridge piers and spread footing

hydroxide and calcium aluminate hydrates to form gypsum and ettringite. This difference in the form of attack means that even concrete containing sulfate-resisting Portland cement and designed in accordance with BRE Digest 363 may be affected.

There are several risk factors in the occurrence of TSA:[2]

- exposure to high levels of sulfate
- very wet, cold conditions
- source of calcium silicate (e.g. Portland cement)
- source of carbonate (e.g. limestone aggregate).

In the case of this bridge, the columns were formed from in-situ reinforced concrete cast in shutters within a cutting through undisturbed Lower Lias Clay. Excavated Lower Lias Clay was stockpiled and later used to backfill around the columns and form a 2 m embankment either side of the carriageway. Oxidation of the pyrite in this reworked clay is thought to have led to high sulfate levels being developed in the ground. In addition, the excavation for the spread footing would have generated a water-logged pit next to the structure, promoting TSA. The concrete used in construction of the bridge included Portland cement and limestone coarse aggregate.

10.4.1 TSA management strategy

The management strategy had two stages, as indicated in Table 10.10. A condition survey was carried out in which softened areas were mapped and the depth of expansion and softening were measured to each face of the columns. Concrete cores and dust samples and fragments were subject to petrographic examination, scanning electron microscope (SEM) microprobe analysis, compressive strength testing, and chloride and sulfate determination.[3]

The attack was principally evident due to softening of the surface of the columns exposed during strengthening work. The distribution of the softening varied, with no attack within 1 m of ground level, partial attack in the form of patches or bands of softening up to 500 mm in size approximately 2 m below ground level, and attack across the full area of all the faces below 4 m depth. The attacked concrete was observed to have a soft, white pasty layer, typically 5 mm in thickness, but no visible cracking (Fig. 10.24). Below this layer, the concrete was found to have cracks running sub-parallel to the surface and the coarse aggregate particles were surrounded by white 'halos' of reaction products. The frequency and size of the halos reduced with depth

Table 10.10 TSA management strategy for the UK highway structure

Action	Purpose
Stage 1 – confirmation of the diagnosis	
Sampling and logging excavation	Confirm soil type and sulfate conditions
Condition survey (mapping area and depth of softening)	Determine the extent of deterioration
Cores for petrographic examination, SEM work and chemical analysis	Confirm TSA from the reaction products Determine sulfate and chloride profiles
Stage 2 – remedial work	
Propping of structure	Ensure safety from dead and live loading, e.g. HGV impact
Cutting and removal of existing piers	Remove deteriorated concrete and provide samples for future testing and examination
Recasting new piers	Replace vulnerable concrete with a concrete mix and additional protective measures meeting the new recommendations

below the exposed surface such that typically at 20 mm depth only occasional white specks of reaction products were present in pores without any physical disruption to the cement matrix; this is described as thaumasite formation rather than TSA (Fig. 10.25).

The reaction products, which were initially very soft, become hard and friable after being exposed for several days. Rust staining was present within some of the attacked areas, and breakouts revealed deep pitting corrosion to the reinforcement, representing 30% loss of section, and black and green reaction products, which turned brown on exposure, characteristic of anaerobic chloride-induced corrosion (Fig. 10.26).

Fig. 10.24 Appearance of concrete at depth increments of 5, 10 and 15 mm (left to right)

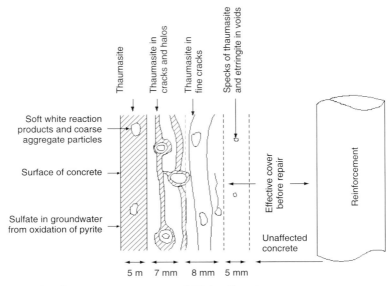

Fig. 10.25 Schematic representation of TSA-affected concrete

Petrographic examination, including point counting, indicated a concrete mix containing 300–400 kg/m³ of Portland cement and a water/cement ratio of 0.48–0.57. This would meet class 1 or 2 sulfate conditions according to BRE Digest 363.[1] The major rock type in the

Fig. 10.26 Deep pitting corrosion to reinforcing bar in TSA-affected concrete

394

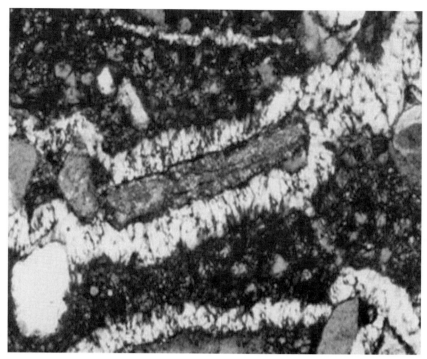

Fig. 10.27 Petrographic examination showing thaumasite around an aggregate particle. (Courtesy of Mike Eden)

coarse aggregate was found to be dolomitic limestone and fine aggregate comprising quartz/metaquartzite with 25% limestone fines. An in-situ cube strength of $60.5–83.0\,N/mm^2$ was estimated from the proportions. The birefringent appearance of reaction products in thin sections under cross-polarisation indicated relatively pure thaumasite around the aggregate particles with little intermixing with other sulfate reaction products such as ettringite (Fig. 10.27). The SEM microprobe analysis confirmed this finding, with the thaumasite being associated with SiO_2/Al_2O_3 ratios of greater than 9, compared with a ratio of less than 0.4 for ettringite.[4]

Sulfate and chloride determinations from 10 mm depth increments of a 100 mm diameter core are shown as profiles in Fig. 10.28. The depth of TSA given by the depth of cracks infilled with thaumasite from the petrographic examination is also indicated. It can be observed that the depth of TSA approximately corresponds to the point at which sulfate values exceed approximately 5% SO_3 by mass of cement,

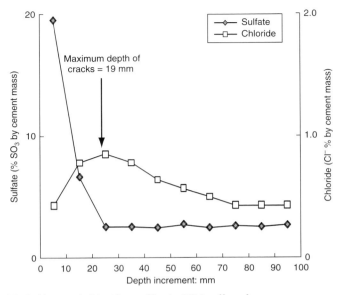

Fig. 10.28 Sulfate and chloride profiles in TSA-affected concrete

equivalent to 6% SO$_4$ by mass of cement. The chloride profile shows a peak of over 0.8% chloride ions by mass of cement 5–10 mm deeper than the depth of TSA. This, together with the evidence of corrosion to the reinforcement, suggests that chloride-induced corrosion may be an underlying problem in areas where TSA is present.

Comparison of the distribution of attack with the results of piezo-metric monitoring over an 18 month period indicate there is no TSA above the maximum ground water level, and full attack to concrete below the minimum water level. Partial attack at one pier coincides with the location of carrier drains through the backfill. The distribution of softening values to the columns indicates there is reduced softening furthest from the oncoming traffic and to faces furthest away from the carriageway. The above indicates that the pattern of attack is closely associated with the presence of water, either as natural groundwater or derived from other sources such as run-off and traffic spray from the carriageway.

The typical pattern of softening and expansion to the columns is indicated in Fig. 10.29. Areas of expansion and softening are closely associated.

The bridge deck was propped as a precaution. The maximum net loss of cross-section for the piers was found to be 33 mm, which is less than

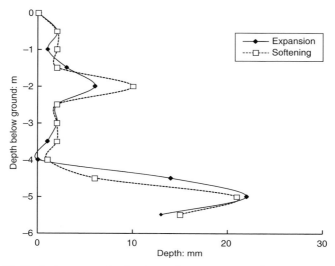

Fig. 10.29 Typical pattern of softening and expansion to columns

the value associated with a risk of buckling failure to the columns. However, in view of the possibility of further deterioration during the remaining service life, it was decided to replace the affected lower portion of the columns and recast a new spread footing above the old base.[5] The cut samples of concrete were removed site for further examination and testing.[6]

10.4.2 Conclusion

Deterioration to piers of a 30-year-old bridge prompted a programme of investigation and repair work. Petrography in conjunction with SEM microprobe analysis confirmed TSA. The pattern of deterioration was expansion and softening of the concrete at the surface, cracking sub-parallel to the surface and white 'halos' of reaction products around the aggregate particles at depth. Attack was thought to have been promoted by the reworking of Lower Lias Clay, leading to oxidation of pyrite. The extent of the attack is closely related to the availability of water.

High chloride levels in the concrete, and staining and pitting to the reinforcement, suggest chloride-induced corrosion may be enhanced by TSA. Uncertainty over long-term deterioration led to removal and replacement of the piers.

References

1. BRE (1996) *Sulfate and acid resistance of concrete in the ground.* BRE Digest 363. Building Research Establishment, Garston, Watford.
2. Thaumasite Expert Group (1999) *The thaumasite form of sulfate attack: risks, diagnosis, remedial works and guidance on new construction.* DETR, London.
3. Floyd, M. and Wimpenny, D.E. (2002) Procedures for assessing thaumasite sulfate attack and adjacent ground conditions at buried concrete structures. *Cement and Concrete Composites*, 25, 1077–1088.
4. Wimpenny, D.E. and Slater, D. (2002) Evidence from the Highways Agency thaumasite investigation in Gloucestershire to support or contradict postulated mechanisms of thaumasite formation (TF) and thaumasite sulfate attack (TSA). *Cement and Concrete Composites*, 25, 879–888.
5. Wallace, J. (1999) Strengthening thaumasite-affected concrete bridges. *Concrete*, 33(8), 28–29.
6. Slater, D., Knights, J.C. and Wimpenny, D.E. (2002) Sulfate and chloride profiles and the visual characteristics with depth from the face in buried concrete subject to TSA. In: *First International Conference on Thaumasite in Cementitious Materials.* Building Research Establishment, Garston, Watford.

10.5 Installation of corrosion protection on the M4 elevated freeway – London

Christopher Atkins, Mott MacDonald, UK

A radical rethink of how to install corrosion protection in reinforced concrete suffering from chloride attack has saved significant time, cost and energy, and offered major health and safety improvements. Cathodic protection has been successfully installed on crosshead beams supporting the M4 elevated section in west London using a novel and highly innovative approach.

10.5.1 Reinforcement corrosion

The M4 is a major elevated freeway into the west of London (Fig. 10.30). It carries 87 000 vehicles per day into and out of the capital of England. It was constructed directly over an existing major highway that carries an additional 51 200 vehicles per day. Both run through a residential area with private dwellings directly adjacent to the twin level route.

The elevated freeway is 3.5 miles long, and is made up of prestressed beams supported on 120 bents using two half joints at each bent. It was constructed in 1967, and has long showed its age. Every winter, huge quantities of de-icing salts are liberally applied to prevent ice forming

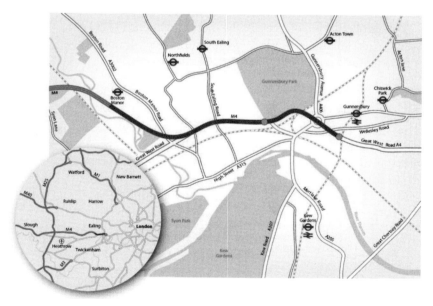

Fig 10.30 The M4 elevated freeway in west London

Fig. 10.31 Twin half joints contribute to the problems

on the freeway. These have penetrated the leaking joints and passed into the concrete.

Levels of chloride of over 5% by mass of cement (17.5 kg/m^3 or 30 lb/yd^3) are regularly recorded, with one value double this. The reinforcement is corroding, and concrete has spalled.

10.5.2   Repair strategy

The shape of the structure means the majority of work has to take place within lane closures. For most of the structure, this can only take place at night when the traffic volumes are lowest. The residential area means that you cannot make any significant noise at night to carry out repairs. The acceptable level of noise is roughly equivalent to loud speech.

The structure owner recognised this was a problem, and started a series of investigations. The first priority was to confirm there was adequate structural capacity and that the public were not at risk from an imminent collapse.

Once this had been demonstrated, the owner went about the task of removing all loose concrete to prevent this falling onto vehicles (Fig. 10.32). Then came the task of repairing the structure, but with

Fig. 10.32 For safety, all loose and delaminated material was removed

one major problem: the area of concrete most likely to be chloride contaminated, and containing the most amount of reinforcement, was the area that could not be accessed due to the half-joints.

A series of trials were undertaken from 1995 onwards where every available repair technique was trialled. This even included replacement of bents. Every currently available cathodic protection system was also trialled. While cathodic protection successfully prevented corrosion without needing to replace the chloride-contaminated concrete, it could only be applied to the accessible areas of the structure (Fig. 10.33).

A large proportion of the pier is accessible, and so could be protected using a standard mesh-and-overlay cathodic protection system. The half-joints needed to be protected, but could not be accessed without shutting the freeway. The only option that was worthy of further pursuit was to drill up and install anodes from the soffit.

To install anodes from the soffit required a total length of drilling of approximately 600 feet. The estimated duration was over 30 days. If reinforcement were encountered, each hole would need to be repositioned and drilled again. Where the crossbeam ran over the pier, holes would need to be drilled in at an angle to position the anodes in the correct place. The holes would then need to have an anode inserted, and be grouted from underneath. Each anode would then have to be individually wired to a primary conductor.

Fig. 10.33 Mesh and overlay cathodic protection can be installed during the day

The works would be difficult to plan and programme (Fig. 10.34). The heavy and persistent traffic during the day meant the drilling would have to be carried out at night, but the strict environmental noise limits in place for the sake of the local residents meant it could not. In addition, there would be the ever-present risk of having to re-drill holes. This would inevitably result in the overall cost of the work being uncertain and highly variable.

In 2006, a repair contract was embarked upon. It was based around installation from the soffit, and it was planned to fix a single bent. A specialist repair contractor was appointed, and after discussions with the designer, a possible alternative approach was put forward that had the potential to revolutionise the repair program.

On reviewing the reinforcement details, it was noted that there was an area running along the crossbeam that contained no reinforcement. This was close to the main inaccessible reinforcement. If it were possible to install anodes in this location, the risk of hitting steel was dramatically reduced, so the works became more manageable.

The lengths of holes required reduced from 600 feet to 60 feet, so it became shorter duration. The shorter duration meant that the works became significantly cheaper. The neighbourhood would be less disrupted. Site safety would improve as there would be fewer people

Fig. 10.34 Work had to be carried out safely without disruption to the traffic flow

working for shorter periods. The only problem was that no one had ever attempted this before.

In order to see if it could be done, the specialist subcontractor obtained 60 feet of temporary concrete barriers and a coring rig. The coring started at one end and remained on course, coming out within 2 inches of where it was predicted. The core drifted down under gravity, but as all the crossbeams to be protected were on a camber, this was not considered a problem. Coring would be started at the highest end, and the natural drift would help keep it on course.

Due to the success of the coring trial, it was decided to use this approach on site (Fig. 10.35). The coring was successfully carried out over three night shifts. A detailed design was produced that split the upper section of the crosshead into two zones. Anodes were fixed to a titanium conductor bar at different locations to vary the current distribution in accordance with the reinforcement.

Reference electrodes were also fixed to the conductor bar to allow for monitoring the system. This was then pulled through the core hole and grouted into place. The manufacturer's recommended grout did not flow sufficiently to enable its use, so a post-tensioned duct grout was employed instead.

Fig. 10.35 A cantilevered gantry allows coring to proceed over the live carriageway

It was still necessary to repair the cantilevered crossbeams, but this was still limited by the noise and traffic problems. The contractor came up with an innovative approach adapting an airport baggage handling truck to provide a mobile soundproofed enclosure that could be raised into position and moved around during night work (Fig. 10.36). This meant that hydro-demolition and shotcrete reinstatement could be carried out at night without disturbing the local residents.

The installation of the first pier was completed in March 2008. The estimated cost savings, assuming drilling from the soffit was possible, were 20%, reducing the cost of repair to approximately $600,000 per crosshead. The estimated cost of crossbeam replacement was $10 million.

Due to the savings involved, another eight piers were able to be repaired within the next 12 months. At present prices, the cost of fixing the entire viaduct reduces from $1.2 billion to $72 million by employing this technique.

The repair system has proved simpler and more efficient to install than anything anyone had developed previously. The approach saves time and money, improves site safety and minimises the duration of the road works and the noise generated during repair. Both environmental and sustainability issues have been addressed in a positive and quantifiable manner.

Fig. 10.36 A modified airport vehicle becomes a mobile soundproof enclosure

Last but not least, the project improved the appearance and feel of the neighbourhood with minimum intervention to the benefit of local residents and London visitors alike.

A truly innovative and sustainable solution to what otherwise was considered an intractable problem (Fig. 10.37).

Fig. 10.37 The first four completed installations

10.5.3 Conclusion

The innovative technique of installing cathodic protection meant that the length of holes drilled was only 60 feet per beam compared with a combined length of 600 feet for a conventional approach.

From an environmental perspective, replacing a pier and crosshead beam would use energy and produce a corresponding volume of CO_2 equivalent to that locked up by 300 acres of forest in a year. Installing cathodic protection using a conventional drilling pattern would have had almost the same carbon footprint. The single hole-through-the-middle approach used 60% less energy. The total time taken to bore each hole was 3 days, compared with 30 days or more for the conventional approach. Also, it dramatically reduced exposure to vibration for the operatives and limited working at height, and there were no day-time road closures.

10.6 Systematic investigation and application of EN 1504 to a concrete repair project

John Broomfield, Broomfield Consultants, UK

10.6.1 Grade II* and Grade II listed structures

The University of East Anglia was founded in the 1960s, when the main campus buildings, including the 'Teaching Wall' and the well-known 'Ziggurat' residential blocks (featured on the English Heritage website) were laid out by Sir Denys Lasdun (Fig. 10.38). The university is proud if its architecture, which has been supplemented by other famous architects.

The Teaching Wall consists of a shallow 'W' of reinforced-concrete buildings approximately 500 m long. The runs of offices, laboratories and lecture rooms are interrupted by lift and stair 'towers' at intervals along its length with water tanks and plant rooms above the main building roof level. The exposed concrete façades are a feature of the Teaching Wall and various parts of the campus, which were given grade II* and grade II listing during the process of the works. Various sections of the Teaching Wall and other campus buildings are linked with elevated walkways.

10.6.2 The condition and situation of the structures

Broomfield Consultants were appointed as corrosion specialist consultants to Jacobs Babtie Consultants to conduct 'forensic structural

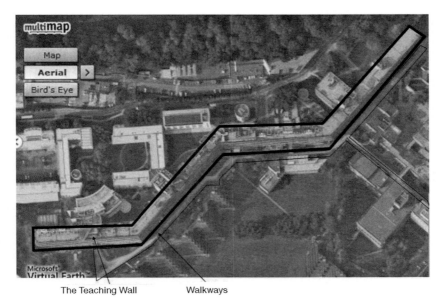

The Teaching Wall   Walkways

Fig. 10.38 The Teaching Wall and adjacent walkways

engineering', initially to the 'Biotower' (Phase 1) and then to all of the reinforced-concrete structures with exposed concrete façades on the campus. Work was conducted in close collaboration with the university departments affected, as well as the estates department, which ran the project, English Heritage and the Norwich City Planning Office, who gave the planning consents for the work.

Phase I work was on the Biotower, a lift and stair tower with an air-conditioning plant room and a water tower above. Detailed investigation showed low cover and carbonation to be prevalent, with some admixed calcium chloride in some 'lifts' of concrete, all leading to reinforcement corrosion. A number of options for repair were investigated, including the possibility of cladding the façade and 'air conditioning' it to remove moisture and stop reinforcement corrosion according to EN 1504, Part 9, Principle 8. However, this was untried technology, and it was considered that no contractor would offer any warranties on such an installation. For that reason, impressed current cathodic protection was applied according to EN 1504-9, Principle 10 (cathodic protection). The specification was according to BS EN 12696:2000, *Cathodic Protection of Steel in Concrete*.

The phase 2 works were on the library walkway, shown in Fig. 10.39, a concrete stairway to another walkway showing severe corrosion damage and two further stair/lift towers in the Teaching Wall. A detailed quantitative condition survey revealed areas of concrete damage due to corrosion from carbonation. This was principally due to low cover, indifferent-quality concrete and the age of the structure. Other areas were deteriorating due to de-icing salt ingress, particularly on the elevated walkways and access stairways. Using the survey data, calculations were made of ongoing chloride and carbonation ingress on a 30 year life projection.[1]

Corrosion modelling was done by using Fick's law of diffusion calculations on cover depth measurements combined with carbonation depths and chloride depth profiles.[2] This indicated that apart from the areas showing immediate damage, few other areas were susceptible to future corrosion.

10.6.3 Applying the principles of EN 1504 to the rehabilitation process

Under EN 1504, Part 9, Section 5.2, the following options are given:

(*a*) do nothing for a certain time
(*b*) reanalysis of the structural capacity

Fig. 10.39 Part of the library walkway showing the Teaching Wall and 'towers' behind. De-icing salts and leachate run-down can be seen where the waterproofing and drainage has failed, allowing corrosion of the slim pier supports

(c) prevention or reduction of further deterioration without improvement of the concrete structure
(d) improvement, strengthening or refurbishment
(e) reconstruction of all or part of the structure
(f) demolition.

Given that the structures are part of a listed site, that further deterioration could lead to health-and-safety problems in some areas, and that the university has set aside a budget for its 'concrete preservation plan' options (c) and (d) were relevant.

The standard options for intervention on a reinforced concrete structure suffering from reinforcement corrosion are:

(a) do nothing for a certain time
(b) complete or partial demolition and rebuild (EN 1504-9, Principle 3.4)
(c) patch repair of local damaged areas (EN 1504-9, Principles 3.1, 3.2 and 3.3)
(d) ingress control via coatings, membranes, sealers, water stops, enclosures or other barriers (EN 1504-9, Principles 1, 2 and 8)

(*e*) impressed current cathodic protection (EN 1504-9, Principle 10 (BS EN 12696))

(*f*) galvanic cathodic protection (EN 1504-9, Principle 10)

(*g*) electrochemical realkalisation (EN 1504-9, Principle 7.3 (CEN/TS 14038-1))

(*h*) electrochemical chloride extraction (EN 1504-9, Principle 7.5 (CEN/TS 14038-2 – in draft))

(*i*) corrosion inhibitors (EN 1504-9 Principle 11.3).

Being part of a listed building and suffering from corrosion damage, options (*a*) and (*b*) were not feasible. Option (*c*) was required in some areas. Option (*d*) was used, but in some areas its use was constrained by the requirement to retain the board-marked finish to the concrete on the listed façades. However, control of ingress of CO_2 and chloride ions was required.

To this end, a proprietary architectural coating was trialled for approval by the university and by the local authority conservation officer. This coating 'tones down' changes in concrete colour and finish, and was considered ideal for minimising the visual impact of patch repairs on the board-marked finish on the concrete façades. The selected coating has anticarbonation properties and is also compatible with a silane for control of moisture and chloride ingress.

In this phase of the works, impressed current cathodic protection, option (*e*) in the list above, was not required on a large enough area to be cost-effective. However, given the presence of active chloride-induced corrosion, an alternative was to use galvanic anodes installed in the patch repairs to minimise incipient anodes (Fig. 10.40). The other electrochemical treatment techniques, options (*g*) and (*h*), were not considered suitable for this project.

10.6.4 Design and specification of the work

Techniques selected therefore included localised galvanic cathodic protection to minimise the incipient anode effect around patches in areas of high chloride (EN 1504-9, Principle 10). An example of such an effect is shown in Fig. 10.40. Penetrating sealers were required to keep out further chloride according to Principles 1.1 and 6.1, and to reduce moisture (Principle 8). Anticarbonation coatings were required to reduce the rate of carbonation (Principles 1.2 and 6.1), and renewal of the waterproofing membrane on the walkway decks was specified to keep moisture and chlorides out of the deck concrete (Principle 1.7). The membrane and improvements of drainage provided reduction in

Fig. 10.40 Incipient anode formation around an old repair on the Biotower plant room prior to phase I repair and impressed current cathodic protection

water leakage, sheltering the walkway substructure from de-icing salt run-down. These techniques were used along with conventional patch repair where required (Principles 3.1, 3.2 and 3.3).

Detailed analysis of the condition survey results allowed the determination of treatments to different elements of the structures, as show in Table 10.11.

The following specifications were written for the job:

(1) A concrete repair specification based on:
   (a) materials according to EN 1504, Part 9 and Part 3 (Class R4 structural grade repair mortar)
   (b) patch repair preparation according to EN 1504, Part 10, Section 7 and Appendix A7
   (c) material application according to EN 1504, Part 10, Section 8, on application, and Appendix A8
   (d) testing on site and of site samples using test methods and values in EN 1504, Part 10 and Appendices A7, A8 and A9.
(2) A coating specification for silane impregnation, based on:
   (a) materials according to EN 1504, Part 9 and Part 2 (1.3C for anticarbonation coating and 1.1(H) and 1.2(l) for silane impregnation for moisture/chloride ingress control)
   (b) the manufacturer's literature for application
   (c) surface preparation according to EN 1504, Part 10, Sections 7 and 8 and Appendix A8

411

Table 10.11 *Treatments used in the work*

EN 1504-9 Principle	Method/principle	EN standard	Elements treated	Materials used
1 – Protection against ingress and 8 – Increasing concrete resistivity	Hydrophobic impregnation: Principle 1.1 and 2.1	EN 1504, Part 2 EN 1062-3 Maximum value: $w = 0.035 \text{ kg/m}^2 \text{ h}^{0.5}$	Walkways below deck level where de-icing salts were applied and the chloride level at rebar is below the threshold for corrosion	Silane compatible with cosmetic coating used to 'tone down' repairs
1 – Protection against ingress	Anticarbonation coating: Principle 1.3c	EN 1504, Part 2 EN 1062-6 Permeability to CO_2: $S_D > 50 \text{ m}$	Parapets on walkways above de-icing salts where chloride levels are very low	Cosmetic coating with anticarbonation properties
1 – Protection against ingress	Waterproofing membrane: Principle 1.7	Not listed in EN 1504	Walkway decks	Waterproofing system
3 – Concrete restoration 7 – Preserving or restoring passivity	Hand-applied mortar: Principle 3.1	EN 1504, Part 3, Class R4 Compressive strength: >40 MPa Adhesive bond: >2 MPa	All damaged elements	Pre-bagged patch repair material
10 – Cathodic protection	Local galvanic anodes: Principle 10	Galvanic anodes not covered yet	Patch repairs with chloride levels in excess of the corrosion threshold	Zinc anodes encapsulated in a proprietary activating mortar

(d) Site testing according to EN 1504, Part 10 and Appendices A8 and A9

(3) An application specification for a waterproofing membrane:
   (a) lifting paving slabs
   (b) conducting repairs
   (c) repairing an improving drainage
   (d) applying waterproofing system
   (e) replacing paving slabs.

10.6.5 Site tests

After applying coatings, cores were taken and sent for testing. Carbon dioxide permeability tests (BS EN 1062-6:2002) gave far better than the 50 m minimum values recommended in the specifications, but they were starting at 23 and 30 m before treatment.

The water permeability test results were:

- Coated   0.03 and 0.04 $kg/m^2 h^{0.5}$
- Partial coated   0.05 $kg/m^2 h^{0.5}$
- Uncoated   0.11 and 0.12 $kg/m^2 h^{0.5}$

BS EN 1062-3:1998, Table 1, states:

I.   High $>0.5\,kg/m^2\,h^{0.5}$
II. Medium   0.1 to 0.5 $kg/m^2\,h^{0.5}$
III. Low $<0.1\,kg/m^2\,h^{0.5}$

Given the requirements for a coating with architectural properties, and the fact that most areas of low cover were repaired, the coated values falling in the II medium range was judged to be an acceptable performance. Also, renewal of the waterproofing and the drainage would reduce the amount of water run-down on the substructure, reducing further the rate of chloride ingress.

Pull-off tests on concrete patch repairs can be conducted according to EN 1542, ISO 4624 and BS 1881, Parts 201 and 207, as described in EN 1504-10 (A9.2 test or observation No. 35). Recommended values are given in EN 1504-10, Table A2. In this project, pull-off tests achieved 0.8 MPa or better.

It should be noted that the specifications were written and the work completed prior to full publication of all parts of EN 1504 and the associated test methods. Not all testing on this project was compliant with the specific CEN test mentioned but used equivalent British standards or other tests in use at the time.

10.6.6 Conclusions

Work on this phase of the project was successfully completed in 2007. There was minimal disruption to campus activities, and both the university and the listing officer were pleased with the final finishes on the listed elevations.

In conclusion, it can be seen that concrete repair systems can be designed, performance specified and applied using the EN 1504 set of documents along with their associated test methods following the principles of corrosion engineering to ensure corrosion prevention before it initiates and corrosion control once damage has initiated.

The first critical part of any repair and refurbishment project is a condition survey which quantifies the type and extent of damage to ensure that:

- only areas in need of treatment are treated
- appropriate treatments are selected
- the current and future requirements of the structure are fully considered in the repair design process.

Appropriate repair systems and materials can then be selected based on the principles of EN 1504, and repair designs and specifications prepared using the product characteristics specified in EN 1504 and the associated test methods. Life cycle costing can also be applied to ensure that the systems used are the most cost-effective ones that meet the client's requirements.

Acknowledgements

The author would like to acknowledge Martin Lovatt, project manager for The University of East Anglia Estates Department, and Andrew Brown, the engineer for Jacobs, for their contributions and for permission to publish this chapter.

References

1. Broomfield, J.P. (2006) A Web based tool for selecting repair options and life cycle costing of corrosion damaged reinforced concrete structures. In: M.G. Grantham, R. Jaubertie and C. Lanos (eds), *Proceedings of the 2nd International Conference on Concrete Solutions, St Malo*. Building Research Establishment, Garston, Watford.
2. Broomfield, J.P. (2007) *Corrosion of steel in concrete – understanding, investigation and repair*. Taylor and Francis, London.

10.7 Improved abrasion resistance of factory floors
R. C. Lewis, Elkem Materials, UK

The textile industry in southern India has been established for many years. During that time, significant damage has been done to the floors in the factories and warehouses. Standard concrete (20 MPa) has been used previously, in conjunction with a surface hardener, to provide the flooring material. The factories are now utilising the high performance of silica fume concrete to gain greater durability – without the use of surface hardeners. The design of the concrete is such that it is very economic to use in terms of increased working lifetime, while still being produced and placed by normal labour-intensive techniques.

10.7.1 Historical perspective
The traditional method used for these industrial large floors – using hand-batched, site-mixed concrete – has been to use the normal 20 MPa mix for ease of production and placement. Whilst still 'green', this concrete is treated with a topping – usually a cement-bound metallic powder – which is then trowelled into the surface.

Any damage to the surface layer will immediately allow attack upon the relatively weak concrete underneath. Deterioration and breakdown can happen very quickly once the surface coating has been broken. Repair and reinstatement of these floor areas can cost not only a lot of money directly, but the downtime and lost revenue for the production unit can also be significant.

New structures are being looked at on a life cycle analysis basis – best durability for the longest working time. The effect of using silica fume concrete has been such that most new units are being built using this material (Figs 10.41 and 10.42).

10.7.2 Improved performance through selection of concrete mix constituents
The benefits of the use of silica fume in concrete have been known for many years in India. This is through use in the nuclear and hydropower industries and its use in road and bridge infrastructures. However, these mixes have been at the top end of the technology scale – requiring very high strength or impermeability characteristics. As such, the design of these concretes has meant a more expensive concrete, even though more economic in the long term.

Fig. 10.41 A new factory unit under construction

The textile industry needed a mix that would give them performance, be economic and be suitable for placement by manual labour (Figs 10.43 and 10.44). A review of the mix normally used, the method of production, placement and curing presented the idea of a simple addition of

Fig. 10.42 The large floor area within the factory unit

416

Fig. 10.43 Hand batching into a tumble mixer on site

about 7% silica fume to the mix. There was no increase in the cement content or other major modifications, with the exception of maintaining the water/binder ratio at 0.40. A few trials found that this mix actually increased the strength from 25 to 30 MPa, the final strength of the

Fig. 10.44 Placing by hand (some larger sites use truckmixers and pumps)

417

Fig. 10.45 The long, single pour, bay after screeding with a razorback

normal mix, up to around 40 MPa. This was acceptable, as it meant no modifications to the bay size or joint cutting.

The fact that the silica fume concrete has no bleed water means that finishing routines can be started within a few hours (Fig. 10.45). This improves the working progress, speeding up the construction rate.

A distinct improvement was noted by the teams doing the power-floating/powertrowelling, as this produced a much tighter surface on the concrete, again making it easier to work (Fig. 10.46). Standard

Fig. 10.46 Powerfloating the surface for a solid, sealed finish

Fig. 10.47 Wet hessian curing. Ponding can also been seen in the top right of the picture

curing – ponding – is frequently used. Wet hessian (sacking) is also used to great effect (Fig. 10.47). Due to the 'sealed' surface of the silica fume concrete, these methods are the optimum for gaining maximum curing effect and thus the best surface durability.

10.7.3 Long-term performance

Cotton-spinning machines are serviced by workers using hand trolleys to collect full bobbins and replace them with fresh ones. These carts weigh up to 150 kg, when full, and run on four small plastic wheels. Each wheel has a central runner about 15 mm across. Thus, the impact area of the 150 kg is focused on four points just 15 mm wide. This is a bad enough abrasion surface when the wheels can actually roll, but once the bearings are clogged with dust and cotton fibres and the wheel drags, the damage can become significant. With the original concrete floors, damage could be seen within a couple of months, as two 'tramlines' gouging into the concrete around the cotton-spinning machines.

Other damage to the floors is caused by the movement of pallet trucks and forklifts, the weight on the wheels of these plus the lifting and dropping of pallets and other equipment and materials. A misjudged

419

Fig. 10.48 Silica fume concrete refurbishment to worn entranceway

pick up with a forklift truck can mean the forks chopping into the concrete surface. Such damage would previously accelerate the breakdown of the concrete below the 'protective' surface hardener.

The new silica fume concrete has been in use for more than a year in a number of factories, and has shown considerably better wear performance (Fig. 10.48). In most cases, the surface of the concrete appears to be even more polished than when first placed. The movement of the trolleys can still be seen, but now the case is that the wheels are cleaning the dust from the surface, hence leaving two highly polished lines. Close inspection of these tracks shows that there has been no damage to the surface (Fig. 10.49).

In all the completed projects the owners are very pleased with the performance of the silica fume concrete. Although designed on the 'low-tech' end of the scale, the durability of this silica fume concrete means that there will be a significant increase in the lifetime of the factory and warehouse floors and thus will give a very high economic value to the owners.

10.7.4 Conclusions
By the end of 2006, some 1 million ft^2 of flooring had been placed using this design of silica fume concrete. Twice that area is already in production. Table 10.12 shows just a selection of the work already completed.

Fig. 10.49 The completed floor in the spinning unit. The polished 'tram tracks' can be seen just in front of the machine. This picture was taken 6 months after production started, and shows no surface wear

Table 10.12 List of projects where silica fume concrete was used for the industrial floors

Factory	Location	Area: ft^2
GHCL Meenakshi Mills	Madurai	125 000
Gangotri Mill	Perundurai	275 000
Jagannata Textile Co.	Coimbatore	100 000
Divyalakshmi Textile Mills	Arupukottai	150 000
Jayajothi Textile Mills	Rajapalayam	85 000
Other projects		>200 000

Acknowledgements

Grateful thanks are due to GHCL, the local management of Meenakshi Mills, Madurai, and Jagannata Textile Co., Coimbatore, for allowing Elkem Materials to visit their sites and use the information to produce this chapter.

10.8   High-strength concrete for the JJ Hospital Flyover, Mumbai, India

R. C. Lewis, *Elkem Materials, UK*

The JJ Hospital Flyover was the first project to combine the strength and durability characteristics in a precast operation where the specified strength is 75 MPa – in a region that used to consider 40 MPa as the highest possible strength obtainable.

10.8.1   Concrete mix design requirements

Due to the location of the flyover, major consideration was given to the aesthetics of the construction. This is a very dense built-up area, and it would not help the traffic situation if the supports of the flyover blocked the street below. It was decided to make the structure as slender and pleasing to the eye as possible, giving maximum room and 'air space' beneath the elevated road, as well as maximising the light to the buildings on either side.

The 'old style' of flyover structure, with three or four large, metre-wide columns every 10–15 m, supporting a thick slab of concrete, would not work in this situation. The idea was to take the design beyond previous building patterns.

Silica fume concrete usage in India was known to the contractor, Gammon India Ltd, through previous projects they had already completed. After discussions on design with the consultant, Dar Consultant Project Management and the approval of Maharashtra State Road Development Corporation, it was decided to utilise the strength potential of silica fume concrete and to set a new precedent for this type of project.

Specifying the concrete at 75 MPa meant that the design could be radically altered to have singular, slender columns and thinner decks. The strength of the structure also meant that the spacing between the columns could be increased, to over 30 m.

10.8.2   Selection and proportioning of mix constituents

Work in the nuclear power sector in India had shown that it was possible to achieve these levels of strength with silica fume concrete. However, the scale of this project was 10 times that of the work previously done. It was necessary to ensure that the silica fume concrete would give the required performance when produced on such a massive precast operation.

Not only was there concern over the durability performance – as well as strength, the values of drying shrinkage, rapid chloride permeability

Table 10.13 Silica fume concrete specifications

Characteristic strength: MPa	75
Target strength: MPa	$83.2(75 + (1.64 \times 5))$
Water/cementitious ratio	0.269
Mix quantities	
Cement: kg	500
Silica fume: kg	50
Fine aggregate: kg	682
Coarse aggregate, 10 mm: kg	384
Coarse aggregate, 20 mm: kg	762
Water: litres	148
Admixture: litres	8.25
Laboratory cube results: MPa	
1 day	30
3 day	46
7 day	71
28 day	86

and the initial surface absorption were to be measured – there was also the plastic properties of the concrete to be taken into consideration. A good surface finish was also required, to keep with the aesthetics of the structure.

With a non-bleed silica fume concrete, precasting in a yard where the ambient temperature may exceed 45°C, curing had to be of paramount importance – as well as workability and slump retention. It was determined that the placing temperature should not exceed 28°C, hence ice-cold water was used for mixing the concrete, and wet hessian curing would begin as soon as the precast elements were stripped and stacked. Concrete for the precast yard had to have a slump value of 80 mm at 90 minutes. For the 'on-site' casting it was necessary to allow for 120 mm at 150 minutes, as the batching plant was some 12 km from the site.

Despite all these considerations, and after extensive trial work, optimised mix proportions were determined (Table 10.13).

The results achieved indicated that a reduction could have been made in the cement content of the mix, but other restraints meant that this was not implemented.

10.8.3 Properties of the finished concrete

Testing of the concrete was as extensive as the trial work. An on-site laboratory was set-up to keep a close track of the concrete quality. The average results are shown in Table 10.14.

Table 10.14 Properties of the finished concrete

Compressive strength: MPa	
1 day	30
3 day	46
7 day	71
28 day	80
90 day	87
365 day	94.5
Drying shrinkage: % (IS 1199)	0.00142
Rapid chloride permeability: coulombs (ASTM C-1202.97)	
Cores from side segments	100
Laboratory samples	240
Initial surface absorption test: ml/m^2/s (BS 1881:1970)	
Precast elements	0.0043
In-situ concrete	0.0161
Compressive strength: MPa	
Cores from spine segments	88.5
Standard deviation	2.1

10.8.4 Conclusions

It can be clearly seen from these results that the concrete has achieved, and easily exceeded, the strength requirements for the construction. It is also clear that the very low permeability of the concrete will result in greatly improved long-term durability. The chloride permeability values alone are the equivalent of latex-modified or polymer-impregnated concrete, and are among the lowest ever seen for a silica fume concrete. The working life-time of the flyover should easily surpass original design expectations.

The use of the silica fume concrete has enabled the building of the JJ Hospital Flyover, using radical design improvements, resulting in a structure that is not only functional but also pleasing to the eye, sympathetic to the environment and an elegant monument to all those involved in its construction.

10.9   High-performance concrete for Tsing Ma Bridge, Hong Kong

R. C. Lewis, *Elkem Materials, UK*

The Tsing Ma Bridge (Fig. 10.50) is the centrepiece of the Lantau fixed crossing in Hong Kong. This is the link across the sea strait from the mainland at Tsing Yi to the island of Lantau, and it carries both the road and rail traffic out to the island – and, more importantly, the international airport of Chek Lap Kok.

10.9.1   Durability requirements
Durability was a major design concern. The bridge would be subject to seawater and high winds and rain, and potentially some seismic activity.

The scale of the bridge – towers over 200 m tall and a main span of over 1400 m – meant that the authorities did not want to face any major maintenance costs for a long time. Hence, the design life of the finished bridge was specified at 120 years.

10.9.2   Selection of constituents and mix proportions
Construction of the towers was to be by slipform; however, the mix design required a high percentage of ground granulated blast furnace

Fig. 10.50 The Tsing Ma Bridge

slag – some 65%. Following much trial work, it was seen that this blend would not gain strength fast enough for placing – even though it met with the temperature criteria for the concrete use. Over 100 mix designs were tested before a suitable one was found. This was a ternary blend containing microsilica in addition to ground granulated blast furnace slag. A small percentage of microsilica (silica fume) gave the early age strengths that were required and also increased the durability character-istics of the concrete. Other concretes for the bridge were also designed to use 5% microsilica in conjunction with 25% fly ash.

With the very tight restrictions on the water/cement ratio, it was noted that adding the microsilica in powder form may mean that the mixing time would be increased and extra superplasticiser required to maintain the slump. It was decided to add the microsilica in a slurry form – an 'instant' mix with the chilled water – so that dispersion of the material would be faster.

As such, the microsilica was delivered to the site in 1 tonne bags which were individually broken into the 'slurry mixer'. Water was added at an equal weight, to achieve a 50:50 blend, and the mixed suspension was pumped into a holding tank before use. Slurry was made up on a day-to-day basis by need, so that excess material did not remain in the tank and settle out.

10.9.3   Conclusion

With a mix design of 390 kg cementitious (30% ordinary Portland cement, 65% ground granulated blast furnace slag and 5% microsilica) the casting of the towers proceeded without problems – only stopping twice for typhoon-strength winds. The long-term results for the concrete returned strengths of over 80 MPa (C65) and chloride diffu-sion coefficients of better than 10^{-14}. The project started in 1992, and was completed in time for the return of Hong Kong to China, in July 1997.

426

10.10 Service life requirement of 100 years: the East-Sea Bridge, Shanghai

R. C. Lewis, Elkem Materials, UK

In the last 20 years, the Shanghai shipping industry has grown with the economy of the surrounding areas. However, the shallow water of the Huangpu River stops the larger vessels from reaching the traditional ports. The construction of a leading international shipping centre, Shanghai Deep Water Harbor, started in 2002. This consists of over 100 container terminals on outlying islands, the East-Sea Bridge and the New Harbor City on the mainland. The East-Sea Bridge is 32.5 km in length, starting at Little Yangshan Island in the Qiqu Archipelago (Zhejiang Province), crossing the north of Hangzhou Bay, to New Harbor City in Luchao, Nanhui District, Shanghai (Fig. 10.51).

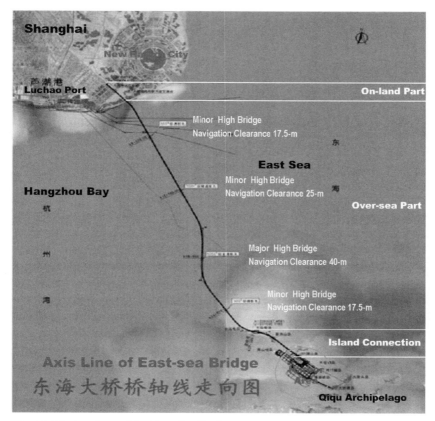

Fig. 10.51 Axis line of the East-Sea Bridge

Fig. 10.52 High bridge

The East-Sea Bridge connects the islands and the mainland for container transportation, so the durability of the bridge is critical. For the first time in China, a service life of 100 years was specified. However, no code or guideline for such durability was available. Extensive studies were carried out at the Shanghai Research Institute of Building Science (SRIBS) to establish scientific, practical and reliable strategies to maximise the life of the bridge.

10.10.1 Profile of the East-Sea Bridge
The East-Sea Bridge has a total length of 32.5 km in three sections: the marine bridge, the island connection and the land-based bridge.

The marine bridge contains one major high bridge, three minor high bridges (Fig. 10.52) and five sections of low bridge (Fig. 10.53):

- the major high bridge is a cable-stayed bridge with a main span of 420 m and a vertical navigation clearance of 40 m
- the minor high bridges are post-tensioned concrete cantilever bridges with main spans of 120–160 m and a navigation clearance of 17.5–25 m
- the total length of the high bridges is 2.15 km
- the low bridges, 22.35 km in total length, are built with precast, post-tensioned box girders, with spans of 50–60 m.

The island connection starts at the south end of the marine bridge at Xiaowugui (Little Turtle) Island, links four islands where the container

428

Fig. 10.53 Low bridge

terminals are located, and is 4.35 km long. The land-based bridge is 2.4 km long, and runs from the north landing point of the marine bridge to the Old Levee of Luchao Port. These sections are built as post-tensioned continuous beams on supports.

The East-Sea Bridge has six lanes plus an emergency lane. With a design life of 100 years,[1] it should resist a Richter scale 7 earthquake and the '100 year' highest probable wind. It also carries a water pipeline and power and communication cables.

10.10.2 Analysis of the environmental load on the structural concrete

10.10.2.1 Meteorology conditions
The East-Sea Bridge is on the south edge of the north subtropics. Influenced by the north-east monsoon, the area is cold in winter and warm in summer. There is high rainfall and an annual average temperature of 15.8°C. The sea water has a salt content in the range 1.0–3.2%.

10.10.2.2 Deterioration mechanisms operating on the structural concrete
Due to the location, freeze–thaw effects and impact and abrasion from floating ice is not relevant. The following durability goals are considered important:

- no expansion caused by the alkali–aggregate reaction (AAR)
- resistance to magnesium sulfate attack

429

- resistance to chloride penetration
- absence of thermally induced cracks
- resistance to carbonation.

An investigation of concrete structures near the East-Sea Bridge site showed that the carbonation of concrete proceeds at a much slower rate than chloride ingress. For medium-quality concrete, the carbonation rate is about 3 mm in a decade.[2] Therefore, the major factor affecting the durability of the East-Sea Bridge is chloride penetration.

10.10.3 Technical measures for enhancing the durability of marine concrete, and evaluation

Various measures to enhance the durability, especially for corrosion control, were reviewed and carefully evaluated:

- high-performance marine concrete (HPMC)[3–7]
- increasing the thickness of the concrete cover
- concrete surface coating
- epoxy-coated rebar[8]
- corrosion inhibitor[8]
- cathodic protection.

10.10.4 The durability strategy for structural concrete used in the East-Sea Bridge

The durability strategy should be reliable, cost-efficient and practical. Based on the evaluations above and on studies using local materials, HPMC was chosen. The thickness of the concrete cover was increased and, for some parts, a concrete surface coating or corrosion inhibitor may be adopted as additional protection.

Based on extensive tests carried out at SRIBS, two blended binders, named marine concretes I and II (MC I and MC II), consisting of Portland cement, silica fume, ground granulated blast furnace slag (GGBS) and fly ash (FA) were formulated. These, with carboxylic-based high-range water-reducing admixture, are used to produce the HPMC. The concrete properties are presented in Tables 10.15, 10.16 and 10.17.

While the mechanical properties are similar to conventional concrete, the durability aspects are substantially improved. The Portland cement content (less than 40% of the total binder) of the HPMC is low, leading to low heat generation. Under adiabatic condition, the temperature increase of the HPMC can be reduced by

Table 10.15 The mechanical properties of the HPMC

Concrete	Compressive strength: MPa, 150 × 150 × 150 mm cube	Split-tensile strength: MPa	Flexural strength: MPa	Axial compressive strength: MPa, 150 × 150 × 300 mm column	Modulus of elasticity: GPa
Grade 35 control	43.3	4.0	7.4	29.4	33.5
Grade 35 MC I	38.7	3.9	7.7	26.7	32.7
Grade 35 MC II	41.0	4.1	7.6	28.9	35.5
Grade 50 control	58.5	4.0	9.0	32.2	36.9
Grade 50 MC I	52.4	3.9	8.7	31.3	36.5
Grade 50 MC II	66.7	4.5	9.9	32.9	41.3

Table 10.16 The carbonation, permeability and frost resistance of the HPMC

Concrete	Carbonation		Water penetration		Freeze–thaw (100 circles)	
	Depth: mm	Strength loss: %	Water pressure: MPa	Penetration depth: mm	Mass loss: %	Loss of dynamic modulus of elasticity: %
Grade 35 control	0.30	0.63	2.5	26.3	0.9	8.1
Grade 35 MC I	0.16	0.42	2.5	7.1	0.6	6.9
Grade 35 MC II	0.16	0.46	2.5	6.5	0.6	7.2
Grade 50 control	0.25	0.50	2.5	20.5	0.7	7.2
Grade 50 MC I	0.17	0.38	2.5	6.6	0.5	6.8
Grade 50 MC II	0.14	0.37	2.5	5.4	0.4	6.4

7–9°C and the peak time of heat release can be delayed for 8–16 hours – beneficial in reducing thermal stress and for control of early age cracking.[2]

The resistance of the HPMC to chloride penetration has been significantly improved. The chloride diffusion coefficient is only 25–35% that of conventional concrete at 90 days (Table 10.18). This will continue to decrease (i.e. increasing the resistance) with time – studies show that the decrease of the chloride diffusion coefficient could continue for 30–50 years.[2,5] Using the structural design and the concrete properties, SRIBS has performed a dynamic analogue computation for chloride

Table 10.17 The resistance of the HPMC to sulfate attack and the ASR

Binder	Sulfate resistance factor	ASR expansion: %
PII cement (100%)	0.92	0.2
PII cement (40%) + MC I mineral admixture (60%)	1.25	0.05
PII cement (40%) + MC II mineral admixture (60%)	1.12	0.03

Table 10.18 The chloride penetration resistance of the HPMC

Concrete	Rapid chloride penetration test (coulomb)	Apparent diffusion coefficient D_a $(\times 10^{-12}\ m^2/s)$[a]
Grade 35 control	1263	4.85
Grade 35 MC I	826	1.28
Grade 35 MC II	741	1.10
Grade 50 control	1112	4.26
Grade 50 MC I	750	1.15
Grade 50 MC II	637	0.95

[a] Here, D_a is tested on the specimens after immersion for 90 days in salt water

ingress, and shown that the service life of 100 years should be ensured with this quality of concrete.

10.10.5 Conclusions
High-performance concrete is specified for many marine bridges with a required service life of 100–120 years. The studies and tests outlined above have shown the required concrete performance, thus confirming HPMC as durable for the marine environment.

Also in the specification for the East-Sea Bridge are concrete quality control and assurance, including full-scale trials and performance confirmation, a testing and inspection scheme, wet curing requirements, etc.

References
1. Shanghai Municipal Engineering Design and Research Institute (2002) The introduction of East-Sea Bridge design. *Science and Technology of Shanghai Construction*, No. 5 (in Chinese).
2. Qiang, X. et al. (2002) Studies on high performance marine concrete applied to Shanghai deep water harbor project, final report for appraisal (in Chinese).
3. Jun, Z. (2001) Corrosion prevention of reinforcing bar of concrete structures exposed to marine environment and de-icing salt. *Highway*, 4, 52–59 (in Chinese).
4. Dinghai, H. (1998) Application examples of high performance concrete with high GGBS content. *Transaction of Building Materials*, 82–86 (in Chinese).

5. Xunjie, C. *et al.* (2000) Successful application of GGBS high performance concrete to a marine structure. *Concrete*, 9, 59–61 (in Chinese).
6. Clear, C. (1991) GGBS in extreme environments, *Concrete*, 27–30.
7. Hassan, K. E. *et al.* (2000) The effect of mineral admixtures on the properties of high-performance. *Cement and Concrete*, 22, 267–271.
8. Federal Highways Administration (2000) *Materials and methods for corrosion control of reinforced and prestressed concrete structures in new construction.* Report No. FHWA-RD-00-081. Federal Highways Administration, Washington, DC.

11

Repair methods

John Broomfield, Broomfield Consultants, UK

11.1   Assessment prior to repair

When approaching a structure in need of repair, it is critically important to understand as much as possible about the structure, its condition and what has caused the deterioration before embarking on repairs. It is also critically important to understand what will be required of the structure after it is repaired so that an appropriate, cost-effective repair is executed.

There is a huge number of destructive, semi-destructive and non-destructive tests that can be used on reinforced concrete structures. In the UK, there is a large and dynamic sector of concrete survey companies which may be independent, part of a construction or concrete repair contracting company, a civil engineering consultant or a university. Some do their own laboratory analysis of samples in house and are National Measurement and Accreditation Service/United Kingdom Accreditation Service (NAMAS/UKAS) accredited. Most of them have specialised teams which carry out the survey using conventional and highly specialised equipment. Using a team with expertise and regular experience with survey and non-destructive testing techniques can be very cost-effective, particularly on large investigations. An extensive list of available investigation techniques are given by the Concrete Bridge Development Group.[1] Broomfield[2] lists the (current in 2007) national and international standards, where available, for many of the techniques discussed, along with a detailed discussion of the merits, limitations and interpretation of techniques.

11.1.1 Condition survey

As stated above, it is important to carry out a sufficiently detailed condition survey to ensure that the present condition of the structure is understood and how it reached that condition. Only then will it be possible to choose the most cost effective repair so that the structure can continue to fulfil its required purpose.

Concrete durability
978-0-7277-3517-1

On most large structures, access will be a major cost of any investigation and survey. Traffic control for highway bridges is frequently the largest single cost of a survey and even of a repair. In some cases, it can be over 50% of the cost. It is therefore important that the survey is well planned with clear objectives and briefing of the survey team and it is adequately resourced. It has been suggested that about 10–20% of the repair budget should be spent on the investigation. It is often better to take more samples and do more tests than initially considered necessary rather than have to go back for more to clear up any ambiguity in the results.

Table 11.1 summarises the major test techniques used in a condition survey and where to find standards and guidance on their application.

11.1.2   Visual survey and desk study

The first part of an investigation is to check what drawings are available, along with any reports of previous investigations, repairs and maintenance relevant to the investigation. It is far easier to mark survey results on elevation drawings if available. If a structural analysis is required, then reinforcement drawings will need to be located or developed, as they will also be for the design and application of any electrochemical techniques such as cathodic protection, electrochemical chloride extraction or realkalisation.

The visual survey is a critical part of the investigation. This may break down into a preliminary survey to understand the extent of the problem and how it has manifested itself, and then a more detailed survey, listing all relevant defects. It is normal practice to use pro formas or elevation drawings or sketches to note down all defects as well as sampling and test locations. A detailed visual survey should be done within touching distance of the surface being surveyed; however, this may not always be possible due to access limitations. In such cases, the areas that can be surveyed by immediate contact should be carried out and then extended to other areas using a binocular survey.

As with all aspects of the condition survey, it is important to decide what will be the result of the visual survey. If it will be used to develop a bill of quantities for repair, it needs to quantify each area of damage to be repaired in terms of depth as well as length and breadth and the number and type of defects.

There are no standards for executing visual surveys that this author is aware of, but good examples and guidance is given by the Concrete Society,[13,14] the Concrete Bridge Development Group,[1] the American

Concrete Institute,[15] the Buildings Research Establishment (BRE)[16] and Broomfield.[2]

11.1.3 Delamination survey

The delamination survey is frequently undertaken alongside the visual survey where it is possible to touch the surface. A hammer or a chain on bridge or car park decks is usually the most cost-effective and accurate method of measurement, although other techniques such as radar, thermography and impact echo have been used, especially where there are overlays such as waterproofing membranes and asphalt wearing courses on bridge decks.

11.1.4 Coring and petrography

It is common practice to extract cores for petrographic analysis and strength testing. Petrographic analysis will reveal internal problems with the concrete such as the alkali–silica reaction (ASR) or sulfate attack, or the presence of high-alumina cement with its risk of chemical conversion and subsequent loss of strength and increase in porosity making it more vulnerable to carbonation (see Section 5.2).

Petrography will also show the extent of voiding and poor compaction and the degree of hydration of the cement. It can be used to estimate the cement content and types and the water/cement ratio. There are many coring companies that can extract cores quickly and efficiently; however, this is a rather destructive technique.

11.1.5 Chloride and carbonation testing

These were discussed in detail in Sections 4.4.5.1 and 4.4.5.2. They are essential in determining the correct repair for structures with corroding reinforcement, as the extent of carbonation and chloride ingress will determine the most cost-effective method of long-term corrosion control. Modelling techniques such as those described in Chapter 6 can be used to predict future rates of deterioration and the effectiveness of applying different repair options.

11.1.6 Corrosion rate, potential mapping and concrete resistivity measurement

These techniques were also discussed earlier, in Sections 4.4.5.3, 4.4.5.4 and 4.4.5.5. Potential mapping is widely used on highway structures and

Table 11.1 Techniques for assessing concrete and relevant international standards

Method	Standards or guidance
Visual	The recognition of ASR is discussed in the SHRP manual on the alkali–silica reaction[3] and in Kaetzel *et al.*[4] The Concrete Bridge Development Group[5] provides a table of 25 examples of cracks and visual defects with photographic examples in an appendix.
Delamination	ASTM D4580-03 describes delamination detection in concrete bridge decks using a chain drag or hammers, the Delamatect system and a rotary percussion device. The standard concludes that the chain drag procedure is the most reliable technique. There is a standard, ASTM D4788, covering infrared thermography of concrete and asphalt-covered concrete bridge decks. This requires a scanner with a minimum thermal resolution of 0.2°C, and says that the temperature difference between a delamination or debonded area should be 0.5°C. The standard is for a vehicle-mounted system, and claims 80–90% of delaminations can be found in concrete decks with or without an asphalt overlay. ASTM D6087-05 covers the use of ground-penetrating radar to evaluate asphalt covered-concrete bridge decks. It is designed for vehicle-mounted or manually driven systems. It claims the system is accurate in detecting delaminations to within +11.2% according to a precision test on 10 bridge decks in New York, Virginia and Vermont.
Cover meter	One of the few standards for cover meters is BS 1881-204. Alldred[5] discusses cover meter accuracy when several rebars are close together. Concrete Bridge Development Group Technical Guide 2[1] gives good coverage of cover meters, their design and performance.
Phenolphthalein	There is a CEN standard on carbonation depth measurement (EN 14629:2006). The Concrete Society report[6] is similar to the CEN standard. BRE Digest 405[7] specifically discusses carbonation and its measurement in Portland cement concrete. BRE Digest 444 Part II[8] and BRE IP 11/98[9] discuss measurement in high alumina cement concrete (HAC).
Chloride content: sampling, analysis and interpretation	CEN has produced a test method for chloride analysis ((Draft pr) BS EN 14629:2007): *Products and Systems for the Protection and Repair of Concrete Structures – Test Methods – Determination of Chloride Content in Hardened Concrete.* There are three relevant ASTM standards: ASTM C1152/C1152M-04e1 (*Standard Test Method for Acid Soluble Chlorides in Concrete and Mortar*), ASTM C1524-02a (*Standard Test Method for Water-extractable Chloride form Aggregate*) and ASTM C1218/C1218M-99 (*Standard Test Method for Water Soluble Chloride in Mortar and Concrete*). ASTM Committee G01.14 is also in the process of developing a laboratory test for chloride thresholds for concretes with different admixtures. ACI 2001 gives extensive discussion on thresholds. BRE Digest 444[9] has a comprehensive table and 'cloud data' graph of chloride content versus risk based on field data.
Reference electrode (half cell) survey	Detailed methodologies for undertaking reference electrode potential surveys and their interpretation can be found in the Concrete Society report,[6] Chess and Grønvold[10] and ASTM C876-09.

Table 11.1 Continued

Method	Standards or guidance
Linear polarisation (LPR)	The macrocell test ASTM G109 is the only standard for measuring corrosion rates of steel in concrete (laboratory tests only). Guidance on LPR comes from the Concrete Society report,[6] and provides a detailed procedure and a pro forma for measuring and recording LPR-based corrosion rates. Detailed instructions are given in the manuals provided by manufacturers and suppliers of the equipment.
Concrete resistivity	Resistivity testing is covered in BS 1881 (*Testing Concrete*, Part 201: *Guide to the Use of Nondestructive Methods of Test for Hardened Concrete*). The Concrete Society report[6] provides a detailed procedure and a pro forma for recording resistivity measurements. The Concrete Bridge Development Group[1] also gives information and guidance. The test is based on soil resistivity, as described in ASTM G57 (*Field Measurement of Soil Resistivity using the Wenner Four-electrode Method*).
Permeability (ISAT)	The ISAT test (BS 1881-208) is particularly useful for checking the effectiveness of coatings. Concrete Society Technical Report 50[11] is a guide to surface treatments for protection and enhancement of concrete. RILEM Report 12[12] goes into great detail on the various tests and transport phenomena in concrete.
Petrography	Standards for petrographic analysis include ASTM C457-98 (*Standard Test Method for Microscopical Determination of Parameters of the Air-void System in Hardened Concrete*), ASTM C856-95e1 (*Standard Practice for Petrographic Examination of Hardened Concrete*) and BS 1881 (*Testing Concrete. Part 124: Methods of Analysis of Hardened Concrete*).
Radar/radiography	ASTM D4788-03 gives a standard test method for its application to bridge decks, which can be used as guidance for its application on other concrete elements.
Ultrasonic pulse velocity	The use of ultrasonic pulse velocity is documented in BS EN 12504-4:2000 (*Testing Concrete. Determination of Ultrasonic Pulse Velocity*).
Impact echo	ASTM C597-02 (*Standard Test method for Pulse Velocity Through Concrete*) and ASTM C1383-04 (*Test Method for Measuring the P-wave Speed and the Thickness of Concrete Plates Using the Impact Echo Method*) cover the application of impact echo to concrete structures.
Radiography	General information regarding radiography of concrete structures can be found in BS 1881-205 (*Concrete Testing. Recommendations for Radiography of Concrete*). National regulations on ionising radiation should also be consulted.

in car park structures, and can be used to map anodic (corroding) areas of reinforcement. The potential survey can be misleading in the presence of excess moisture, galvanised steel or other embedded metal (such as window frames). Local corrosion rate measurements can be used to confirm that corrosion is occurring where mapped, and can be used to predict rates of section loss and time to cracking and spalling.

11.1.7   Other tests

A comprehensive list of tests and their relevance is given by the Concrete Bridge Development Group[1] and Broomfield[2] as well as in Table 11.1.

11.2   The repair strategy

As stated in Section 11.1, successful repair requires understanding of the following:

- the structure, including critical locations for its ongoing performance
- its present condition
- the deterioration mechanisms
- the future condition if unrepaired
- repair options, their suitability and their limitations.

The new European standard BS EN 1504 *Products and Systems for the Protection and Repair of Concrete Structures* (Parts 1–10) gives specifications for concrete repair materials in Parts 2 to 7. These cover

- coatings, sealers and impregnation (Part 2)
- concrete and mortar repair materials, structural and non-structural (Part 3)
- structural bonding (Part 4)
- concrete injection materials (Part 5)
- anchoring of reinforcing bars (Part 6)
- reinforcement corrosion protection (Part 7).

Part 1 is definitions and Part 8 is quality control and conformity, of interest principally to manufacturers to ensure their products comply with the directive.

Part 9 gives general principles for the use of products and systems, while Part 10 gives information on how to specify the materials in a concrete repair performance specification. It also describes how to apply the materials and what site tests and values should be specified in a non-mandatory appendix.

Part 9 states that the basic options for protection and repair are:

(1) do nothing for a certain time
(2) reanalyse the structural capacity possibly downgrading the capacity
(3) prevention or reduction of the deterioration without improvement
(4) improvement, strengthening or refurbishment
(5) partial or full reconstruction
(6) partial or full demolition.

Item 4 will include provisions such as corrosion control for the reinforcement or control of factors such as moisture that influence the ASR or sulfate attack, or surface protection to stop damaging agents reaching the concrete surface.

11.2.1   Structural issues

A detailed review of structural issues is beyond the scope of this book, but they must be considered both in terms of any reduction of structural capacity due to deterioration and also reduction of structural capacity during repairs, for instance when concrete is removed. As stated in EN 1504, Part 9, the structural impact of the repair process and the means by which loads will be carried both during and after repair must be considered.

11.2.2   During assessment

A condition evaluation as described here is not necessarily a structural survey. A structural engineer must be consulted if there are concerns about the capacity of the structure either due to corrosion damage or for any other reasons. Any excessive deflection of structural elements, misalignment, impact damage, excessive cracking, loss of concrete or loss of steel section will require a structural evaluation before repairing corrosion damage. Structural issues for concrete bridges are discussed by the Concrete Bridge Development Group.[1] For car parking structures, they are discussed by the Institution of Civil Engineers,[17] and for older buildings by Macdonald.[18] If there are structural issues, then the existence of reinforcement drawings becomes more important. In their absence, a careful cover meter survey with a scanning cover meter may supply sufficient information. However, it is very difficult to get detailed information about hinges, joints and other structurally important locations from cover surveys, as steel is often very congested, multilayered and sometimes may be buried deeply below accessible surfaces.

11.2.3 During repair

Removal of concrete during repair can compromise the integrity of the structure. This may lead to limiting the amount of break-out permitted during repair, allowing adequate time for concrete strength gain after repair and a requirement for propping during the repair process itself. If repairs are to be structural, then a structural engineer must design them. It is usually necessary to support the live and dead loads on the structure during repair, so that the loads will go through the repairs after the support is removed. In some cases, structural support is needed to remove significant amounts of concrete. A structural support system for concrete repair is shown in Fig. 11.1. The above-ground structure has been tied into the foundations. BS EN 1504, Part 3, clearly differentiates between structural and non structural repair materials.

It should also be noted that if there is extensive delamination of concrete, the bond between steel and concrete may have already been compromised.

Fig. 11.1 Structural support system on the Midland Links Motorway elevated section (Junction 9, M6) prior to repair. The support system can be jacked up to take some of the load while concrete repair is undertaken. Conductive coating cathodic protection systems can be seen on cross heads in the background. (Courtesy of Kevin Davies, taken December 2005)

11.3   Repair due to concrete attack

There are a number of mechanisms for attack of concrete, as covered in Chapters 3–9. These can be internal reactions of the constituents such as the ACR or sulfate attack, or it can be external attack by acids, freeze–thaw or fire, erosion or impact. Since the aim of all repairs is to repair existing damage and control future damage, it is important to understand how far the damage has gone and how much further it may go.

11.3.1   Sulfate attack

This topic is covered more extensively in Section 4.1. There are several forms of sulfate attack, including external sulfate attack, internal delayed ettringite formation and thaumasite attack. Repairs may be structural or may require replacement of lost cover to reinforcement or protection from ingress of sulfates and water.

11.3.2   Alkali–silica reaction

This topic is covered more extensively in Section 4.3. The ASR causes unsightly staining on the concrete surface. By the time it is investigated, the aggregates may have completely reacted, in which case repairs can be carried out without further concern. Repairs may therefore be mainly cosmetic. However, ASR-induced expansion may move bearings off pads, and damage and distort other elements or adjacent structures. Repairs may be structural, requiring jacking or structural propping. Testing and analysis of cores may be required to determine the extent to which further expansion could occur and the likelihood of treatments in successfully controlling it.

As water is a necessary reactant, controlling moisture access by protecting the concrete surface with coatings, penetrating sealers or better drainage may reduce the reaction rate. This subject is dealt with in greater detail in Section 4.3.

11.3.3   Acid attack

Section 4.2 discusses the mechanism of acid attack on concrete. Acids with a pH of less than 6.5 will attack concrete. Concrete repair, fairing coats, and cementitious and other coatings may be required to repair existing damage and prevent future damage to the concrete. In some cases, the residue after attack may be washed away, exposing reinforcement and leaving large aggregate residues in the system, in other cases it may leave a softened layer at the surface that needs to be removed back to sound concrete before repair.

11.3.4   Fire damage

The subject of fire damage is discussed in Section 4.6. Fire may weaken concrete and therefore require requiring jacking or structural propping during the repairs. The repairs may need to be structural or the replacement of compromised cover concrete.

11.4   Physical concrete repair

11.4.1   Structural repairs

This may include the complete or partial removal of elements of a structure or of damaged concrete while leaving the reinforcing network in place. Alternatively, it may use structural plate bonding or structural bonding of glass or carbon fibres.[14,19]

11.4.2   Patch repairs

Where concrete cover is damaged due to reinforcement corrosion or attack from external sources, patch repairing is conventionally applied. A good-quality patch repair will use a proprietary low-shrinkage repair mortar or concrete. The material may be applied by hand, by spray or pumped into suitable shutters. The patch area will have clean, dust- and debris-free surfaces and squared edges.

When bidding for concrete patch repair work, the form of contract and the form and accuracy of the bill of quantities will have an influence on the way the bidding contractor perceives the risk to itself and therefore how it prices the bill of quantities. As stated in Section 11.1.2, the investigation should provide an estimate of quantities broken down into depth and area of concrete removal which is as accurate as possible. It must allow for ongoing deterioration between the survey and the day the contractor carries out repair, the fact that break-outs are larger than the area that is delaminated and the shortcomings of the delamination measuring technology.

There are a number of methods for removing concrete, and the choice depends on the number, size and depth of the patches, the reinforcement layout, the specification, the budget and the contractor's preferences. If concrete is just starting to spall due to carbonation or if an electrochemical treatment such as cathodic protection is planned, then a simple repair of removing unsound concrete, cleaning the rebar surface, squaring the edges and putting in a sound, cementitious, non-shrinking repair material may suffice. This may also be sufficient for 'one off' repairs such as fire damage, an ASR that has gone to completion or external attack or damage that will be prevented from recurring.

Before patching carbonated concrete, the cracked and spalled concrete must be removed from around the rebar or as far as the carbonation front goes, whichever is the greater. The cementitious patch material is chosen to ensure that the steel is back in a high-pH, alkaline environment. This will encourage the reformation of the passive layer, to stop reinforcement corrosion.

Concrete removal may be by hand-held percussion tools, hydro-demolition or, if large areas of concrete are being removed, by milling, vehicle or robot-mounted hydrojets. With increased health and safety concerns about vibration-induced injuries, hydrodemolition is increasingly popular for reasonably extensive repairs. Hydrodemolition has several advantages over percussion tools:

- it reduces noise and vibration, so there is less impact on users of the structures or the neighbours
- it cleans the bars as it goes
- it leaves the exposed concrete surface less damaged than percussion tools, and ready for concrete application
- it eliminates the risk of vibration injuries to operatives.

However, it is not always suitable, because of the need to deal with the water effluent, the access requirements and the logistics for smaller repairs. It may need follow-up to remove enough concrete from behind the reinforcement due to 'shadowing'.

If patching is required due to chloride-induced corrosion, then the usual specification is to remove concrete to a depth of about 25 mm behind the rebar, ensuring that all corroded steel is exposed around delaminated areas and the rebar is cleaned to a near-white finish to remove all rust and chlorides. Removing concrete behind the reinforcement and then cleaning the rear surface of the bar can be extremely difficult, especially if the reinforcing bars are highly congested. It requires careful specification and selection of concrete removal techniques.

Figure 11.2 shows three types of repair: a bad repair, a good chloride repair and a patch suitable for cathodic protection, where protection

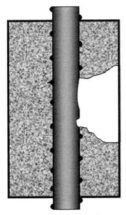 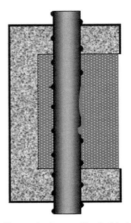 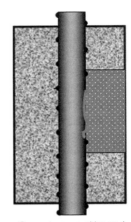

| Feathered edges and poor preparation allow breakaway at edges and poor keying | Squared edges cutting behind the bar and removal beyond the corroded area restores passivity and removes contamination | Concrete removal beyond the corroded area is not required for electrochemical treatment |

Fig. 11.2 Patch repairs: bad, good and compatible with electrochemical treatment

will not be provided by the patch itself. The cut edges and faces of the concrete must be square and clean of all dust and debris in all cases.

The choice of general type of patch repair material is discussed below. Wetting or bonding agents may be used to promote adhesion between the repair and the original concrete. Bonding agents should not be used where electrochemical treatment is required, as it could create an electrically insulating layer. If an electrochemical technique is being used, then repairs only need to reinstate damaged concrete. A clean rebar surface is needed, and a simple patch to rebar depth in cracked, spalled and delaminated areas will prepare the structure for the application of the anode. In some cases, the anode and overlay will fill the excavated areas as a single operation. Materials used must be compatible with the electrochemical treatment. These issues are more fully discussed in Sections 11.6–11.8.

In a recent evaluation of 230 case histories of repairs under the CONREPNET programme, it was found that 20% of repairs failed within 5 years, 60% failed within 10 years and 90% within 25 years.[20] The most common forms of failure were cracking and debonding of patch repairs. Corrosion and the ASR were the most difficult problems to repair.

One of the reasons why corrosion is one of the most difficult problems to repair, particularly with patch repairs, is the incipient or ring anode effect.

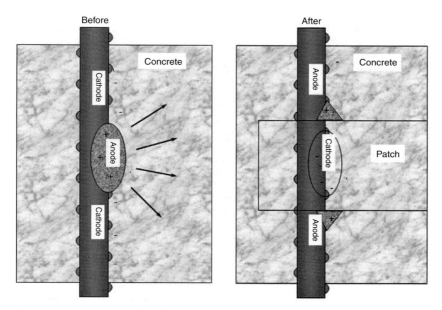

Fig. 11.3 Incipient anode schematic showing how the anode is displaced to the edge of the repair by the formation of a new cathode in the patch repair

In Section 4.4 and Figure 4.3, the corrosion process was explained in terms of the formation of anodes where corrosion occurs, and adjacent areas where a benign cathodic reaction occurs. Figure 11.3 shows what can happen in the adjacent concrete when an anode is patch repaired.

The formation of an anode forces the creation of adjacent cathodes around it, limiting the active area of corrosion. If the anode is suppressed and is turned into a cathode by patch repairing, the previously cathodic areas will become anodic, and corrosion will initiate around the repair. This problem is particularly prevalent for chloride-contaminated structures, as the concrete is generally wetter than for carbonated structures and therefore the lower electrical resistance of the concrete means that anodes and cathodes are well defined and separated from each other. Examples are illustrated in Figs 11.4 and 11.5.

It is therefore important to realise that patch repairs will only be cosmetic and short term unless the underlying cause of deterioration is fully addressed.

The UK Concrete Society has a standard specification and bill of quantities for concrete repair.[21] A more up-to-date specification and method of measurement is available on the UK Concrete Repair

447

Fig. 11.4 A poorly executed patch repair with surrounding spalling due to incipient anodes on a building

Association website. (In the USA, there is *Concrete Repair Manual*, published jointly by the American Concrete Institute and the International Concrete Repair Institute.[22]) A quantitative visual and delamination survey will be required to develop the bill of quantities.

Fig. 11.5 Incipient anode showing the damaged repair (left side) and corrosion in the original concrete due to incipient anode formation on an integral beam under a car park deck exposed to chlorides. The original repair appears to be well executed but could not stop incipient anode formation

11.5   Surface protection systems

One of the easiest ways of addressing the underlying causes of corrosion may be to apply a coating or other form of surface protection. If there is ingress of carbon dioxide, chlorides, water or oxygen, then, in principle, such barriers can stop further ingress.

Obviously, a coating or barrier applied during the initial construction is more likely to be successful in keeping out contaminants than one applied once problems come to light. It is important to determine whether application of surface protection is going to be cost-effective once deterioration has started.

11.5.1   Coatings

The construction products industry has produced a range of coatings, including penetrating sealers and impregnation. The new European standard EN 1504, Part 2, refers to hydrophobic impregnation, impregnation and coatings. These can be used for protection against ingress, moisture control, increasing electrical resistivity by limiting moisture content, and for physical resistance and surface improvement.

In terms of corrosion control, it has been demonstrated experimentally that coatings can control corrosion in carbonated structures in outdoor exposure in the UK, but not where corrosion is due to chlorides.[23] EN 1504, Part 2, lists test methods for measuring the permeability to carbon dioxide, water and water vapour. There is no agreed European standard yet for measuring chloride ion diffusion through concrete.

In general terms, anticarbonation coatings are high-build, pigmented coatings. Penetrating sealers such as silanes or siloxy silanes are hydrophobic penetrating sealers which line the concrete pores with a hydrophobic coating while leaving the pores open. This will minimise water, and hence chloride, ingress while allowing the concrete to 'breathe out' water vapour.

It is important that coatings are applied properly with the specified surface preparation, coverage rate, wet and dry film thickness and adhesion. It is interesting to note that Sergi *et al.*[23] found that the best coatings for controlling carbonation-induced corrosion did not have a very high pull off strength. They suggested that this allowed better crack bridging, accommodating minor crack movements.

It is important that the pull-off strength of the substrate is tested prior to coating as well as after coating. A very poor-quality substrate may be unsuitable for coating and may require a penetrating sealer or other measures to prepare a suitable surface for coating.

11.5.2   Concrete overlays

Concrete overlays are widely used in North America for rehabilitating corrosion damaged bridge decks. The problems of shrinkage crack control, delamination and 'curling' due to differential shrinkage have been dealt with by the development of considerable practical expertise by American highway contractors. Such techniques are rarely used in Europe, where waterproofing membranes and asphalt overlays are more commonly applied to bridges.

Decks to be overlaid may be prepared by removing all concrete cover behind the reinforcement or selective area, or removal by milling, hydrodemolition percussion tools or a combination. It may be necessary to replace damaged reinforcement or the complete top mat.

11.5.3   Waterproofing

Waterproofing is used extensively on new bridge decks in many western countries where de-icer use is extensive. Two notable exceptions are the USA and Canada. A range of materials, in the form of either sheets or liquids, are approved for use in different countries. A synthesis of practice of waterproofing membranes in North America was published by Manning.[24] Approximately 50 systems were tested by the UK Transport Research Laboratory in a trial of membrane systems.[25]

11.5.4   Barriers and enclosures

Barriers and enclosures can keep out chlorides. They can also be used to reduce the humidity level at the reinforcement surface below the threshold of corrosion. It is not clear from the literature what that level is, but concrete within buildings rarely corrodes even though it is frequently carbonated to rebar depth. Also, buildings constructed with calcium chloride added as a set accelerator show corrosion on external façades but rarely, if ever, internally, unless there is moisture ingress. It is therefore feasible to construct cladding over a building façade and ventilate it to the interior of the building to control reinforcement corrosion.

11.6   Chemical corrosion inhibitors

ISO 8044:1989 defines a corrosion inhibitor as 'A chemical substance that decreases the corrosion rate when present in the corrosion system at suitable concentration, without significantly changing the concentration of any other corrosion agent.' This definition excludes coatings, pore blockers and other materials that act on the water, oxygen and chloride concentrations, and clarifies what we are talking about.

In the case of steel in concrete, we are talking about chemicals that are either admixed into fresh concrete or that are applied to the concrete surface, usually long after the concrete has hardened. They must then transport through the hardened concrete to react on the reinforcing steel surface to slow down the rate of corrosion. This difference between admixed and surface-applied inhibitors is fundamental. By admixing during the construction process, the precise dosage is controlled, and conditions can be readily tested. The protective mechanism is in place before corrosion initiates.

Applying inhibitors to hardened concrete as part of a rehabilitation system is far more problematical in ensuring effective dosage at the rebar, and in measuring, testing and predicting long-term performance. Also, if corrosion has started, the inhibitor must penetrate the layer of corrosion product and stop corrosion that has started, not just prevent its initiation.

There are a number of ways of subdividing inhibitors:

(*a*) By their action. The corrosion reaction in concrete occurs by the formation of anodes and cathodes. Corrosion inhibitors can therefore be:
   - anodic inhibitors – suppressing the anodic corrosion reaction
   - cathodic inhibitors – suppressing the cathodic reaction
   - ambiodic inhibitors – suppressing both anodes and cathodes.
(*b*) By their chemistry and function:
   - inorganic inhibitors – nitrites, phosphates and other inorganic chemicals
   - organic inhibitors – amines and other organic chemicals
   - vapour phase or volatile inhibitors – a subgroup of the organic inhibitors (generally amino alcohols) that have a high vapour pressure.

All of these inhibitors have been widely used by corrosion engineers for many years to protect steel and other metals such as electronic components and in the chemical process and power generation industries. In some cases, their chemistry is well understood, in other cases less so.

11.6.1   Admixed corrosion inhibitors

Of the available materials, the most widely applied is calcium nitrite admixture in new concrete. This inhibitor is of proven effectiveness as long as the chloride:nitrite ratio stays less than about 1.8:1.[26] A dosage rate of 10–30 litres per cubic metre of concrete is generally specified, depending on the expected maximum chloride level at the rebar. This has proven efficacy, and is usually the benchmark against which other inhibitors are tested. Nitrites act as set accelerators (their original use as a concrete additive was as a non-chloride set accelerators for use in cold climates). They are usually therefore sold in proprietary formulations, including a set retarder to avoid flash setting of the mix, adjusted to local conditions.

Competitor companies offer other proprietary materials. They are generally organic inhibitors with active amine and ester groups. The main problem with most organic and inorganic inhibitors is that they act as severe set retarders: these formulations of amines, esters and nitrites do not.

The US Federal Highways Administration has expressed some concern that inhibitors may be leached out of concrete, particularly in marine exposure conditions. Some preliminary work on calcium nitrite has been undertaken.[27] The author understands that further long-term tests are underway.

11.6.2   Applying corrosion inhibitors to hardened concrete

Vapour phase inhibitors or migrating corrosion inhibitors have been used to impregnate packaging, greases and waxes for many years to protect steel machinery and components, particularly before use. An American company realised in the 1980s that they might be effective in diffusing through concrete pores and protecting reinforcing steel. Several companies now offer proprietary formulations, which are again amine and ester based, with amino alcohols as the main volatile component.

Amino alcohols are ambiodic, forming a film on the steel surface, blocking both anodic and cathodic reactions. They can be applied as a coating on the surface of the concrete, a 'plug' of material in a hole, admixed into repairs. The suppliers claim that these materials will move very rapidly through the air voids in the concrete, through pores and microcracks, to reach the steel and protect it.

A migrating inhibitor with no pretensions of volatility is MFP or mono-fluorophosphate. This relies on capillary action and other transport mechanisms to get the material down to the steel through the concrete cover. A paper has showed that it could be quite successful in carbonated concrete.[28] Calcium nitrite has also been offered in a formulation for patch repairs or overlays, with the intention that the nitrite will diffuse out of the patch and protect the steel.

11.6.3 Track record, case histories, monitoring

One issue with corrosion inhibitors is to determine how effective they are. There are two aspects to this: Can they be detected and levels measured quantitatively in the concrete and at the rebar surface? Are they effectively depressing the corrosion rate?

Detection is obviously specific to a particular active inhibiting chemical. Some tests are expensive and difficult. Much work has gone on to develop a test for amino alcohols, with mixed results so far. As stated earlier, it is far easier to carry out a test on admixed inhibitors than for those applied to hardened concrete. There can be dangers with under-dosing inhibitors, in that corrosion may occur very aggressively in small anodic pits. It is therefore important that the risk of pitting and under-dosing is fully understood by those using these materials.

The most effective method of assessing corrosion inhibitors is to measure the corrosion rate. If a corrosion inhibitor is effective, it should reduce the active corrosion rate by at least one and preferably two orders of magnitude. The author has reported long-term monitoring of structures which had corrosion inhibitors applied in 1995.[29] Amino-alcohol-based inhibitors were applied by surface application, down holes drilled in the concrete and in repair patches to support structures with over 1% chloride by weight of cement cast into the concrete as calcium chloride set accelerator. Over the years, a large number of patch repairs have been carried out. This trial on over 60 structures showed only a slight reduction in the corrosion rate and the onset of cracking of treated structures at approximately the same time as untreated controls. In a very small-scale trial with another proprietary amino-alcohol-based inhibitor, short-term results were more encouraging, with a 91% reduction in the corrosion rate measured with a linear polarisation device in an area showing active corrosion rates and corrosion potentials. However, there has been no follow-up to this trial on a car park deck with low cover, which would maximise the likelihood of inhibitor reaching the steel.

The UK Transport Research Laboratory has recently set up laboratory tests on a range of inhibitors available in the UK. It is also doing field tests on bridge beams on the Midland Links Motorway near Birmingham for the Highways Agency. The BRE has had long-term outdoor exposure trials going for several years. BRE also did tests many years ago on calcium nitrite.[30]

In the USA, corrosion inhibitor testing goes back many years. The calcium nitrite inhibitor cast into concrete has been widely tested in the laboratory and the field. Amino-alcohol-based inhibitors were first trialled in hardened concrete in the 1980s, with some of the most comprehensive testing done by the Strategic Highway Research Program.[31,32] A recent review of the field trials on bridge decks suffering from chloride ingress showed no beneficial effect of inhibitor treatments.[33] Similarly, a large-scale exposure trial of inhibitors on slabs also found no beneficial effects of a number of inhibitors.[34] The exception was calcium nitrite cast into the concrete.

If coatings are applied after the inhibitor is introduced, then corrosion monitoring is made harder. In order to take measurements of the corrosion condition, either the coating must be removed locally or (preferably) corrosion rate probes embedded in the concrete, as has been done on some car parks in the UK.[35] Ideally, this should be done before the inhibitor is applied, so that 'control' measurements can be taken prior to treatment. Permanent monitoring probes will effectively monitor the changes in corrosion rate without disturbing the structure, and ensuring that readings are made in consistent ways in consistent locations.

The biggest issues with corrosion inhibitors are for those wishing to use them for rehabilitation rather than for durable construction.

The research and field trials on corrosion inhibitors reveal the following issues:

- there is very little independent field data on the performance of corrosion inhibitors
- the available field data are often poor, with no clear evidence of the amount of inhibitor applied, whether it reached the rebar and if it is reducing corrosion rates and extending time to cracking
- many claims have been made about the transport of inhibitors through hardened concrete – these claims need to be independently assessed
- we will need definitive evidence of the dosage versus chloride level to achieve a given (low) corrosion rate
- for application to hardened concrete, we need quantitative data on its penetration versus concrete cover and concrete permeability

- we need more information on the performance of inhibitors, particularly well-controlled field trials, and long-term corrosion monitoring.

Corrosion inhibitors are inexpensive materials that can be applied simply. However, until there is clear evidence of their envelope of effectiveness in terms of chloride level, corrosion rate reduction, dosage, cover and longevity, it will be difficult to do comparative whole-life costing of inhibitor application versus proven alternatives such as cathodic protection. This is particularly so for inhibitors applied to hardened concrete. In the case of inhibitors for admixing into fresh concrete, whole-life costing programmes are available from the manufacturers that show the cost-effectiveness of their inhibitor versus some of their other products or alternative methods of improving durability in aggressive environments.

11.7   Cathodic protection

Cathodic protection is a way of providing comprehensive corrosion protection to the reinforcing steel in a concrete structure. It is combined with patch repairs, as described in Section 11.4.2. The repairs need to be compatible with the passage of electricity through the concrete and should exclude insulating bond coats on the steel and metallic fibres in repair materials which could short circuit the anode to the steel.

11.7.1   Impressed current cathodic protection

The best known form of cathodic protection for steel in concrete uses an impressed current from a mains-powered transformer/rectifier to deliver direct current along wires to a distributed anode system mounted on the concrete surface or embedded in the concrete. The level of protection is determined by measuring the steel potential against an embedded reference electrode (see Sections 4.4.5.3 and 4.4.6.1). The arrangement is illustrated in Fig. 11.6.

11.7.1.1   Cathodic protection theory

There are a number of ways of explaining how cathodic protection works. The simplest is to consider Fig. 4.13. If corrosion is occurring at the anode and generating electrons while a benign cathodic reaction is consuming electrons nearby, then installing an external anode and

455

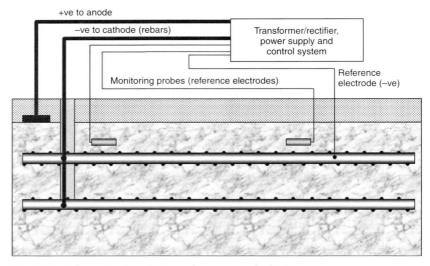

+ve to anode
−ve to cathode (rebars)
Transformer/rectifier, power supply and control system
Monitoring probes (reference electrodes)
Reference electrode (−ve)

Fig. 11.6 Schematic of an impressed current cathodic protection system

applying a direct current can be used to force all the steel to become anodic or cathodic. By connecting the reinforcing steel cage to the negative terminal, excess electrons are injected into the steel preventing Fe^{2+} ions from being generated and encouraging the formation of hydroxide ions, and therefore corrosion stops. Another definition of cathodic protection is to depress the electrochemical potential of the anodes (see Chapter 9) to below that of the cathodes, so that current no longer flows between them, and corrosion is suppressed.

11.7.1.2 Impressed current anodes

A number of electrochemical cathodic reactions can occur on the anode surface. Most of them lead to the formation of acidic products at the anode, which must therefore have the following characteristics:

- it must be electrically conductive, to get the direct current into the concrete
- it must be compatible with being embedded in or mounted on concrete long term (at least 10 years and ideally 40 years or more)
- it must have a low interfacial electrical resistance at the steel–concrete interface
- it must be resistant to the acids formed at the anode–concrete interface

The primary materials used for impressed current anodes for steel in concrete are either mixed-metal oxide-coated titanium (MMO/Ti) or carbon-loaded composite materials. There is also a proprietary conductive ceramic which is manufactured in various configurations by one anode supplier.

The mixed-metal oxide coating on the titanium anode lowers the interfacial resistance, allowing it to conduct electricity into the concrete. It also resists the acidic reactions. The thickness and formulation of the coating determines the life of the anode in ampere-hours per square metre of anode surface area. There is a standard test for anodes in concrete designed to test the embedded anodes, originally designed for mixed-metal oxide-coated titanium anodes.[36]

Mixed-metal oxide-coated titanium anodes frequently come in the form of an expanded mesh to maximise the surface area in contact with the concrete. This may be in the form of ribbons, 11–15 mm wide, which can be placed in saw cuts in concrete or spaced off from rebars when applied to new construction. Its most common application is as a sheet of mesh typically 1.2 m wide, fixed to the concrete surface with a concrete overlay applied.

Another popular application is mesh rolled into a tube and inserted into holes in the concrete. Strings of individual probe or discrete anodes are used to form a zone. In one proprietary design, each anode has a resistor, which helps to balance current flow in the anode zone. Examples are shown in Figs 11.7 and 11.8.

Carbon-based anodes are either conductive organic paint coatings with a graphite 'pigment' or two proprietary materials which use nickel-coated carbon fibres as the conductor and are encapsulated in a wet-sprayed cementitious overlay or in a conductive paint coating, as illustrated in Figs 11.9–11.11. There is a NACE test method for qualifying these anodes.[37]

Another anode system can be used as an impressed current or a galvanic cathodic protection anode (see Section 11.7.2). This is thermal sprayed zinc. It has the attributes of a coating (no change in load, cross-section or clearances), but is more moisture-tolerant than carbon-loaded organic coatings. It is widely used in the USA but less so in Europe. One example is shown in Fig. 11.12.

Different anode systems have different advantages and limitations.

(a) Conductive organic coatings are easy to apply and do not change the loads or dimensions of the structure but are moisture-sensitive and have limited resistance to wear and tear. A cosmetic top coat

457

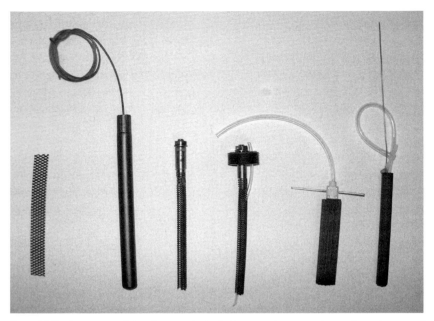

Fig. 11.7 A range of 'probe anodes'. Left to right: mixed-metal oxide-coated titanium ribbon, mixed-metal oxide-coated titanium tube anode, mixed-metal oxide-coated titanium mesh probe anode with resistor, mixed-metal oxide-coated titanium probe anodes with a resistor and cap for retaining grout when installed in soffits, and two shapes of conductive ceramic anodes. All anodes come in varying diameters and/or lengths

can provide an attractive finish that hides any patch repairs. The system has a limited life compared with mixed-metal oxide-based anodes. They have modest current outputs.

(b) Arc-sprayed zinc does not change the loads or dimensions of the structure and is more moisture-tolerant than organic carbon-loaded coatings but is more difficult to apply. It also has modest current outputs.

(c) Probe anodes are easy to install, do not change the loads or dimensions of the structure but require the use of percussion drills, so there are health-and-safety implications. They are often installed unobtrusively in surface details in buildings. They can short circuit to the reinforcing steel unless the steel location can be accurately mapped. The anode zone current output can be varied by the size and spacing of anodes.

(d) Ribbon anodes do not change the loads or dimensions of the structure, but they require chases in the concrete and adequate concrete cover to avoid short circuiting directly to the steel. Like probe

458

Fig. 11.8 A mixed-metal oxide-coated titanium mesh anode fixed to a concrete surface prior to applying a sprayed concrete overlay. (Courtesy of WS Atkins and the Highways Agency)

anodes, they can be installed unobtrusively in surface details in buildings. They also can short circuit to the reinforcing steel unless the steel location can be accurately mapped. The current output can be varied by the spacing of the anodes.

(e) Mesh anodes with their cementitious overlays increase the loads and dimensions of the structure. They can be extremely durable, and take wear and tear very well. Patch repairs will be hidden under the overlay. However, they require excellent quality assurance and control procedures to ensure a good bond between the original surface and the overlay. They can have very high current outputs, using different meshes and even multiple layers of mesh.

459

Fig. 11.9 A conductive paint coating applied to a building suffering from calcium chloride admixed into the concrete on the lowest two floors during construction. (Courtesy of Taylor Woodrow Construction)

Fig. 11.10 The same building after application of a cosmetic/protective top coat. (Courtesy of Taylor Woodrow Construction)

Fig. 11.11 Applying the conductive cementitious overlay to a bridge. (Courtesy of BASF – Construction Chemicals (UK) Ltd)

(*f*) The sprayed conductive mortar anode also increases loads and dimensions of the structure. It is very durable and has excellent adhesion to the original concrete surface. Its output is not as great as a mesh but this can be increased by increasing the anode thickness.

Thus, it can be seen that there is a wide range of anodes suitable for different applications with different installation requirements, different durability, wide-ranging appearance and different current output.

11.7.1.3 Control and monitoring systems

Impressed current cathodic protection systems are powered by a series of small power supplies, usually applying one or two amps per zone at a few volts direct current. Each zone will have one or more embedded reference electrodes (see Sections 4.4.5.3 and 4.4.6.1 and Fig. 11.6). These are used to measure the electrochemical potential of the steel to determine whether there is sufficient protection applied to the steel in the zone.

Fig. 11.12 Arc-sprayed zinc applied to a bridge on the M6 in Cumbria. The very top and the cantilevered ends of the leaf piers have probe anodes installed due to the higher moisture exposure

Measurements may be made manually on a simple system of 1–4 small zones, but is more commonly linked to a logger which records the steel to reference electrode potential automatically, switching off the current briefly to make 'iR-free' measurements. They will also take depolarisation measurements, typically once a month, where they will switch off the current for 24 hours logging of the potential decay.

The European standard for cathodic protection of steel in concrete BS EN 12696:2000 has a number of optional criteria for achieving corrosion control. The most commonly applied to atmospherically exposed reinforced concrete is that the steel to reference electrode potential should decay by 100 mV in less than 24 hours from the instant after switching off the system.

Control and monitoring systems are generally modular, with power supplies and voltage and current monitors slotting into a rack to power the system and measure anode zone currents, voltages and the steel to reference electrode potentials. If it is automated, the logging system will record values on a regular basis, and conduct depolarisation tests. Data can be collected by going to site, but usually it is done with a suitable telecommunications link to an office-based computer, which can also be used to adjust the current and voltage outputs once the data have been analysed.

11.7.1.4 Feasibility and conceptual design

When considering cathodic protection, a repair designer must first ensure that the cathodic protection system address the deterioration problem, either fully or in combination with other procedures, such as patch repairs, applying coatings and controlling water ingress. It is essential that any adverse effects are considered. The primary one is that there is no risk of hydrogen embrittlement of any prestressing steel. Other issues include any coatings, sealants or other insulating materials that might impede the current flow from the anode to the cathode.

As discussed in Section 11.7.1.2, there is a wide range of anode types. More than one anode can be used on a structure, as shown in Fig. 11.12.

The reinforcement dimensions and layout must be determined in order to calculate the steel surface area and hence the required current density for the anode and the total current for the power supplies. The structure will generally be broken down into a series of independently powered and monitored anode zones, depending on the size and geometry of the structure, the steel densities within it and its exposure (e.g. submerged, tidal and splash zones on a marine bridge substructure).

The location and design of connections to the reinforcement and to the anodes will be undertaken, along with the number and location of reference electrodes and possibly other monitoring probes. Cable dimensions, cable runs and cable management systems will be determined. The number, capacity and performance requirements of the control and monitoring system will also be set down in the design. Some of these issues may need adjustment once on site, but they should all be set down in design documents with suitable drawings, a quality management system, method statements, and a materials and equipment list before work starts, along with a full health-and-safety plan and procedures (in the UK, as required under the Construction (Design and Management) Regulations 2007).

11.7.1.5 Installation and operation

The system will be installed by a competent and experienced contractor to the designs and method statements provided. For surface-mounted

463

anodes, surface preparation is the key to a successful, durable installation. This may require trials to achieve the required bond and finish, as well as adhesion tests during anode application. Cable runs may be embedded in overlays, recessed into the concrete or mounted the surface with clips, conduit or cable trays.

In the UK, the usual requirement is for the contractor to install and commission the system, and then, after commissioning (which usually constitutes contract completion), the contractor operates the system for 11 months, undertaking the quarterly monitoring and then providing a final 11 month full review of the system performance. The system is then handed over to the client, which must then either be trained in the operation of the system or must place a contract with a suitably qualified cathodic protection engineer to undertake the ongoing monitoring. There is a new European standard, BS EN 15257:2006, which defines the levels of competence and certification of cathodic protection personnel: the design, installation and monitoring shall be under the supervision of a level 3 certified engineer, but work can be carried out by level 1 and level 2 technicians to method statements provided by the level 3 engineer. However, it will take several years for there to be a reasonable reservoir of engineers and technicians with suitable certification to be appointed.

11.7.2 *Galvanic or sacrificial anode cathodic protection systems*

11.7.2.1 *Theory*
Galvanic cathodic protection relies on the fact that metals such as aluminium and zinc are higher in the electrochemical series than iron or steel. This means that when coupled together in a corrosive electrolyte, they will corrode preferentially, and can protect the steel from corrosion. If we think of a simple zinc carbon single cell (battery), the zinc case is the anode which corrodes; the carbon rod in the centre does not corrode. In reinforced concrete, we use a cruder cell of steel and zinc, where the zinc corrodes, producing a small current at up to about 1 V DC, which protects the reinforcing steel.

11.7.2.2 *Galvanic anodes*
Anodes for steel in concrete are generally zinc, although there are presently two made of aluminium/zinc/indium alloy on the market as

Fig. 11.13 Thermal spraying zinc on a bridge substructure in the Florida Keys. (Reproduced with permission from the Florida Department of Transportation)

proprietary galvanic anodes. Anodes come in the following forms:

- Coatings on the surface, generally electric arc sprayed (zinc or aluminium/zinc/indium alloy). The thermal spraying process of applying a galvanic zinc anode system to a bridge substructure is shown in Fig. 11.13. The coating thickness is a few tenths of a millimetre, to avoid adhesion problems due to thermal expansion during application. The thickness achieved will determine the life of the system. Pure zinc can be used on marine structures, as the concrete stays wet and therefore electrically conductive. A 'humectant' spray can be applied to decrease the electrical resistance of the concrete if applied inland in dryer conditions. The proprietary aluminium/zinc/ indium systems are designed to work in marine and non-marine conditions, as the alloy has a higher output than pure zinc.
- Zinc sheets fixed to the surface with a conductive hydrogel adhesive are shown in Fig. 11.14. The amount of zinc in the system can give lives of 25 years or more for this anode. It is important that moisture cannot get at the hydrogel, which is soluble and can leak out of inadequately sealed systems in wet conditions.
- Expanded zinc mesh in a permanent form clamped to a column and filled with cementitious grout, as shown in Fig. 11.15. The amount of zinc in the system can give lives of 25 years or more for this

465

Fig. 11.14 Sheets of zinc hydrogel applied to a bridge substructure. (Reproduced with permission from BAC Corrosion Control Ltd)

Fig. 11.15 Expanded zinc mesh anodes in a permanent form filled with cementitious grout and clamped to columns on a jetty in St Helier, Jersey

466

Fig. 11.16 A disk of zinc embedded in a proprietary mortar wired to the reinforcement exposed during repairs. (Reproduced with permission from Fosroc (UK) Ltd)

anode. It is generally used on piles or columns in marine conditions, but has been applied in a flat form to jetty soffits.

- A disk of zinc embedded in a proprietary mortar wired to the reinforcement exposed during repairs, as shown in Fig. 11.16. This anode is particularly aimed at the 'incipient anode' problem (see Section 11.4.2), extending the life of patch repairs and reducing the ongoing risk of corrosion around the repair rather than providing structure-wide corrosion control.

- A rod of zinc embedded in a proprietary mortar inserted in a core hole in the concrete and connected in a string of other similar anodes to the steel, as shown in Fig. 11.17. This is a 'structure-wide' development of the anode shown in Fig. 11.16.

11.7.2.3 Control and monitoring systems

Generally, there is not much to control or monitor in galvanic cathodic protection systems compared with an impressed current system. Systems for measuring current flows can be inserted in trial systems or in

467

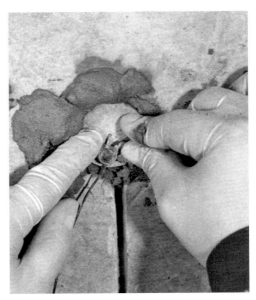

Fig. 11.17 A rod of zinc embedded in a proprietary mortar being inserted in a core hole in the concrete and connected in a string of other similar anodes to the steel to form an anode string or array. (Reproduced with permission from Fosroc[19] Ltd)

representative locations by separating an anode or group of anodes and measuring the current flow to the steel, usually via a resistor in the circuit. It is also possible to insert a switch and use an embedded or portable reference electrode to measure instant off and depolarisation (see Section 11.7.1.3). However, the current cannot be adjusted, except by the use of resistors if the current is too high.

Since galvanic anodes have finite lives and since they cannot be adjusted, structures containing galvanic anodes should be checked on a regular basis. Monitoring can be performed by:

- visual survey of the structure and the cathodic protection system to check for corrosion of the steel or any effects on the anodes
- potential survey of the structure to determine the activity of embedded anodes
- potential shifts of the steel (if a switch between anode and cathode is installed)
- current flow from the anodes to the steel (if a switch between anode and cathode is installed)
- corrosion rate measurements (if a switch between anode and cathode is installed).

11.7.2.4   Feasibility and conceptual design

As with an impressed current design, the choice of anode is critical. It is, then, important to ensure there is sufficient anode material available to provide the required design life. This will be based on calculations of the steel surface area and the likely current demand (measured in trials or estimated from comparable exposure conditions). For the two embedded anodes, the anode supplier has tables of anode density versus the steel surface area.

11.7.2.5   Installation and operation

It can be seen that most anodes can be installed using fairly conventional techniques known to the civil engineer. The exception is if arc spraying zinc or aluminium/zinc/indium onto concrete. A detailed specification for such an application has been written by the American Welding Society,[38] based on its extensive use on bridges in the Florida Keys and other structures on the Florida coast and elsewhere in the USA.

11.8   Electrochemical chloride extraction and realkalisation

Electrochemical chloride extraction and electrochemical realkalisation are two closely related techniques that use a similar arrangement to that shown in Fig. 11.6, but use a temporary anode and a high current to re-establish a passive, alkaline, chloride-free environment around the steel.

11.8.1   Theory

As established in Section 4.4, the chloride ion Cl^- is effectively a catalyst for corrosion of steel when it has accumulated at a sufficient concentration at the steel surface. We can remove chloride ions by applying a negative charge to the steel and a positive charge to an external temporary anode, repelling the chloride ions from the surface and removing a certain proportion from the concrete. In addition, hydroxide ions are formed at the steel surface by the cathodic reaction, as shown in Fig. 4.13. The hydroxide ions repassivate the steel surface and, in particular, the actively corroding pits, suppressing corrosion. As the hydroxide ion concentration rises and the chloride concentration declines, the chloride/hydroxide ion ratio drops below the threshold for corrosion. Additional effects include the infilling of voids at the steel surface by the migration of the ions under the electric field, leading to the precipitation of calcium hydroxide. This means that there is a reduction in initiation sites for future corrosion.

For carbonated concrete, the generation of hydroxide ions realkalises the concrete. Also, an alkaline electrolyte is used, which moves into the concrete under normal absorption capillary and diffusion processes and, in some cases, by electro-osmosis (discussed in the next section).

11.8.2 Anodes

The mixed-metal oxide expanded titanium mesh and mesh ribbon anode described in Section 11.7.1.2 is widely used in tanks or bunded systems on decks. Plain steel mesh is also used. The steel is consumed, and can stain the concrete surface. However, its consumption avoids the generation of acid at the steel surface, which can etch the concrete surface. A proprietary system uses a shredded paper (papier mâché) spray using a suitable electrolyte. Figures 11.18 and 11.19 show examples of these anode systems.

Electrolytes for electrochemical chloride extraction include tap water, calcium or other hydroxide solutions to maintain alkalinity, or

Fig. 11.18 Electrochemical chloride extraction steel mesh plus cellulose fibre overlay being applied. (Reproduced with permission from Makers Ltd)

Fig. 11.19 Cassette-tank system using a mixed-metal oxide-coated titanium anode. (Reproduced with permission from Makers Ltd)

lithium hydroxide or lithium borate, where there are concerns about alkali silica reaction.

For realkalisation, potassium carbonate is recommended, as it chemically increases the pH of the concrete at the surface while the electrochemical realkalisation is occurring at the steel.

11.8.3 Control and monitoring systems

Power supplies range from simple mains rectified direct current supplies to highly sophisticated monitoring and logging systems that measure and control the current flow to the systems. Voltages are kept below $50\,V$ DC for health-and-safety reasons, and current below $4\,A/m^2$ of the steel surface area, to minimise effects on the concrete. The charge passed in amps per square metre of steel surface area should be logged.

471

11.8.4 Feasibility and conceptual design and specification

There is a US NACE Standard Practice for electrochemical chloride extraction and realkalisation.[39] In Europe, the first part of DD CEN/ TS 14038-1:2004 on realkalisation has been published, but Part 2 on electrochemical chloride extraction is still in preparation. The issues of electrical continuity, short circuits from steel to anodes, hydrogen embrittlement of susceptible steels and ASR are even more critical for these techniques than they are for impressed current cathodic protection. Like an impressed current cathodic protection, the areas to be protected will be divided into independently controlled zones. It is important to measure the steel surface area in each zone and that zones reflect differences in steel densities.

One major issue for application of these systems is end point determination. For realkalisation it is fairly straightforward, as phenolphthalein can be used to measure the elevation of pH above the level where corrosion can occur (see Section 4.4.2). However, there are risks and therefore cost implications if a contractor has to achieve a given level of realkalisation without some limitation on the total number of amp-hours passed and therefore the time spent realkalising.

For electrochemical chloride extraction, a number of criteria are given in NACE SP0107-2007.[39] These are the achievement of a specified level of chloride in the vicinity of the reinforcement, such as less than 0.4% chloride by mass of cement within 25 mm of the bar, or the passing of a given amount of current per square metre of steel, usually in the region of 600–1500 Ah/m^2, and achieving a chloride/ hydroxide ratio of less than a defined level such as 0.6. Contractors can of course guarantee to achieve a criterion based on the charge passed. There are risks associated with requiring the contractor to achieve given chloride concentrations or chloride/hydroxide ratios without giving an upper amp-hour limit. If treatment goes on longer than expected to achieve the end point criteria, this could have cost and programming implications for a project unless current density limits are used in conjunction with a concentration criterion.

11.8.5 Installation and operation

The installation process is similar to that for an impressed current cathodic protection system, with checks of electrical continuity of the steel, connections to the steel, installation of the anode, checks for short circuits before and after applying the electrolyte, and then application of the current with regular monitoring, testing and sampling. There should be detailed records of all results of testing before, during and on completion of the treatment.

11.9   Electro-osmosis

The phenomenon of electro-osmosis has already been mentioned in connection with electrochemical realkalisation. When a porous medium such as concrete contains a solution, then an electric current applied between an anode and a cathode will move the water, usually from the anode to the cathode. This leads to drying of cathodic protection anodes for pipelines in soils. The basis of the phenomenon is that when a compound dissolves and forms charged ions, water molecules attach themselves to the ions. This happens more for positively charged metal ions than for negatively charged ions. Therefore, more water is carried by the positive ions towards the negative cathode.

There is proprietary technology for applying this methodology to reinforced concrete. The proprietary part is the use of pulses to reduce the build up of charged ions at the electrodes and the reinforcement, which would increase the electrical resistance. The system works by the installation of cathodes in areas where the water can be discharged, and anodes in areas where water must be removed.

The technology is described by McInery *et al.*[40,41] Several systems have been installed in the UK on a range of structures. The author was the consultant for a trial system with horizontal mixed-metal oxide-coated titanium wire anodes and cathodes across the leaf piers and cathodes in the ground to draw water away. The aim of the system was to reduce the relative humidity in the concrete to a level which would not sustain corrosion below, say, 60% relative humidity (see Section 4.11.4 and Lambert[42]).

The main problem that the author is aware of with such systems is that the anode is heavily stressed if there is significant water ingress or run-down. Very high current densities can occur on the anodes. This can lead to them failing. It is therefore likely that anodes are in need of further development before the technology can be considered fully developed for external applications where random wetting events cannot be avoided.

In addition, the pulsing process requires sophisticated electronics, which must prove durable over many years. They are likely to be less reliable than simpler transformer rectifier systems providing straight direct current for impressed current cathodic protection.

Although electro-osmosis systems for drying building materials (e.g. concrete or masonry block basements) are well-known technology, at the time of writing, this technology is in its infancy for the corrosion control of reinforced-concrete structures. The issue of stray-current-induced corrosion of the reinforcement should also be considered. Some proprietary electro-osmosis systems are designed to work with the reinforcing cage. Others are not, and, as such, could accelerate reinforcement corrosion.

11.10 Selection of repair techniques, achieving the design life and life cycle costing

Achieving an effective repair requires the balancing of first cost, life cycle cost, aesthetics, effectiveness of the repair and disruption during repair and the future requirements for the structure.

11.10.1 Costs

The aim of any repair is to be cost-effective for the remaining life of the structure. Costs should include the first cost and the life cycle cost. Effectiveness will include what is feasible, what is aesthetically acceptable and any other constraints such as access. Such determinations should be carried out by an engineer who is fully up to date with all the technologies being considered and with the relevant standards for their application. For instance, when considering impressed current cathodic protection, the wide range of anodes available now make it applicable to far more structures than previously when only expanded mixed-metal oxide-coated titanium mesh and conductive paint were available.

Obtaining cost data is difficult, particularly to obtain up-to-date prices and rates for more specialised techniques. There are comparative 2000 price data for the UK for most of the techniques described in this chapter on a BRE projects website.[43] These can be used to generate first cost and life cycle cost comparisons. The website also gives guidance on the constraints and limitations of the techniques. More up-to-date information on impressed current cathodic protection costs for highway structures can be found in Broomfield.[44] However, these costs were collected from UK Highways Agency projects with more complex contractual, access and health-and-safety issues than many other projects, so the reported costs may be twice those of, say, a parking structure.

474

11.10.2 Treating the causes as well as the symptoms

A well-planned investigation of the structure and the damage on it is critical to the correct choice of repair. There is little guidance from published standards on site investigations, but guidance documents such as those by the Concrete Society[45] and the Concrete Bridge Development Group[1] cover the techniques to be used and their application. There is also more background in books on the subject, such as that by Broomfield.[2]

11.10.3 Planning investigations and conducting repairs using EN 1504

The new European standard EN 1504 (Parts 1 to 10) contains a lot of valuable information on concrete repair materials, and systems and their application. It is important to understand that this standard was written under the Construction Products Directive, and therefore it is designed to inform manufacturers of the requirements for their products to achieve CE marking and how to specify the materials properties so that comparable materials that meet the performance requirements for the project can be offered by materials suppliers. However, Parts 9 and 10 give guidance on the range of repair options available, how to specify the materials and how they should be applied on site.

A systematic method of using EN 1504 would be as follows:

- EN 1504-9 summarises the evaluation process covering both structural and reinforcement corrosion issues. The corrosion evaluation is covered in Section 4.4.5 of this book.
- EN 1504-9 summarises the options. These are covered in this chapter. EN 1504-9 discusses:
  - the 'do nothing' option, with or without monitoring
  - structural issues
  - prevention or reduction of further deterioration
  - improvement, strengthening or refurbishment of all or part of the structure
  - full or partial reconstruction.
- EN 1504-9 then lists a series of factors to be considered when choosing a rehabilitation option:
  - intended use of the structure
  - performance characteristics (e.g. water tightness, fire resistance)
  - repair life time
  - opportunities for additional repair and monitoring

- life cycle costing throughout the residual life of the structure
- performance and properties of the existing substrate
- final appearance.
- There are additional issues, such as:
  - health and safety
  - environmental impact
  - appropriateness of the selected option(s)
  - conformity with the standard.
- The concrete rehabilitation materials for non-electrochemical techniques can then be listed, and the required properties selected from Table A-1 in the informative annex to EN 1504-9. This lists tests to achieve the required characteristics (bond strength, shrinkage, elasticity, etc.) from EN 1504, Parts 2 to 7.
- The production of a specification and application conditions and methods are discussed in EN 1504-10. The most useful information being in the informative annexes.
- Finally, EN 1504-9 requires the compilation of a record of the works and an inspection and maintenance procedure.

A basic procedure for conducting a repair would be as follows:

(a) Determine the cause and extent of damage by a quantitative survey using suitable techniques that measure how much repair is needed and what causes of deterioration must be addressed.
(b) If the cause is structural, then look at strengthening options such as plate bonding and other reinforcement, structural concrete repairs, partial or full replacement, etc.
(c) If the cause is concrete deterioration (ASR, sulfate attack, etc.), address the causes and conduct conventional concrete repairs using first cost and life cycle cost analysis to ensure the most cost-effective solution.
(d) If the cause is reinforcement corrosion, again we must address the cause of deterioration. The first requirement is to determine whether it is chloride or carbonation or both, to shortlist options.
(e) For all repair options, consider the following:
  - required remaining life of the structure
  - required life of the treatment or the time to the next intervention
  - possible side-effects of treatments
  - aesthetics
  - access issues
  - other constraints

 – available expertise and familiarity with the systems
 – ongoing maintenance and monitoring requirements.
(*f*) Once a repair option is selected, produce suitable designs and specifications using the relevant national and international standards.

11.11   Conclusion

There is a wide range of repair options available to repair and rehabilitate damaged reinforced-concrete structures, as detailed in this chapter. No one technique is necessarily superior to the others, although some (such as corrosion inhibitors applied after corrosion initiates and electro-osmosis) are less tried and tested than others.

Selection of the optimum combination of repair techniques requires a well-designed and executed condition survey. This is followed by proper evaluation of the available options by an engineer fully conversant with the techniques and experienced in their application.

While EN 1504 provides a useful systematic framework for concrete repair, it is no substitute for a knowledgeable and experienced engineer.

Design and specification of repairs is likely to be most cost-effective if the contractor is fully aware of the information available as well as any gaps in that information. The equitable sharing of risk between the contractor and client can be critical to the pricing and success of any concrete repair project.

Whichever repair and corrosion control system is applied, it is important to fully and accurately record the work done and the future maintenance requirements in an operation and maintenance manual or the health and safety file for the structure.

References

1. Concrete Bridge Development Group (2002) *Guide to testing and monitoring the durability of concrete structures.* Technical Guide 2. Concrete Society, Camberley.
2. Broomfield, J.P. (2007) *Corrosion of steel in concrete, understanding, investigation and repair.* Taylor and Francis, London.
3. Stark, D. (1991) *Handbook for the identification of alkali silica reactivity in highway structures,* SHRP-C-315. Strategic Highway Research Program, National Research Council, Washington, DC.
4. Kaetzel, L., Clifton, J., Snyder, K. and Kleiger, P. (1994) *Users guide to the Highway Concrete (HWYCON) Expert System.* SHRP Report and

Computer Program SHRP-C-406. National Research Council, Washington, DC.

5. Alldred, J.C. (1993) Quantifying the losses in cover-meter accuracy due to congestion of reinforcement. In: *Proceedings of the 5th International Conference on Structural Faults and Repair*. Edinburgh, vol. 2, 125–130.

6. Concrete Society (2004) *Electrochemical tests for reinforced concrete*. Technical Report 60. Concrete Society, Camberley.

7. BRE (1995) *Carbonation of concrete and its effects on durability*. BRE Digest 405. Building Research Establishment, Garston, Watford.

8. BRE (2000) *Corrosion of steel in concrete, part 2. Investigation and assessment*. BRE Digest 444. Building Research Establishment, Garston, Watford.

9. BRE (1998) *Assessing carbonation depth in ageing high alumina cement concrete*. BRE Information Paper IP 11/98. Building Research Establishment, Garston, Watford.

10. Chess, P. and Grønvold, F. (1996) *Corrosion investigation – a guide to half cell mapping*. Thomas Telford, London.

11. Concrete Society (1997) *Guide to surface treatments for protection and enhancement of concrete*. Technical Report 50. Concrete Society, Camberley.

12. Kropp, J. and Hilsdorf, H.K. (1995) *Performance criteria for concrete durability*. RILEM Report 12. ERepair of concrete damaged by reinforcement corrosion – report of a working party. Technical Report 26. Concrete Society, Camberley.

14. Concrete Society (2003) *Strengthening concrete structures with fibre composite materials: acceptance, inspection and monitoring*. Technical Report 57. Concrete Society, Camberley.

15. American Concrete Institute (2001) *Protection of metals in concrete against corrosion*, ACI 222R-01. American Concrete Institute, Farmington Hills, MI.

16. BRE (2000) *Corrosion of steel in concrete, Part 2. Investigation and assessment*. BRE Digest 444. Building Research Establishment, Garston, Watford.

17. Institution of Civil Engineers (2002) *National Steering Committee for the Inspection of Multistorey Car Parks. Recommendations for the inspection, maintenance and management of car park structures*. Thomas Telford, London.

18. Macdonald, S. (2003) *Concrete – building pathology*. Blackwells, London.

19. Concrete Society (2004) *Design guidance for strengthening concrete structures using fibre composite materials*. Technical report 55. Concrete Society, Camberley, 2nd edn.

20. Tilly, G.P. (2006) Past performance of concrete repairs. In: M.G. Grantham, R. Jauberthie and C. Lanos (eds), *Concrete Solutions: Proceedings of the 2nd International Conference, St Malo*. BRE Press, Watford, paper 2.

21. Concrete Society (1991) *Patch repair of reinforced concrete – subject to reinforcement corrosion: model specification and method of measurement*. Technical Report 38. Concrete Society, Camberley.

22. American Concrete Institute (2008) *Concrete repair manual*. American Concrete Institute, Farmington Hills, MI, 3rd edn.

23. Sergi, G., Seneviratne, A.M.G., Maleki, M.T., Sadegzadeh, M. and Page, C.L. (2000) Control of reinforcement corrosion by surface treatment of concrete. *Proceedings of the Institution of Civil Engineers: Structures and Buildings*, 140(1), 85–100.

24. Manning, D.G. (1995) Waterproofing membranes for concrete bridge decks: a synthesis of highway practice. In: *NCHRP Synthesis 220, National Cooperative Highway Research Program*. Transportation Research Board, National Research Council, Washington, DC.

25. Price, A.RC. (1989) *A field trial of waterproofing systems for concrete bridge decks*, TRRL Research Report 185. Transport Research Laboratory, Crowthorne.

26. Virmani, Y.P. and Clemena, G.G. (1998) *Corrosion protection – concrete bridges*. Technical REPORT RD-98-088. Federal Highway Administration, Washington, DC.

27. Li, L., Sagüés, A.A. and Poor, N. (1999) In-situ leaching investigation of pH and nitrite concentration in concrete pore solution. *Cement and Concrete Research*, 29(3), 315–321.

28. Raharinaivo, A. and Malric, B. (1998) Performance of monofluorophosphate for inhibiting corrosion of steel in reinforced concrete structures. In: *Proceedings of the International Conference on Corrosion and Rehabilitation of Reinforced Concrete Structures*. Orlando.

29. Broomfield, J.P. (2000) Results of long term monitoring of corrosion inhibitors applied to corroding reinforced concrete structures. In: *Corrosion 2000*. Houston, paper 791.

30. Treadaway, K.W.J. and Russell, A.D. (1968) The inhibition of the corrosion of steel in concrete. *Highways Public Works*, 36(19), 40–41.

31. Al-Qadi, I.L., Prowell, B.D., Weyers, R.E., Dutta, T., Gouru, H. and Berke, N. (1993) *Concrete bridge protection and rehabilitation: chemical and physical techniques – corrosion inhibitors and polymers*. SHRP-S-666, Strategic Highway Research Program, National Research Council, Washington, DC.

32. Prowell, B.D., Weyers, R.E. and Al-Qadi, I.L. (1993) *Bridge protection and rehabilitation: chemical and physical techniques – field validation*. SHRP-S-658, Strategic Highway Research Program, National Research Council, Washington, DC.

33. Sohangpurwala, A.A., Islam, M. and Scannell, W.T. (1997) Performance and long term monitoring of various corrosion protection systems used in reinforced concrete bridge structures in North America. *Proceedings of the International Conference on Repair of concrete structures*. Additional papers: 9 to 19. Svolvaer Norway Publ, Norwegian Road Research Laboratory, Oslo.

34. Sprinkel, M. and Ozyildirim, C. (1998) Evaluation of exposure slabs with corrosion inhibitors. *Proceedings of the International Conference on*

Corrosion and rehabilitation of reinforced concrete structures, Florida, FHWA-SA-99-014, Federal Highway Administration, Washington, DC.

35. Broomfield, J.P., Davies, K. and Hladky, K. (1999) The use of permanent corrosion monitoring in new and existing reinforced concrete structures. In: *NACE Corrosion/99.* Houston, paper 99559.

36. NACE (2007) *Testing of embeddable impressed current anodes for use in cathodic protection of atmospherically exposed steel-reinforced concrete,* TM0294-2007. NACE International, Houston, TX.

37. NACE (2007) *Test procedures for organic based conductive coatings anodes for use on concrete structures,* TM 0105-2005. NACE International, Houston, TX.

38. American Welding Society (2002) *Specification for thermal spraying zinc anodes on steel reinforced concrete,* AWS C2.20/C2.20M:2002 (ANSI). American Welding Society, Miami, FL.

39. NACE (2007) *Electrochemical realkalization and chloride extraction for reinforced concrete,* SP 0107-2007. NACE International, Houston, TX.

40. McInerney, M., Morefield, S., Cooper, S., Malone, P., Weiss, C.A., Brady, C., Taylor, J. and Hock, V.F. (2002) *Electro-osmotic pulse (EOP) technology for control of water seepage in concrete structures,* ERDC/CERL TR-02. US Army Engineer Research and Development Center, Construction Engineering Research Laboratory, Champaign, IL.

41. McInerney, M.K., Cooper, S.C., Hock, V.F. and Morefield, S.W. (2004) Measurements Of water and ion transport in concrete via electro-osmosis. In: *Proceedings of Corrosion/2004.* NACE International, Houston, TX, paper 04351.

42. Lambert, P. (1997) Controlling moisture. *Construction Repair,* 11(2), 29–32.

43. Broomfield, J.P. (2006) A web based tool for selecting repair options and life cycle costing of corrosion damaged reinforced concrete structures. In: M.G. Grantham, R. Jauberthie and C. Lanos (eds), *Concrete Solutions: Proceedings of the Second International Conference, St. Malo.* BRE Press, Walford, 505–513.

44. Broomfield, J.P. (2008) *Budget cost and anode performance information for impressed current cathodic protection of reinforced concrete highway bridges.* Technical Note 11. Corrosion Prevention Association, Aldershot.

45. Concrete Society (2004) *Electrochemical tests for reinforcement corrosion.* Technical Report 60. Concrete Society, Camberley.

12

Issues related to performance-based specifications for concrete

John Bickley, J. A. Bickley Associates, Toronto, Canada
Douglas Hooton, University of Toronto, Canada
Kenneth C. Hover, Cornell University, Ithaca, USA

12.1   Introduction

As part of its initiative to facilitate a construction industry change from prescription specifications to performance specifications (P2P), the US National Ready Mixed Concrete Association (NRMCA) commissioned a review of the international state-of-the-art. A literature search was made (Bickley, Hooton and Hover, 2006),[1] and from this, a draft performance specification was produced (Bickley, Hooton and Hover, 2008).[2] This draft specification is being used as one of the inputs by a current special American Concrete Institute (ACI) task group on performance specifications. While the international literature was reviewed, this chapter deals with performance specifications from the North American perspective.

From the review of current standards and the literature, it became clear that while there was an almost universal interest in performance, primarily for durability, there were few specifications that contained any pure performance criteria. Most defined exposure conditions that pertained to each country, and then tabulated concrete mixture contents and limits that studies had shown would result in the desired durability. These include maximum limits for water/cement (w/c) or water/cementitious materials (w/cm) ratio, minimum cement contents and an acceptable range of air contents. There is an almost universal use of supplementary cementitious materials, such as fly ash, ground granulated blast furnace slag and silica fume, either as additions or in blended cements. All the specification documents assumed the use of statistical quality control to assure consistent strength conformity.

It also became clear that the term 'performance specification' means many things to many different people. This is not necessarily because of any misinterpretation. This is because there is such a wide array of

Concrete durability
978-0-7277-3517-1

options and valid interpretations, making it imperative that the term be carefully defined in any given context. Parties could agree in principle to execute work under the performance specification umbrella and yet have widely differing views about mutual expectations.

A lack of reliable, consistent and standardised test procedures for evaluating concrete performance is frequently cited as a major barrier to the adoption of performance specifications. Some of the available tests can be expensive, take a long time to run and may not be as precise as desired. Short bid times and quick construction starts create a difficult situation for a concrete supplier faced with the need to develop a performance mixture and to perform prequalification testing. In a number of jurisdictions, such as state highway departments, some advanced tests have been site-proven and then specified in subsequent years for pay items in contracts.

On the other hand, in the face of an international mindset that says that testing technology has not yet caught up with performance philosophy, there are a wide range of tests that are available today, and have been used successfully on important concrete projects. These tests methods can be called into action to support performance-based specifications. While some may complain that current tests are not ideal or are insufficiently accurate or precise, which of our everyday concrete quality tests are ideal? If a new test only has to be as accurate, as precise or as meaningful as the slump test, there may be many new developments to choose from. While perhaps not used for acceptance, even a non-standard test can provide useful information, if it provides timely information for construction decisions and the site-engineer has experience with interpreting the results.

The advent of performance specifications could significantly change the distribution and sharing of responsibility among the owner, contractor and concrete supplier. It would be up to the owner (through design professionals) to clearly specify performance requirements together with the test procedures used for acceptance. In the case of true end-result specifications based on hardened, in-place concrete properties, the execution of these requirements would be the joint responsibility of contractor and concrete supplier. They would assume the risk involved and would have to work closely to determine the appropriate concrete mixture. Quality management programmes would also be required from both, since the successful installation of a concrete mixture would be imperative to achieving acceptance by the owner.

The transition to performance specifications as another, complementary way of doing business will require a dedicated educational effort, and advantages and disadvantages will have to be made concrete, so to speak. The motivation will have to come from clear benefits that can be shared at many levels of the industry and not just because it is time for a change.

12.2   What is a performance specification?

It is useful to begin the discussion by sampling a range of definitions of the term 'performance specification'.

12.2.1   Definitions

As documented on its website, the NRMCA discusses a performance specification as follows:

> A performance specification is a set of instructions that outlines the functional requirements for hardened concrete depending on the application. The instructions should be clear, achievable, measurable and enforceable. For example, the performance criteria for interior columns in a building might be compressive strength and weight since durability is not a concern. Conversely performance criteria for a bridge deck might include strength, permeability, scaling, cracking and other criteria related to durability since the concrete will be subjected to a harsh environment.
>
> Performance specifications should also clearly specify the test methods and acceptance criteria that will be used to enforce the requirements. Some testing may be required for prequalification and some for jobsite acceptance. The specifications should provide flexibility to the contractor and producer to provide a mix that meets the performance criteria in the way that they choose. The contractor and producer will also work together to develop a mix design for the plastic concrete that meets additional requirements for placing and finishing, such as flow and set time, while ensuring that the performance requirements are not compromised.
>
> Performance specifications should avoid requirements for means and methods and should avoid limitations on the ingredients or proportions of the concrete mixture.
>
> The general concept of how a performance-based specification for concrete would work is as follows:
> - There would be a qualification/certification system that establishes the requirements for a quality control management system, qualification of personnel and requirements for concrete production facilities.

- The specification would have provisions that clearly define the functional requirements of the hardened concrete.
- Producers and contractors will partner to ensure the right mix is developed, delivered and installed.
- The submittal would not be a detailed list of mixture ingredients but rather a certification that the mix will meet the specification requirements, including prequalification test results.
- After the concrete is placed, a series of field acceptance tests would be conducted to determine if the concrete meets the performance criteria.
- A clear set of instructions outlining what happens when concrete does not conform to the performance criteria.

12.2.2   US Federal Highway Administration (FHWA)

Part of the FHWA Performance Specifications Strategic Roadmap is devoted to a detailed description of performance specifications. It states: 'A performance specification defines the performance characteristics of the final product and links them to construction, materials and other items under contractor control'.

12.2.3   Canadian standard CSA A23.1

In Canadian standard CSA A23.1, a performance concrete specification is defined as follows:

> A performance concrete specification is a method of specifying a construction product in which a final outcome is given in mandatory language, in a manner that the performance requirements can be measured by accepted industry standards and methods. The processes, materials, or activities used by the contractors, subcontractors, manufacturers, and materials suppliers are then left to their discretion. In some cases, performance requirements can be referenced to this Standard, or other commonly used standards and specifications, such as those covering cementing materials, admixtures, aggregates or construction practices.

12.2.4   UK Highways Agency

The UK Highways Agency defines performance specifications as follows:

- *Output measures* define the end product of works carried out on the network. This is usually in the form of a series of outputs that will deliver the desired outcome. For example meeting road surface skid resistance requirements is one output that will help enable the safety outcome to be realised.
- *Outcome measures* define the benefits that should be delivered as a consequence of the works carried out on the network. This will usually take the form of the level of service required. For example journey time reliability or level of safety.

12.2.5   Cement and Concrete Association of New Zealand

Cement and Concrete Association of New Zealand – 'A performance-based specification prescribes the required properties of the concrete but does not say how they are to be achieved.'

12.3   Why have performance specifications become an issue now?

The competing or sometimes complimentary philosophies of prescriptive versus performance specifications have been around as long as there have been concrete specifications. In 1928, ACI Committee E-1 and the Concrete Reinforcing Steel Institute Committee on Engineering Practice proposed their 'Joint Code': *Building Regulations for Reinforced Concrete*. This document was the precursor to what many consider to be the 'first' ACI code, published in 1936 and also called *Building Regulations For Reinforced Concrete* (ACI 501-36T).[*] These early documents permitted use of concrete mixtures without 'preliminary tests of the materials to be used', as long as w/c met the prescriptive requirements.

The 1929, the Joint Code went on to require tests of at least one specimen per 100 cubic yards (CY) of concrete placed. Interestingly, this early set of regulations permitted a performance-based alternative to the prescriptive w/c requirements for concrete strength, by allowing prequalification of a mixture on the basis of test data correlating strength to w/c. Four different w/c values had to be tested with four specimens each (the forerunner of today's 'three-point curve'), and w/c approved for production was that value corresponding to a compressive strength 15% higher than specified (the forerunner of today's so-called 'overdesign'

---

[*] Proposed by Committee 501, the new code was presented as revised and tentatively adopted at the Annual Convention of the American Concrete Institute, Feb. 25, 1936.

strength). Once the proposed mixture was approved, no substitutions in materials were permitted without additional tests. A bias towards the prescriptive specification of w/c was apparent, however, as the frequency of testing had to increase to one specimen per 50 CY (40 m^3) placed if w/c had been established on the basis of contractor testing. There is no mention of durability or permeability in the 1928 Joint Code.

The state of the technology in those early days of the industry can be gauged from this statement from A.R. Lord's *Handbook of Reinforced Concrete Building Design*,[3] published in 1928: 'Engineers are so accustomed to thinking of concrete for buildings in terms of 2,000-lb [per in^2] strength at 28 days that it may be novel to consider using a 3,000-lb [per in^2] concrete as the basic mix' (Lord, 1928, p. 186). But changes were soon to be in the works.

S.C. Hollister, visionary engineer and educator, was president of the ACI in 1933–1934. In his outgoing address, he predicted the advent of chemical admixtures and high-strength concrete by saying:

> One may grow so accustomed to the surrounding conditions that they are accepted as a sort of status not subject to review... Who may say, for example, whether it is possible to achieve mobility or workability with an agent other than water... We see the many varied and intriguing avenues of development that present them-selves... Imagine, for example, the concrete with an available strength of 10,000 pounds per square inch. Smaller columns, thinner and lighter beams and slabs would at once result. Present limitations... would at least double. A new basis of design would be required. The achievement of today was the goal of yesterday.

Seventy-one years after Hollister's predictions, today's concrete has become a complex and truly 'engineered material', and this develop-ment has intensified the debate surrounding performance specifications.

Over this same time period of concrete's transition to an engineered material, many concrete producers have likewise transitioned from being merely 'truckers' who deliver concrete mixed in accordance with a specified recipe to being well informed on concrete materials, including complex aggregate grading, chemical admixtures and a wide range of cementitious materials. Similarly, when the 1928 code was published, the design professionals had the responsibility for preparing detailed prescriptive specifications and conducting careful inspections of the mixing process. More recently, fewer specifications require pre-determined concrete recipes or materials and production inspections.

Likewise, there has been a shift in the responsibility for concrete ingredients and mix proportions towards the concrete producer and away from the design professional. Today's review of concrete mixture submittals 'for general conformance with the contract documents' is a significant evolution from the fully specified mixture proportions of only a few years ago.

Other changes that have swept the industry include recognition that for many modern concrete applications, strength is no longer the only, or even the most important, issue. Portland cement is not the only cementitious material; water content and aggregate size are not the only factors that influence slump, and w/c is not the only factor influencing permeability. Air content is easily specified and readily measured in the field, but freeze–thaw durability and scaling resistance are more dependent on air void size and distribution in the paste than on the total air volume in the concrete. Chemical and mineral admixtures affect air, workability, setting time, bleeding, rate of strength gain, and early and later age strength. These same admixtures may or may not be mutually compatible. At the same time, it has become more difficult to write a prescriptive specification that can take advantage of these developments and avoid their pitfalls. It has become evident that evaluating the durability of concrete is more difficult than evaluating strength. It is more difficult to predict or assure the long-term service life of concrete than it is to predict or assure the short-term load capacity.

Thus, the simultaneously increasing demand for improved concrete durability and the growing complexity of concrete mixture design and proportioning lead us back to the prescription-to-performance debate. Interest is further fuelled by the changes in construction technique that have accompanied these newer concrete materials developments. An example of all of these factors is a high-performance, high-density, low-permeability concrete. Such mixtures often blend Portland cement with one or more cementitious materials such as silica fume and fly ash or slag, use up to two performance grades of water-reducing admixture, incorporate at least three sizes of aggregate and may have set-retarders and/or corrosion inhibitors plus an air-entraining admixture. Proportioning such a mixture requires experience with these specific materials, including recognition that the normal relationships between workability and water content, and between strength and w/cm, need to be re-calibrated. Handling, placing, finishing and curing such a mix requires experience as well to accommodate rapid surface drying, rapid setting and a high shrinkage potential. So, if the question

487

is, 'Why discuss performance specifications now when prescriptive specs have been used since the early ACI codes in 1928?' One answer is that we are now demanding more of the concrete, and that it may be difficult to take full advantage of the wide range of material and construction combinations and options under a strictly prescriptive specification.

12.4   The essence of prescription versus performance

The P2P initiative is directed towards a shift in the focus of concrete specifications. A prescriptive specification focuses on the properties of the raw materials, mixture proportions, the batching, mixing and transport of the fresh concrete, and the full range of construction operations from placing to curing. Prescriptive specifications rely on observed or implied relationships between the details specified and the final, in-place, or 'end-product' or 'end-result', performance of the concrete. Under a prescriptive specification, the end-product performance may or may not be described. In contrast, a pure performance specification 'starts with the end in mind', fully describing the required performance characteristics of the end-product, leaving materials selection, proportioning and construction means and methods up to the party contractually bound to comply with the specifications. Under a pure performance specification, it is the responsibility of the concrete producer–contractor team to select materials and conduct construction operations that will produce the required concrete performance. Proponents of prescriptive specifications say, 'Here is how we want you to proportion and install the concrete, and if done in accordance with these instructions, the results will be satisfactory'. Proponents of performance specifications say, 'Just tell me what you want done; don't tell me how to do it'.

Keys to the concept of performance specifications include:

(*a*) The ability of the specifications writer to discern the performance characteristics appropriate to the owner's intended use of the concrete.
(*b*) The ability of the specifications writer to describe these performance characteristics clearly, unambiguously and quantitatively so that performance can be evaluated.
(*c*) The availability of reliable, repeatable test methods that evaluate the required performance characteristics (along with performance compliance limits that take into account the inherent variability of each test method).

(*d*) The ability of the concrete producer–contractor team to correlate choices of materials, mixtures and construction techniques to the required characteristics so that projects can be planned and bid, risks and costs can be assessed, and materials and construction operations adjusted to comply with performance requirements.

These four keys present at least the following challenges:

(*a*) Under current, predominantly prescriptive specifications, end-product performance is not always comprehensively spelled out at the specification stage. For example, prescriptive specifications may not explicitly include requirements for abrasion resistance, scaling resistance or limitations on concrete cracking. Nevertheless, unsatisfactory performance in any of these categories is often pointed out after the concrete has been installed. The rationale for finding the concrete unsatisfactory may be that these common end-result requirements are generally implied and that the concrete would have been satisfactory if the prescriptive requirements would have been met. In contrast, performance specifications require an 'up front' description of owner expectations. In most cases, this can take significant additional effort and expertise beyond that required for prescriptive specifications by design professionals working on the owner's behalf.

(*b*) Some commonly expected (although uncommonly specified) performance characteristics are not readily clearly definable or readily quantified. In-place cracking, movements due to shrinkage, scaling, pop-outs, colour variations or local incidents of abrasion are easy to spot, but more difficult to describe in an unambiguous way.

(*c*) Despite an explosion of research and development into new concrete test methods, the industry does not yet have a comprehensive suite of test methods or the predictive models to allow their use to reliably predict service life in general.

(*d*) Some contractors and concrete producers will need additional training to be able to select materials and construction operations that will produce the required concrete. Design professionals will also need additional training or special expertise to develop the reliable performance requirements.

12.5   Advantages and disadvantages

The primary advantage to specifying end-product performance is that a knowledgeable concrete producer–contractor team has the flexibility to develop a unique combination of materials and construction methods that will achieve the owner–designer's stated objectives. Performance specifications will therefore work well when the producer–contractor team has the necessary expertise; the owner–designer can clearly articulate the requirements, and appropriate and sufficiently precise test methods are available for documenting the specified performance. A prescriptive specification may work best when there is a reliable connection between the specified materials, means and methods and the desired outcomes. The simpler the concrete mixture and the less restrictive the required outcomes, the more likely a prescriptive specification will be an efficient and reliable way to specify concrete. As the P2P initiative is implemented, it will be necessary to identify those opportunities for which performance specifications offer the greatest advantage, as well as to identify situations where a more conventional approach is more appropriate. Further comments on the advantages and disadvantages for both prescriptive and performance specifications are listed in Tables 12.1 and 12.2.

12.6   Available options

Given that any method for specifying concrete materials and construction services will have both advantages and disadvantages, the challenge (and the opportunity) is to develop appropriate specifications that maximise the advantages and minimise the disadvantages. This also means having a range of available specification-types that may be most appropriate in any given situation. For example, the British standard BS 8500 defines five approaches to specifying concrete:

- designed concrete – (similar to current practice under ACI 318)
- designated concrete – (a specific and certified mix that meets requirements of designed concrete)
- prescribed concrete – (fully prescriptive, 'recipe' specification)
- standardised prescribed concrete – (a 'standard mix' as with many public works-type standard mixes)
- proprietary concrete – (full performance).

The official definitions of these options are a bit difficult for the North American audience to understand, but they cover the range from pure performance to pure prescriptive to calling for a specific, pre-approved, proprietary concrete mixture. As evidence that 'there is nothing new under the sun', Elwyn Seelye's classic civil engineering reference

Table 12.1 Advantages and disadvantages of prescriptive specifications

Advantages	Disadvantages
Some designers and producers may have more confidence in, and be more comfortable with, traditional prescriptive approach	Some designers and producers may not be confident that prescriptive specifications lead to desired end performance
Expertise required at the spec-writing stage	Some specification writers may not have such expertise, especially with modern materials combinations
Value and effectiveness of the product is 'designed-in' by the specifier	Limited opportunity for optimisation of the concrete beyond the specification-writing stage
Newer materials and methods can be implemented if the specifier has remained technologically current	Limited opportunity to take advantage of the producer's unique access to materials, material combinations, plant, equipment, technology, expertise or knowledge of local materials and conditions
The specification reflects the specification writer's understanding of the relationship between the desired properties of the concrete and the specified materials, means and methods	The relationships implied in the prescriptive requirements may not be as reliable or the same as the relationships assumed for the specific materials or project conditions
The specification writer has the opportunity to control any aspects of the process, from concrete materials selection and proportioning to batching, mixing, transporting, placing, consolidating, finishing and curing the concrete. This control is exercised through prescriptive specification requirements	The interests of all parties may not be represented in the prescribed specification, i.e. raw materials suppliers, concrete producer, concrete-placing contractor, concrete-finishing contractor, general contractor, construction manager, owner, investor, end-user
Basic specification compliance tests are inexpensive, generally accepted and commonly available	Basic specification compliance tests may not tell us as much as we would like to know. Test results may be more variable (less reliable) than supposed. Conventional results may reflect the material as delivered and as subsequently cured under standard conditions, rather than as installed and as cured in actual field conditions. Standard tests may report results at concrete ages other than are critical for assurance of quality. Standard tests may not enable accurate prediction of longer-term concrete performance in the actual environment

Table 12.1 Continued

Advantages	Disadvantages
Prescriptive specifications could be interpreted to limit the concrete producer's responsibility to adhere to the prescribed requirements, and could be interpreted to limit concrete producer's liability for post-chute influences on concrete behaviour	The producer often ends up being liable for post-chute concrete behaviour anyway, at least until expensive tests demonstrate placing, consolidation, finishing or curing problems
The concrete producer only need batch required materials in the required manner. Concrete materials and mixture expertise may not be required for typical applications	Since limited expertise is required, limited expertise is applied; a lowest common denominator industry emerges. Concrete producers have limited technical, economic or creative control of the product
In the absence of explicitly defined end-results, contractual performance requirements can be implied in addition to the explicitly stated prescriptive requirements. The owner–designer may object to the subsequent concrete performance regardless of compliance with prescriptive specifications	In the absence of explicitly defined end-results, and if no desired end-results can be implied, the owner–designer may be dissatisfied with the end-product but have limited recourse if all the prescriptive requirements were met
Prescriptive specifications can be written to clearly separate the concrete producer's responsibility from the concrete contractor's responsibility	Even with prescriptive specifications, lines of responsibility can be blurred, especially when testing is conducted on concrete sampled anywhere beyond the truck chute or anytime after job site addition of water
Prescriptive specifications 'level the playing field', allowing concrete producers with a wide range of levels of expertise to compete	Prescriptive specifications diminish the incentive for a given concrete producer to optimise a mixture or to exercise quality control beyond the level of competitors

book, *Specifications and Costs* (first published in 1946), has model specifications for both prescriptive or 'fixed-ratio' concrete mixtures and for performance-oriented 'controlled concrete'. In the first case, 'Concrete shall be, by dry volume, of those proportions that are shown on the drawings'. In the second case, 'Controlled concrete shall conform to the following requirements', followed by a table showing 'class of concrete', and '28-day compressive strength' (Seelye, 1946).[4]

Table 12.2 Advantages and disadvantages of performance specifications

Advantages	Disadvantages
Designers can focus on what is needed rather than how to get it (not all designers are familiar with *how* to best achieve end results)	Specifying how to achieve satisfactory concrete has been a traditional design responsibility. Engineers may be concerned over a perceived reduction in control
Opportunity to focus on the concrete behaviours and characteristics that really matter	The specifier may not be sure about what those characteristics are, nor about how to measure them. Reliable tests may not be available to quantify the desired outcomes. Performance tests may be more expensive, more time-consuming or require more special expertise compared to conventional tests
The concrete producer–contractor team has technical, logistical, economic 'creative control' or influence on the product Opportunity to take advantage of unique materials, material combinations, plant, equipment, technology, expertise, knowledge of local materials and conditions Flexibility in mix proportioning can be opportunity to produce a better overall mixture, or a more economical mixture that meets all performance requirements, or both A more durable product leads to a lower life-cycle cost	End-product properties are influenced by materials, concrete production, concrete delivery, mix adjustments by the contractor, placing, consolidation, finishing, adjustments to mix properties at surface, ambient conditions, moisture control, temperature control. There are many parties involved, and each party has a unique influence on the product. It may be difficult to separate those influences and responsibilities There may be increased cost in the prequalification stage and durability-related or in-place testing may be more expensive It may be difficult to take advantage of life-cycle economic benefit in a low bid (first-cost) contract
Assumed relationships between concrete performance and mix characteristics can be augmented or replaced with tests of concrete properties	Tests beyond the routine slump, air and cylinders are likely to be more expensive and more complicated, and their precision must be taken into account in the specification
In those cases where prescriptive specifications are interpreted to give the producer–contractor full responsibility for end-result concrete behaviour, even when prescriptive requirements have been met, a switch to performance specifications does not necessarily result in any additional responsibility	In those cases where prescriptive specifications clearly limit the producer's responsibility to comply with instructions, the switch to performance specs and the accompanying responsibility for end results is a considerable increase in responsibility

Blends of philosophies within a given specification are frequently encountered. ACI 318, for example, allows for acceptance for strength on the basis of strength test results (performance), but, for durability, limits are placed on the maximum w/cm (prescription.) Likewise, the Canadian CSA A23.1–04 (2004) 'performance' specification includes prescriptive w/cm ratio limits. The NRMCA has already developed an example of a 'minimally prescriptive' specification that recognises the current prescriptive limits of ACI 318, but allows for maximum flexibility via end-product performance requirements.

Further options exist in the distinction between performance characteristics that are used for acceptance of the concrete, in contrast to those used to adjust the amount paid. Payment schemes can be developed to provide incentives for good performance and to exact penalties for marginal performance, as long as the concrete that remains in place has met minimum requirements. A related issue, however, is that given the joint responsibility generally inherent in performance specifications, fairly distributing cash bonuses or penalties among the parties who contributed to the concrete quality can be a difficult problem.

12.7   Concept of 'point of performance'

Given the multiple stages of concrete production and installation, and that 'custody' of the concrete changes hands multiple times before the properties of the end-product are fully developed, one question is, 'At what point in the process do we specify and evaluate the concrete?' This is further complicated by the fact that concrete performance is to some extent in the eye of the beholder. Contractually speaking, performance is in the eye of the party who wrote the specification, and thus defined the required performance characteristics. To a concrete producer buying raw materials, 'end-product performance' applied to aggregates may be defined by density, aggregate grading or uniformity of the FM (fineness modulus) of the sand, or compliance with a specification such as ASTM C33 with no further stipulation as to where or how to extract the rock or how to process it. Similar examples could be given for requirements for the performance of cement or cementitious materials or admixtures. A purely performance-minded concrete producer might not ask for cement mill test reports, and might not even demand that cement meet specific chemical or fineness requirements of EN 197 or ASTM C150, specifying instead that the cement meet requirements for strength, rate of strength gain, soundness, shrinkage, setting time, water demand, uniformity, and limited expansion in alkali–silica reaction (ASR) tests. Likewise a performance-minded buyer of chemical and mineral admixtures might say, 'Don't tell me what they are made of, just guarantee that they will perform in the concrete and that they are mutually compatible.' While some of these performance characteristics for raw materials are easily specified and readily evaluated with long-established test methods, and the responsibility for meeting them is rather clear, the situation gets complicated when the raw ingredients are combined.

To a contractor buying concrete from the producer, 'performance' might be defined in terms of concrete quantity delivered per hour, workability, pumpability, finishability, setting time or early age strength for formwork or shoring removal. If performance is evaluated at the point of discharge from the concrete truck, responsibility is fairly clear, but as soon as evaluation moves to the point of discharge from the pump, responsibility starts to get fuzzy. If the concrete is not 'pumpable', does the concrete producer have to redesign the mixture or does the pumping service have to change equipment? The project owner, on the other hand, may not be concerned with any of these raw materials, fresh concrete or construction issues. To the owner, performance is defined by having sufficient in-place load-carrying capacity to allow the safe operation of the facility and sufficient in-place durability to withstand the service environment for the financially intended life of the facility. The idea of a performance specification can therefore imply different things to different players, and achieving that required performance can (and must) become the joint responsibility of more than one party. Since the owner's chief interests are the in-place, long-term properties of the concrete, meeting such requirements will necessarily be the joint responsibility of the raw materials suppliers, concrete producer, formwork subcontractor, pumping subcontractor, placing and finishing subcontractor, and the general contractor or construction manager that is managing the entire process. It is therefore necessary to be more definitive, and to talk about the 'point of performance', i.e. when and where in the multiple processes of concrete making to curing and protecting are we going to define the required performance of the concrete? The related question is, 'Who bears the responsibility for achieving the specified performance?'

The following example focuses on specifying and installing a superflat industrial floor, and demonstrates that the terms 'concrete performance', 'performance specification' and 'point of performance' can take on different scope and meaning for various parties to the overall project (Table 12.3). (This example considers only the flatness aspects

495

Table 12.3 Example of multiple responsibilities and multiple performance criteria for a superflat industrial floor

Example performance requirement	Party setting the performance requirement	Party obligated to achieve the specified performance	Point of performance	Is performance measurable?
Superflat floor	Owner–specifier	General contractor	Hardened concrete floor, in service	Flatness can be measured at any time. Conventional flatness criteria are intended to apply prior to shrinkage and curling
Finishing to proper tolerances	General contractor	Concrete floor placing and finishing contractor	Hard concrete floor, day after placing	Place/finish must be evaluated prior to curling
Consistent rate of concrete delivery	Concrete floor placing and finishing contractor	Concrete producer	At concrete delivery	Observed on site
Consistent timing and rate of concrete placement	Concrete floor placing and finishing contractor	Concrete producer, pumping contractor, place-finish crew	As observed during placing	Observed on site
Consistent concrete finishability	Concrete floor placing and finishing contractor	Concrete producer	As observed during finishing	Somewhat subjective. Influenced by crew, equipment and weather
Consistent concrete bleeding	Concrete floor placing and finishing contractor	Concrete producer	As observed during finishing	Can be, but is rarely measured
Consistent concrete setting characteristics	Concrete floor placing and finishing contractor	Concrete producer	As observed during finishing	Test for concrete setting not yet standardised

Table 12.3 Continued

Example performance requirement	Party setting the performance requirement	Party obligated to achieve the specified performance	Point of performance	Is performance measurable?
Cement with consistent setting behaviour	Concrete producer	Cement manufacturer	As evaluated at cement delivery to concrete batch plant	Is typically reported by cement producer, but rarely measured later
Cementitious materials with consistent setting behaviour	Concrete producer	Cementitious materials supplier	As evaluated at delivery to concrete batch plant	Can be, but is rarely measured
Chemical admixtures with consistent setting behaviour	Concrete producer	Admixture supplier	As evaluated at delivery to concrete batch plant	Can be, but is rarely measured
Aggregate with consistent grading	Concrete producer	Aggregate producer	As evaluated at aggregate delivery to concrete batch plant	Can be measured, but rarely done during concrete production
Concrete with controlled shrinkage (and related curling)	General contractor	Concrete producer	Prequalify materials? As evaluated prior to construction Sample at time of construction to verify	ASTM C157 for samples taken on site, but no reliable in-situ test
Contribution to shrinkage (and related curling) of the entire range of concrete ingredients from aggregates to admixtures	Concrete producer	Cement, cementitious materials, aggregate and admixture suppliers	Prequalify materials? As evaluated prior to construction Sample at time of construction to verify	ASTM C157 for samples taken on site, but no reliable in-situ test

Table 12.3 Continued

Example performance requirement	Party setting the performance requirement	Party obligated to achieve the specified performance	Point of performance	Is performance measurable?
Contribution to shrinkage and related curling from timing of finishing and curing, type and duration of curing, and job site microclimate	General contractor	Concrete floor placing and finishing contractor	After construction but before service loading	ASTM C157 for samples taken on site, but no reliable in-situ test
Influence of subgrade preparation specification and compliance, installation of reinforcing or dowels	Owner–specifier	General contractor and parties other than the concrete placing and finishing contractor or the concrete producer	Fundamental quality is defined prior to concrete placement, but effect on floor performance not evident until after construction	Compaction tests and inspection of reinforcing and dowels prior to concrete placement
Accuracy of anticipated loads, floor thickness, joint spacing, and detailing, reinforcement	Owner	Designer–specifier	Fundamental quality is defined during design, but effect on floor performance not evident until after construction	End-results can be measured, but can be complicated to attribute effects to specific sources

of the floor and ignores other critical performance characteristics such as strength, cracking and abrasion resistance.)

As shown in this example, the performance required by the owner is a consequence of the design, materials and construction performance. Performance specifications can thus be efficient for the owner, quickly zeroing-in on the key operational characteristics of the installed concrete, but they become equally efficient for the concrete producer only when the concrete materials performance aspects that contribute to the owner's required performance have been identified and can be controlled. The contractor must likewise control construction operations to achieve the owner's requirements. Satisfying the owner's performance requirement specification requires that the various parties influencing performance accept their mutual responsibility.

From the owner's overall project perspective, there is no question that the most meaningful point of performance is the hardened concrete, in place, at an age of concrete that is indicative of the service-life capacity and longevity. However, the frequently acceptable state of the practice is to evaluate concrete properties as sampled at either the point of discharge from the concrete truck, sampled at the point of placement or both. From the results of these tests, the in-place capacity and durability are projected or assumed on the basis of known or implied relationships. (In many cases, the in-place concrete properties can be measured using standard tests developed for that purpose, but such test programmes are not necessarily the norm.) Thus, one could consider a less comprehensive but more conventional performance specification that targets the performance of concrete as sampled at the time of casting. For example, the concrete strength value that is most meaningful in determining structural capacity is the in-place strength value, as influenced by materials, mixing, consolidation, curing and in-place time–temperature history. However, most frequently this in-place value is inferred on the basis of standard cylinders sampled at the time of placement and cured under laboratory conditions. Thus, one could specify the strength of standard laboratory-cured cylinders as a performance criterion. Similarly, air content in the fresh concrete, (or even air bubble size in the fresh concrete using the air void analyser) are viable performance criteria. Even if the performance-minded specifier chose to ignore slump as a performance criterion, the concrete contractor might demand that concrete arrive at the site at a particular slump, thus re-introducing slump as a performance requirement for the concrete producer.

12.8 Potential performance, prequalification and identity testing

It is difficult, time-consuming and expensive to deal with hardened concrete that has been determined to be unsatisfactory. This reality heavily influences decisions about how to specify and evaluate concrete, and how to control and assure its quality. In contrast, consider a typical industrial example of manufacturing steel bolts for construction purposes.

The bolts can be made and tested at the factory and only shipped if found to be satisfactory. Alternatively, the bolts could be sampled and tested upon delivery on site, and only used if proven to be satisfactory. If the bolts are no good, they can be scrapped or shipped back, and replaced with a new batch. Even in the worst case, unsatisfactory bolts in place can still be removed and replaced without destroying the structure.

Switching from steel bolts to concrete, the quality control environment is far more complex. The properties of concrete are not developed at the time of shipment, and in almost all cases the concrete must be installed long before its properties can be reliably measured (even though it can be sampled before installation). Further, the installation process itself affects the concrete properties. Further still, if the concrete is found to be unsatisfactory, it is not a simple matter to replace it with a new batch. For this reason, the industry has developed a number of intermediate checks such as review and approval of proposed concrete mixtures, prescriptive specifications for raw materials and mix proportions, and fresh concrete tests of temperature, slump and air content, and accelerated strength tests to limit (but not eliminate) the chances of ending up with an unsatisfactory material in place.

Under a performance specification, owners and specifiers may minimise the chances of ending up with unsatisfactory, hardened concrete in-place by first demanding evidence that the proposed concrete materials, mixture and methods have the potential to meet specification requirements. It may therefore be necessary to 'prequalify' the materials and or methods on the basis of historical records of performance or by providing laboratory test data. It makes sense to approve the use of a 'known' winning combination of materials and construction technique, especially when a fresh set of tests to demonstrate performance requires more time than normally is available before the concrete is needed on site. However, given the inherent variability in concrete materials, batching and mixing, even if a mixture has been prequalified, it will still be necessary to prove that the concrete actually delivered, placed and finished is in fact the same material that was demonstrated to have been satisfactory during the prequalification process. 'Identity testing' is the term used in BS 8500 to describe such on-site testing to validate the identity of the mixture. Identity testing seeks to verify some key characteristics of the concrete that relate to the desired performance, and could take the form of typical slump, air and strength tests, water content, fresh concrete air void analysis, or some non-destructive or in-place method. For example, consider a performance requirement for an in-place value of 1000 coulombs for the ASTM C1202 rapid

Table 12.4 *Concrete properties of interest*

Fresh concrete	Transition	Hardened state
Workability	Rate of slump loss	Compressive strength
Slump	Time to initial set	Tensile strength
Response to vibrator	Time to final set	Flexural strength
Pumpability	Rate of strength gain	Shear strength
Finishability	(compression)	Fatigue strength
Segregation	Rate of strength gain	Fracture toughness
Bleeding	(tension)	Elastic properties
Air content	Rate of stiffness gain	Shrinkage
Stability of air bubbles	Time to frost resistance	Creep
Uniformity of mixing	Tolerable rate of	Porosity
Consistency of properties	evaporation	Pore size distribution
Temperature	Plastic shrinkage	Permeability
Yield	Drying shrinkage	Air void system
	Temperature changes	Frost resistance
		Abrasion resistance
		Sulfate resistance
		Acid resistance
		Alkali resistance
		Thermal volume change
		Heat capacity
		Thermal conductivity
		Electrical conductivity
		Density
		Radiation absorption
		Colour
		Texture
		Cost

chloride permeability test, and assume that a mix had been prequalified based on pre-construction C1202 testing. During actual construction, the challenge is to perform a suite of tests on concrete sampled at the time of placement that can be used to verify that the concrete as delivered is substantially the same as the concrete that had previously been shown to meet the 1000 coulomb requirement. This could be some combination of water content, fresh unit weight, compressive strength or field calorimetry, for example. There may also be periodic C1202 tests on samples as delivered, accelerated strength testing and/ or of cores extracted from the structure. Further, differences in sampling, numbers of samples or test conditions can make it appropriate to set different acceptance criteria for in-place or job site tests compared with controlled laboratory tests used for prequalifying a concrete mixture.

While prequalification or pre-certification provides evidence to the owner–specifier that the producer–contractor *can* install a product that meets the performance specifications, it does not prove that such has actually been done. It is likely, therefore, that performance specifications will include requirements for pre-construction demonstrations of suitability that could range from the simple submittal of historical data as evidence of past performance all the way to the casting of demonstrations or sample slabs that could be evaluated by in-place methods. But, it is also likely that additional testing during or after construction would supplement such prequalification. This is because of the significant potential for batch-to-batch variation of the concrete due to variations in the raw materials themselves and the fundamental difficulties in precisely controlling water and air.

12.9   Concrete performance characteristics

One clear advantage of performance specifications is that they focus attention on the concrete properties that are the most important for a given situation. Conventional testing often concentrates on slump, air content and 28 day cylinder strength, even though one or more of these properties may not be relevant to the owner's desired performance, while more relevant performance requirements may not be tested at all. As presented in the National Highway Institute's highway materials course manual (Hover, 2003),[5] one way to look at a broader range of concrete properties is shown in Table 12.4, where concrete is evaluated as it transitions from the fresh to the hardened state. While the 'fresh concrete' properties are rather conventional and the hardened concrete list is expanded well beyond the typical cylinder break, the transitional properties are frequently not specified but are nevertheless critical to the safe and economical progress of a concrete construction project. It is instructive to note that within this list (which could be expanded), relatively few properties are commonly specified and tested even though owner satisfaction is commonly based on a far larger set of criteria.

The NRMCA-sponsored P2P contractors' joint task group developed a list of performance characteristics oriented around various applications. For example, concrete for exterior pavements should place easily, be finishable, have no pockmarks, set in a reasonable time, be freeze–thaw and de-icer salt scaling resistant, and have low permeability and low shrinkage. For tilt-up concrete construction, early age flexural strength is critical along with setting time, consistent colour and low shrinkage. Tilt-up finishability was considered important, but not as critical as for a floor. Indoor slabs-on-grade were said to require low to minimum shrinkage, consistent set times, good finishability and good workability (in that order).

12.10   Exposures and exposure classes

One effective way of specifying concrete durability is to require that the concrete remain serviceable for a minimum period of time in a specified environment. Prescriptive specifications usually approach durability by requiring particular ingredients (such as fly ash or air-entraining admixtures), proportions (such as minimum cementitious materials content or maximum w/cm) or requiring construction operations (such as wet curing for a specified duration). Each of these factors is a means to an end, where the required 'end' is durable in-place concrete. Conceptually, requiring that the concrete remain serviceable for a given period of time when exposed to a particular set of environmental conditions specifies the 'end-result' itself. (The term 'remaining serviceable' would require further quantitative definition.) Nevertheless, clearly and unambiguously specifying the service environment that the concrete must endure puts all bidders on an equal footing in regard to the expectations for durability and challenges each prospective supplier-contractor team to jointly figure out ways to economically blend concrete materials technology with construction practice to achieve the required endurance.

Standardised 'exposure classes' can be developed to serve as descriptions of common environmental exposures. This has been done effectively in many codes, including EN 206 and the Canadian A23.1 standard. It should be noted that until ACI 318 was revised in 2008, it addresses many of these issues in various tables, but it was not focused to make it prominent. Thus, it was rare when a specification based on ACI 318-05 deliberately and clearly points out the required exposure. Based on the recommendation of the authors, in the 2008 revision, exposure classes were finally adopted, similar to most international codes.

12.11   Prescription and performance elements in the ACI 318 building code

12.11.1   Durability requirements

ACI 318 contains elements of both prescriptive and performance specifications. Chapter 4 of ACI 318 contains tables of durability requirements in terms of maximum w/cm, total air content, limitations on supplementary cementitious materials, limits on chloride content and requirements for cement types, each as a function of exposure. Since the code was only revised in 2008 to explicitly require that exposures be defined in the drawings and specifications, prior to this there was confusion about whose responsibility it was to determine durability requirements. In some cases, project specifications imply that it is the contractor's responsibility to consult the code and to determine the appropriate concrete mixture requirements. This approach is unambiguous only when the specifier explicitly defines the expected exposure conditions.

While maximum w/cm, minimum air content, limitations on supplementary cementitious materials, limits on chloride content and requirements for cement types are clearly prescriptive in nature, some flexibility for materials selection and proportioning remains, in that there are no maximum or minimum limits on the total weight of cement or supplementary cementitious materials nor are there maximum or minimum water contents. The code imposes no requirement on aggregate content except that which derives from the limits on nominal maximum size of coarse aggregate and good mixture-proportioning practice. For resistance to freezing and thawing cycles, even air content is given as a function of coarse aggregate size, which is an approximate (if indirect) way to account for the need for more air as paste content increases. (The increased surface area of smaller aggregates demands more paste – thus higher air content is required for mixtures with smaller aggregates.)

12.11.1.1 w/cm ratio limits

As discussed earlier, prescriptive requirements for w/c were incorporated in codes as far back as 1928 (Table 12.5), and a prescriptive option for w/c based on strength was built into every code through 1986. (The modern versions of the code use the term 'water/cementitious materials ratio' (w/cm).) The prescriptive association between w/c (w/cm) and strength was dropped in 1989, but the more performance-oriented approval of mixtures on the basis of strength test results plotted as a function of w/c (the three-point curve) remains to the current (2008) edition. Although w/c had been limited by the code for freeze–thaw durability since 1947 (6 gallons/sack = 0.53), the current more comprehensive tables describe maximum w/cm limits and minimum f'_c values for permeability control, de-icer salt scaling resistance, and corrosion protection. This table first appeared in 1989.

Table 12.5 Assumed strength of concrete mixtures for plastic concrete (from the 1928 Joint Code – Building Regulations for Reinforced Concrete)

Assumed compressive strength[a] at 28-days in pounds per square inch: MPa	w/c in US gallons per sack of cement	w/c in lb water per lb cement (by mass)	Approximate ratio of cement volume to dry total aggregate volume
1500 (10.3)	$8\frac{1}{4}$	0.73	1:7
2000 (13.8)	$7\frac{1}{2}$	0.67	1:6
2500 (17.2)	$6\frac{3}{4}$	0.60	$1:5\frac{1}{4}$
3000 (20.7)	6	0.53	$1:4\frac{1}{2}$

[a] Strength value assumed in structural design.

Further, the table of 'Requirements for concrete exposed to sulfate-containing solutions', includes maximum limits on w/cm, depending on sulfate exposure.

The concept behind these tables is fundamental to Portland cement concrete behaviour. The porosity and permeability of hardened cement paste is intimately connected to the volume of mix water, as much of the volume initially occupied by mix water in the fresh paste remains as pore space in the hardened paste. A more porous paste implies a more porous mortar, and a more porous mortar implies a more porous concrete. This was conclusively demonstrated in a large number of tests, ranging from pioneering work on pastes to the classic permeability studies on mortars and concretes conducted by the Bureau of Reclamation in association with dam construction in the western USA (Ruettgers et al.).[6] The conclusion is always the same: for any given mixture, permeability decreases as w/c (or w/cm) decreases. But it is not true that permeability is uniquely or absolutely defined by w/c or w/cm across all mixtures when aggregate size and content, total paste content, total water content, paste composition (different types of cementitious materials), age or method of test are allowed to vary.

The high-pressure permeability tests conducted by Ruettgers et al. showed that pastes are far more permeable than mortars or concretes at the same w/c, and that for concrete at any given w/c, the permeability can vary by a factor of 10–100 as other mixture ingredients change. (The scatter was considerably narrowed when the results were recomputed as permeability per pound of cement per cubic foot). Nevertheless, these results were based on Portland cement as the only binder, and therefore do not begin to reflect the differing capacity of various supplementary cementitious materials to affect permeability. Further, if the issue is in-place permeability, the significant influences of consolidation, finishing and curing need to be taken in account as well.

It is clear, then, that the w/cm table values do not define a particular level of *concrete* permeability (although for a pure cement binder, the limiting values may come closer to defining a level of *cement paste* permeability). It may be that these requirements could evolve along the lines of ACI's older default requirements for w/c based on specified strength that were used only when no other strength versus w/c ratio data were available. Prescriptive w/c values could be overridden on the basis of permeability index test data, and proposals along these lines are currently before ACI 318 subcommittee A. If a permeability index value were specified (or some similar property that relates to the transport of fluids and dissolved solids through hardened concrete), a value of w/cm could be determined that would meet that specified requirement for a given set of concrete materials.

Allowing for an increase in the limiting w/cm values when supported by test data can be advantageous to all parties. This is because any given mixture requires a basic water content to achieve the necessary workability (pumpability, compactability and finishabilty), and the requisite total cementitious materials content is determined by dividing water content by w/cm. For any given level of workability and water demand, the lower the w/cm, the higher the total cementitious material content and the greater the paste content. More paste requires more air, and as the paste content goes up, the total aggregate content must go down. Higher paste and lower aggregate generally lead to increased shrinkage and creep, and the combination of high paste content and low w/cm increases the risk of plastic shrinkage cracking for any given rate of evaporation. Increased paste content also increases the total heat of hydration, with a greater temperature rise and risk of thermal cracking and strength reduction. Controlling these heat effects requires changing the blend of cementitious materials or the use of other mixture- and construction-related cooling techniques. Thus, requiring a w/cm that is lower than it needs to be to develop the desired permeability can, therefore, increase the cracking potential of the concrete if the paste content is not limited. Specifying a minimum cement content that is higher than needed to meet strength and/or durability criteria can have the same effect.

As a final comment on w/cm, the ACI 318 tables also require a minimum f'_c with each maximum value of w/cm. As stated in the code commentary, the minimum strength requirements 'will ensure the use of a high quality cement paste', and help to guard against mismatched specifications such as 'Max. w/cm shall be 0.40, and f'_c shall be 3000 psi' (20 MPa). The strength requirements are a pragmatic recognition that since w/cm and permeability are not normally evaluated, strength is often the only indicator of concrete quality. Concrete that truly has a w/cm of about 0.40 is likely to have a 28 day compressive strength in excess of 5000 psi (35 MPa). Exceptions are possible, of course, given the wide range of materials available, but the current code has no specific provisions for accommodating such exceptions. If a clearly defined end performance requirement were available and specified in lieu of w/cm, perhaps the strength associated with that characteristic could be determined for a specific mixture.

12.11.1.2   Freeze–thaw durability and scaling resistance

The prescriptive requirements for w/cm when the concrete is to be exposed to freezing and thawing while in moist condition and for de-icer salt scaling have been addressed in Section 1.11.1. It is further noted that ACI 318 only has limits for total fresh air content 'as delivered', with no further distinction between so-called entrained and entrapped air. ACI 318 makes no mention of air bubble size in the fresh concrete or air content, air void size or air void distribution and spacing in the hardened concrete.

If the total air content is seen as a durability characteristic by itself, then one could consider the ACI 318 table values to be performance criteria. If the total air content is seen as one of the factors leading to freeze–thaw resistance, with the needed air content varying with paste content and air void size, then the table limits are prescriptive. Of course, if freeze–thaw resistance is the desired end-result, then demonstrated performance in a freeze–thaw test would be the most useful measure, and it is entirely possible that fully successful performance could be achieved in some mixtures at total air contents significantly lower than those required by the table. (This can be the case with low paste content and a stable system made up predominantly of microscopically small air voids.) On the other hand, when the air voids are predominantly large, it is possible that the total air contents required by the code will not necessarily lead to the desired durability. The code-mandated total air content is therefore only part of the story, and either freeze–thaw tests or air void analyses are needed to increase confidence (but still not guarantee) freeze–thaw durability. This is why use of the air void analyser to test fresh concrete or ASTM C457 (the microscopic analysis of hardened concrete) are of great interest, especially if specifications move from the prescriptive air content to a more performance-oriented criterion. Note also that since freeze–thaw durability or de-icer salt scaling resistance are dependent on both materials and construction procedures (such as finishing and curing), prequalification of a concrete mixture based on freeze–thaw or scaling tests demonstrates only the potential of the material to achieve the required durability (Hover, 1994).[7] It may be necessary to demonstrate this potential, however, for certain combinations of cementitious materials and admixtures, as proof of compatibility, especially when the producer wants to prove that durability can be achieved at air contents or spacing factors that are outside the values recommended for conventional mixtures. However, if standard freeze–thaw or scaling tests are being contemplated for prequalification testing, the multi-month duration of these procedures has to be kept in mind.

If in-place freeze–thaw resistance were the objective, there would be little question that the most meaningful point of sampling would be as handled, pumped, squeezed, dropped, pressurised, depressurised, consolidated, finished and cured in-place. While current code provisions can be interpreted to imply testing fresh concrete at the truck chute, sampling at the point of placement is increasingly common, yet there are no clearly required values for the total air content at the point of placement. Some specifiers automatically invoke the ACI 318 air content values at placement, which can be too conservative when the air bubbles remaining after handling and consolidation are predominantly small. This leads to the need to batch the concrete at higher than normal air contents to accommodate the normal losses that occur during handling. This can in turn lead to significant drops in strength. This confusion is avoided if in-place hardened air void system criteria or freeze–thaw testing are specified as performance requirements in lieu of reliance on total air.

There is also a table in ACI 318 listing 'Requirements for concrete exposed to deicing chemicals', that sets limits on the maximum percentage of total cementitious materials by weight for fly ash, slag, silica fume and other pozzolans. Under a more performance-oriented approach, a concrete producer might be permitted to demonstrate that resistance to freeze–thaw damage and de-icer scaling can be achieved with a particular combination of cementitious materials, when finished and cured as the contractor intends. With appropriate lead time, this might be an opportunity for effective prequalification via actual freeze–thaw or scaling tests. However, scaling tests need to relate to real life exposure conditions, and the ASTM C672 test is known to be overly harsh to concretes containing fly ash or slag, partly due to the inadequate time to achieve the concrete's potential before initiation of freeze–thaw cycles, and in part due to different times of set (affecting the time of finishing) (Boyd and Hooton, 2007).[8]

12.11.1.3 Corrosion protection

The first line of defence against the corrosion of embedded metals is to inhibit the penetration of water, oxygen, carbon dioxide and salts from the concrete surface to the level of the embedded metal. ACI 318 sets minimum requirements for the depth of cover, with the requirement that in a corrosive environment, concrete protection shall be 'suitably increased, and denseness and nonporosity of the protecting concrete shall be considered'. This provides for the interesting interplay of a structural design feature (bar cover) and the material property of the concrete. Given that deeper bar cover is often associated with wider crack widths at the concrete surface, overall performance of the completed structure might be improved with the coordination of bar cover and concrete materials properties.

Permeability-related properties of the concrete have already been addressed in regard to w/cm requirements, but it may be reiterated that the use of supplementary cementitious materials such as fly ash, slag, silica fume or other pozzolans can be an effective way of decreasing the permeability of the concrete. It is expected that within a given set of materials and proportions, permeability will decrease with w/cm. Across a wide range of material combinations and proportions, the value of w/cm that leads to a particular and desired value of permeability is expected to vary. Adjusting the code limit on w/cm on the basis of mixture composition is not currently permitted within the ACI 318 code provisions.

To reduce the amount of potentially corrosive chloride in concrete, the ACI 318 sets limits on the 'maximum water soluble chloride ion concentration'. These limits vary with exposure conditions and between reinforced and prestressed concrete, but there are no options for differences in concrete composition, permeability or w/cm. ACI 318 also limits minimum depths of cover to reinforcements for different conditions.

12.11.1.4   Sulfate durability

In addition to the w/cm limits mentioned earlier, minimum strengths and requirements for allowable types of cementing materials are included in the table requirements. These requirements are based on the nature and severity of the sulfate exposure. Limits on w/cm and associated values of f'_c are intended to limit permeability to reduce the ingress of sulfates, and the earlier comments about a mix-specific correlation among w/cm, strength and permeability apply. Given the wide range of cementitious materials available and the wide range of types and concentrations of sulfate exposures, it is difficult to make a one-size-fits-all approach in the code, but it is equally difficult to allow multiple exceptions and adjustments in the absence of performance tests that would demonstrate the sulfate resistance of a given mixture.

12.11.2   Concrete not exposed to aggressive exposures

It is important to note that when durability is not a concern, the code imposes no prescriptive limits of any kind on w/c, w/cm, air, cement content or percentages of supplementary cementitious materials. Thus, the ACI 318 door is open for performance specifications of concrete that will not be exposed to aggressive conditions. With the exception of a chloride limit of 1.0%, the code imposes no prescriptive limits in the absence of freezing and thawing or risk of corrosion or sulfate attack, unless the finished structure is intended to have a low permeability.

12.11.3   Strength requirements

Chapter 5 of ACI 318 unfortunately makes concrete strength requirements appear far more complex than they actually are. As is common in any rational quality control system, the code provisions recognise that accepting concrete whose average strength equals the specified strength is unreasonable since about half of the concrete accepted would have below-average strength and therefore below-specified strength. Code provisions also recognise that rejection of all concrete with strength results lower than a specified 'minimum' is just as unreasonable, as the cost and frustration associated with a zero-tolerance policy would be prohibitive. The code therefore requires that concrete mixtures be selected that demonstrate at least a 99%

chance of meeting the two principal strength requirements:

- ACI 318-08, Section 5.6.3.3(a). Every arithmetic average of any three consecutive strength tests equals or exceeds f'_c.
- ACI 318-08, Section 5.6.3.3(b). No individual strength test (average of two cylinders) falls below f'_c by more than 500 psi when f'_c is 5000 psi or less; or by more than $0.10f'_c$ when f'_c is more than 5000 psi.

Working backwards from these reasonable requirements, any concrete mixture that meets both of the acceptance criteria 99% of the time will have a readily predictable average strength that will always be higher than the specified strength, and here the code imposes a strictly performance-oriented requirement. The amount by which the required average concrete strength must be higher than the specified strength depends on the level of precision of the concrete producer-test laboratory team. (This level of precision is indicated by the value known as the 'standard deviation'.) The lower the demonstrated variability in cylinder test results (lower standard deviation), the lower is the required difference between specified and average strength.

These provisions contain no arbitrary 'safety factors' and merely express the reality of everyday 'normal' variability. Unfortunately, the ACI uses the misleading term 'overdesign' to refer to the difference between specified and required average strength. Thus, a fully rational procedure that incorporates tolerance for occasional low breaks, which provides the owner with 99% confidence regardless of producer and is self-adjusting across concrete producers with varying levels of sophistication and quality control, is made to look like an arbitrary strength requirement above and beyond that which is really needed.

There is a common and perhaps dangerous misconception that the code requirements for concrete strength 'overdesign' are an extra layer of conservatism in addition to other factors of safety applied in structural design. In actuality, the code has an integrated, three-part approach to establishing structural reliability. First, 'load factors' take into account the variability of service loads and the likelihood that the structure will experience an overload during its service life. The value of these load factors takes the nature and predictability of various loads and load combinations into account. Second, the code applies a strength-reduction factor that accounts for a difference between computed and actual strength of a member, based on factors such as variable dimensions and rebar placing tolerances, and reflects the nature and consequences of structural failure. Third, the variability in the strength of the concrete itself is accounted for only by the statistical quality control requirements on concrete strength. The load and strength reduction factors are based on the assumption that the concrete strength meets the two strength acceptance requirements 99% of the time. Thus, the code provides a comprehensive and inter-dependent approach to reliability: load factors, strength reduction factors and the statistical quality control requirements for concrete strength. The load and strength reduction factors are not intended to make up for a shortfall in concrete quality, and once the structure is designed, concrete that fails to meet the code requirements for strength will reduce the design load-carrying capacity. This is why the situation has to be investigated when concrete strength test results drop below currently specified values.

These strength requirements began to take their current format with the introduction of the 1971 code, but the bottom line is that for most concrete producers and laboratories, compliance with the code require-ments for compressive strength requires that the average concrete strength will be about 10–15% greater than the specified strength. The 1928 ACI code required 15% greater than f'_c, and this requirement was carried all the way to the 1963 code that immediately preceded the 1971 code, at which time today's slightly more rigorous statistical approach appeared.

A final comment on strength is that a fully performance-based, in-place strength requirement is embedded in ACI 318, Section 5.6.5.4:

> 5.6.5.4 – Concrete in an area represented by core tests shall be considered structurally adequate if the average of three cores is equal to at least 85 percent of f'_c and if no single core is less than 75 percent of f'_c. Additional testing of cores extracted from locations represented by erratic core strength results shall be permitted.

This code provision is an interesting example of an in-place requirement that is different from the requirements for the corresponding standard laboratory-cured test. In this case, in-place strength of 85% f'_c is accep-table compared with approximately 1.10–1.15f'_c for the average of laboratory-cured test results (recall that the statistical quality control provisions require an average strength that often turns out to be 10–15% greater than f'_c). However, some portion of this difference has also been attributed to the differences between cores and cylinders, and to conditioning of cores.

12.12   Changing role of testing

A transition to performance specifications literally means a transition to performance testing as well. But, if we can only specify those properties that we can reliably test, our current range of specifiable performance criteria are limited to our current battery of approved, standardised tests.

From the owner's perspective the most meaningful tests are those that evaluate the concrete properties as influenced by materials, proportions, mixing and transport, placing, consolidating, finishing, curing and concrete temperature, i.e., in-place testing of the hardened concrete. While many proven options are available to evaluate in-place strength, fewer alternatives are ready to go for evaluating durability, although ASTM C457 (microscopic analysis) and C1202 (rapid chloride permeability) are viable candidates in most cases. In general, we have few means of reliably predicting durability without testing samples extracted from the hardened structure. It is also likely that since laboratory-cured cylinders have been traditionally considered an acceptable basis for evaluating concrete strength, they will continue to play a significant role in performance testing. Where reliable mixture-specific correlations can be demonstrated between various properties of interest and the results of more conventional tests such as density, cylinder strength or beam strength, such common tests might be considered an acceptable surrogate.

Since the fully hardened, in-place properties are the most desirable but also the most difficult to get and are often at least 28 days too late, there is an intense need to evaluate the concrete earlier to get an early warning of problems or to gain early confidence that all will be well. The sooner information is obtained about the early hardened properties of any given load of concrete, the sooner any adjustments can be made to the materials, proportions or processes for subsequent concrete placements; and the sooner remedial measures can be initiated on the concrete already installed or the sooner construction practices can be altered (i.e. longer form or shoring removal times, or extended curing). Early age or accelerated strength testing is useful in this regard, and becomes absolutely essential as the consequences of later-age discovery of unsatisfactory concrete become more expensive. Likewise, the sooner that durability-related test results such as ACI C1202 (rapid chloride permeability) or ACI C457 (microscopic analysis) become available, the sooner remedial action such as requiring a sealer on marginal concrete or the sooner mix proportions or admixtures can be changed.

From a logistical perspective, tests are needed that can be performed on the fresh concrete at the time of delivery (when we can still accept, reject or adjust the product) that will provide data for predicting the likelihood that required hardened properties will be achieved. But aside from slump, air and unit weight, we have no tests that are fast enough to allow an accept/reject decision on the truck at hand. The challenge of predicting performance from air or unit weight has already been discussed, and predicting performance on the basis of slump may be impossible. The water content test and air void analyser are examples of 'fresh concrete' tests that have the potential to deliver truly useful information for predicting long-term performance, but this information can at best be used to influence subsequent loads as the concrete evaluated by these tests will already be in the structure by the time the results are available. As a final comment on tests of fresh concrete, given the influence of subsequent construction operations it remains entirely possible that concrete that is satisfactory at the point of delivery will not yield the desired performance characteristics in place. Alternatively, if concrete is found to be unsatisfactory at the point of delivery, it is unlikely that subsequent construction operations will bring substantial improvement. Thus, there remains a need for some type of screening tests for the fresh concrete, if only to serve as an early warning for material that may be unsatisfactory. Screening tests could include air or water content, or density (unit weight of fresh concrete). Cylinders are undeniably useful, but cannot return immediate data (except for their as-cast weight).

As discussed earlier in this chapter, given the variables that affect concrete performance and the consequences of installing unacceptable concrete, it makes sense to pre-certify concrete production facilities as having the capability of making the kind of concrete that is required, and to prequalify specific mixtures and materials based on their demonstrated potential to meet all performance requirements. Under normal circumstances this prequalification would only be step one, followed by screening tests at delivery and the programme of subsequent tests of hardened concrete. However, given that the ability to meet performance requirements starts with appropriate plant and equipment, concrete-making materials and knowledgeable personnel, it also makes sense to develop certification programmes or expand existing ones to be able to identify those operations that are broadly capable of meeting the specifications. Likewise, personnel certification programmes should be expanded, especially into the skill areas of mix design with modern materials, effect of construction operations on concrete quality, and use of advanced quality control and quality assurance testing. Furthermore, the certification of plants and personnel not only elevates knowledge level but also establishes pride and credibility.

12.13   Risk and responsibility

In the 2004 edition of CSA A23.1, *Concrete Materials and Methods of Concrete Construction*, the main Canadian standard on concrete, the owner is offered two options for the specification of concrete: performance or prescription. Each option delineates what the owner will specify and what the contractor and supplier shall do. It is stated that the performance requirements apply

> when the owner requires the concrete supplier to assume responsibility for the performance of the concrete as delivered and the contractor to assume responsibility for the concrete in place.

It is thus clear that the responsibility of the concrete supplier ends with the discharge of the appropriate concrete mix from the mixer or delivery unit. The contractor, on the other hand, will be responsible to place, compact and cure the concrete so that it matures to have the strength and durability characteristics required by the owner. The text of the options is given in Table 5 of the standard, as shown below as Table 12.6.

The ACI's Committee on Responsibility in Concrete Construction (ACI RCC-05)[9] reports that

> Construction has now reached a level of complexity that makes design input from constructors and subcontractors desirable and sometimes essential. This input, whether submitted as value engineering proposals, responses to performance requirements or design alternatives, has a legitimate place in concrete construction.

But, in the case of a conversion to performance specifications, taking this 'legitimate place' will entail some redistribution of risk and responsibility. RCC-05 goes on to cite the 'over-riding principle... that responsibility and authority must be congruent'. This suggests that if the concrete producer and contractor are to be held ultimately responsible for concrete performance, then it is reasonable to give them freedom of action to provide a product that meets the demanding performance requirements developed by the design professional. In one sense this means that the producer–contractor will take on additional responsibility in return for the added freedom to select and

Table 12.6 Prescriptive vs performance specification responsibilities *(adapted from CSA A23.1-04)*

Alternative	The owner shall specify:	The contractor shall:	The supplier shall:
(1) Performance: when the owner requires the concrete supplier to assume responsibility for performance of the concrete as delivered and the contractor to assume responsibility for the concrete in place.	(a) required structural criteria including strength at age (b) required durability criteria including class of exposure (c) additional criteria for durability, volume stability, architectural requirements, sustainability, and any additional owner performance, prequalification or verification criteria (d) quality management requirements (e) whether the concrete supplier shall meet certification requirements of concrete industry certification programmes and (f) any other properties they may be required to meet the owner's performance requirements	(a) work with the supplier to establish the concrete mix properties to meet performance criteria for plastic and hardened concrete, considering the contractor's criteria for construction and placement and the owner's performance criteria (b) submit documentation demonstrating the owner's preperformance requirements have been met and (c) prepare and implement a quality control plan to ensure that the owner's performance criteria will be met and submit documentation demonstrating the owner's performance requirements have been met	(a) certify that the plant, equipment, and all materials to be used in the concrete comply with the requirements of this Standard (b) certify that the mix design satisfies the requirements of this Standard (c) certify that production and delivery of concrete will meet the requirements of this Standard (d) certify that the concrete complies with the performance criteria specified (e) prepare and implement a quality control plan to ensure that the owner's and contractor's performance requirements will be met if required (f) provide documentation verifying that the concrete supplier meets industry certification requirements, if specified and

516

(g) at the request of the owner, submit documentation to the satisfaction of the owner demonstrating that the proposed mix design will achieve the required strength, durability, and performance requirements

(a) provide verification that the plant, equipment, and all materials to be used in the concrete comply with the requirements of this Standard

(b) demonstrate that the concrete complies with the prescriptive criteria as supplied by the owner and

(c) identify to the contractor any anticipated problems or deficiencies with the mix parameters related to construction

(a) plan the construction methods based on the owner's mix proportions and parameters

(b) obtain approval from the owner for any deviation from the specified mix design or parameters and

(c) identify to the owner any anticipated problems or deficiencies with the mix parameters related to construction

(2) Prescription: when the owner assumes responsibility for the concrete.

(a) mix proportions, including the quantities of any or all materials (admixtures, aggregates, cementing materials, and water) by mass per cubic metre of concrete

(b) the range of air content

(c) the slump range

(d) use of a concrete quality plan, if required and

(e) other requirements

proportion materials, and to plan means and methods of construction. Viewed another way, on many current projects the concrete producer and contractor are already heavily responsible for concrete behaviour, even when following prescriptive specifications. For many producer–contractors, the principle of 'congruency of authority and responsibility' under a performance specification might simply mean getting the freedom that comes with the *current* level of responsibility. However, under a prescriptive specification based on sampling at the truck chute, responsibility for end-result concrete problems frequently tends to gravitate towards the concrete producer. Under a clear performance specification with at least some in-place testing of hardened concrete, the contractor's joint responsibility for in-place concrete performance may become more visible.

Under a performance specification, design professionals will have a clearer responsibility to articulate exposure conditions and performance characteristics. However, committing to a discrete list of explicitly required concrete performance criteria introduces the risk that an unspecified, long-term performance problem may develop. For example, if the stated performance criteria included strength, shrinkage, permeability and frost resistance, but did not specifically require immunity to the ASR, who is responsible if the ASR develops sometime later? This is not to suggest, however, that the responsibility for a problem like the ASR is necessarily clear under a conventional prescriptive specification that does not include explicitly stated requirements for special ASR-related testing. Meeting the requirements for aggregates in ASTM C33 does include a general requirement for being non-deleterious due to the ASR, but the specific test methods and limits to be used are not, so specifications need to clearly state any ASR requirements for clarification.

Another interesting issue is raised in the RCC-05 document by first stating that:

> it can be appropriate [for the design professional] to delegate certain aspects of engineering design to specialty engineers working for the constructor or subcontractors. When any of this design work involves engineering (as opposed to simply detailing), it should be done under the control of an engineer who is licensed in the state of the project and who takes responsibility for such work.

Many concrete materials engineers would agree that concrete mixture design and proportioning is in fact 'engineering'. RCC-05's report also raises the issue of the design professional's responsibility to review mix submittals:

> The Design Professional should review the mixture proportions and submittals concerning materials, procedures and testing data, but the Constructor remains responsible for compliance with the requirements of the Contract Documents. If approval is required, the Contract Documents should state so specifically.

At the top of the performance specification pyramid, the owner gains the opportunity for clarity of expected performance, but along with the design professional the owner has to accept the risk associated with a finite list of those quantifiable objectives. Under current specifications there is a tendency to seek relief from the contractor for a wide range of longer-term performance problems, some of which may come from unstated exposures or unspecified service conditions. Pinning these down puts most of the cards on the table. In a related issue, the performance criteria selected have to be indicative of the concrete's ability to meet the owner's functional needs. When the UK Highways Agency studied the risks of adopting performance specifications, one of its concerns as owners of the highway network and as buyers of construction services was 'inappropriate application of performance measures resulting in a situation whereby suppliers can meet targets without achieving the desired outcome'. A related concern was 'Mismatches between contract performance requirements and client objectives'. If the performance criteria are overspecified relative to the owner's needs, the product is unduly expensive, and if the criteria are underspecified, the result can be 'poor operational performance, excessive maintenance and premature replacement'.

12.14 A few comments about the current general state of practice

When considering new or alternative approaches to specifying concrete construction, it can be helpful to be reminded of the current state of practice. The following outline has been developed for that purpose, serving as a background for further thinking about specifications in general.

(1) Qualifications:
    (*a*) There is a broad range of expertise and experience in the marketplace.

(b) It can be difficult to qualify contractors or producers on basis of quality record or expertise.

(c) Experience often works against a knowledgeable producer and contractor in a low-bid situation, especially when specifications are subject to interpretation.

(2) Concrete mixture pre-placement approval (prequalification) is based on one or more of the following:

(a) Historical records for strength performance.

(b) Laboratory tests to document strength record.

(c) Use of the ACI 318 'three-point-curve' to demonstrate strength as a function of w/c (w/cm).

(d) The detailed ACI 318 Chapter 5 on the statistical method for mix approval based on strength.

(e) Producers do not frequently prequalify on the basis of air, freeze–thaw testing, scaling resistance testing, corrosion protection or permeability (or its surrogate tests). (The lead time for these tests can be excessive, and there is no guarantee that the pre-placement mixture is same as the mixture actually installed.)

(3) We often specify:

(a) Minimum cement or minimum total cementitious contents:

(i) ACI 318 is silent on this.

(ii) ACI 318 does set limits of proportions of supplementary cementitious materials (SCMs) based on percentage by mass of total cementitious materials for various exposures.

(iii) ACI 318 dictates the cement type or pozzolan or slag use for various sulfate exposures.

(b) Maximum aggregate size:

(i) ACI 318, Chapter 3, matches ACI 211.1.

(c) w/c (w/cm):

(i) w/c is either directly incorporated into project specifications for special exposure conditions or included by reference to ACI 318, Chapter 4, for freeze–thaw, de-icers, corrosion and sulfate exposures.

(d) Temperature:

(i) ACI 301 has limits for maximum and minimum temperatures.

(ii) ACI 305 recommends a mix-specific waiver based on tests for maximum temperature.

(iii) No specific temperature guidance is provided by ACI 318, but there are caveats to avoid harmful effects.

(e) Slump:
  (i) ACI 318 gives no guidance on slump.
  (ii) ACI 211 on mix design has some outdated, pre-admixture recommendations.
  (iii) ACI 301 has some outdated recommendations.

(f) Air:
  (i) The air content is either directly incorporated into project specifications for special exposure conditions or included by reference to ACI 318, Chapter 4. ACI 318 values are for 'total air, as delivered'. No ASTM C457 hardened air values are implied. Code values are tied to old ACI 211 assumptions of mortar content based on water contents as a function of maximum aggregate size, but with no adjustment for admixtures or aggregate grading.

(g) Chloride penetration resistance for corrosion-sensitive structures:
  (i) ACI 318 gives no guidance for ASTM C1202 'coulomb' values, but the code does specify cover depth and allows adjustments of cover depth based on concrete quality.

(h) Compressive strength:
  (i) Test specimens are laboratory cured.
  (ii) Statistical quality control is used.
  (iii) Effectively, there is as 'option' for in-place testing via cores.

(4) We do not often specify:
  (a) Air void system parameters (we check them after the fact if we have a problem, and then often act as though ASTM C457 was specified).
  (b) Shrinkage (we check after the fact if we have a problem, and then often act as though ASTM C157 was specified).
  (c) Actual concrete material transport properties such as actual permeability, sorptivity or diffusivity.
  (d) Curing measures required to achieve a durable surface (CSA A23.1-04 has addressed this issue).
  (e) Measures of curing effectiveness.

(5) We often measure:
  (a) Temperature of fresh concrete.
  (b) Slump of fresh concrete.
  (c) Air content of fresh concrete:
    (i) At the truck chute.
    (ii) Sometimes at the point of placement (although there are no standard procedures for doing so, and no clear

521

acceptance criteria for air content beyond the point of discharge from the concrete truck).

(d) Strength under laboratory-cured conditions.

(e) Chlorides with the rapid chloride test (ASTM C1202), when specified:

    (i) Specimen commonly cut from a 4 inch (100 mm) cylinder as sampled at the truck chute.

    (ii) Specimen sometimes cut from a core extracted from the structure.

(6) We do not often inspect or measure:

(a) w/c or various quantities of water:

    (i) Water content in the drum when the truck is loaded.

    (ii) Changes in aggregate moisture content during production.

    (iii) Water added on site.

    (iv) Retempering water or water applied to the concrete surface or finishing tools.

(b) Quality of aggregate (including grading) during production.

(c) Quantity of aggregate during production.

(d) Quality of cementitious materials during production.

(e) Quantity of cementitious materials during production.

(f) Unit weight (density) or yield of fresh concrete.

(g) Air bubble size in fresh concrete (air void analyser).

(h) In-place strength of concrete (field-cured cylinders are *not* the same as in-place testing). (However, the in-place conditions must be clearly provided along with test results, as cases have occurred where the field-cured cylinders were taken into the heated construction office the day after casting in the winter. One site superintendent stated: 'If I leave them on the structure the strengths will be too low and the structural engineer will not let me remove the forms.')

(i) In-place chloride penetration resistance of the concrete.

(j) In-place temperature of the concrete.

(k) In-place curing effectiveness.

(l) In-place degree of consolidation.

(m) In-place air void system parameters.

(n) In-place air void system parameters at the concrete surface.

(7) Tests, variability and precision. (The values given below are approximate (half the D2s (acceptable range of two results) between laboratory precision limits as defined by ASTM) and simplified. For complete and accurate information on variability and precision, refer to the relevant ASTM standard.)

(*a*) Slump:
   (i) Time dependency is not normally taken into account.
   (ii) Precision $= \pm 11$ mm.
(*b*) Air:
   (i) One pressure air meter air pot $= 0.1\%$ volume of a $6\,m^3$ truck.
   (ii) Precision $= \pm 0.4\%$
   (iii) One core examined by ASTM C457 examines about 1000 of the 1 trillion air voids in a $80\,m^3$ placement. (At \$500 per test we can't afford statistical significance.)
   (iv) ASTM C457 precision $= \pm 1.16\%$.
   (v) Fresh concrete air void analyser spacing factor $= \pm 25\%$.
(*c*) Strength:
   (i) Reported cylinder strength is also a function of the cylinder maker and cylinder breaker:
   Precision $= \pm 4\%$.
   (ii) Cores are sensitive to moisture conditioning:
   Precision $= \pm 6.5\%$.
(*d*) ASTM C1202 rapid chloride penetration resistance test:
   (i) Precision $= \pm 21\%$.
(*e*) ASTM C157 length change:
   (i) precision $= \pm 0.0037\%$ if cured in water and 0.0069% if cured in air.
(*f*) AASHTO microwave oven test for w/c in fresh concrete:
   (i) ± 0.03–0.05.
(8) We have reasonably reliable relationships between:
   (*a*) w/c (w/cm) and the strength of a given concrete mixture at a given maturity.
   (*b*) w/c and the permeability of a given hardened, Portland cement paste (no SCMs) at a given maturity.
   (*c*) Strength of a given hardened Portland cement paste and maturity.
   (*d*) Total paste content and shrinkage.
   (*e*) Total water content and shrinkage.
(9) We do not have reliable, mix-independent relationships between:
   (*a*) Slump and water content.
   (*b*) Slump and w/c or w/cm.
   (*c*) Slump and strength.
   (*d*) Slump and risk of segregation (in the presence of admixtures).
   (*e*) Slump and pumpability.
   (*f*) Slump and finishability.

(g) Slump and response to vibrator.

(h) Air content (alone) and frost resistance.

(i) Air content (alone) and de-icer scaling resistance.

(j) Strength and durability.

(k) Strength and w/c or w/cm.

(l) w/c or w/cm and permeability over a broad range of mixtures and binders.

(m) Strength as represented by cylinders cured in the laboratory and cores extracted from the structure.

(n) Strength indicated by either cylinders or cores and the strength of the concrete surface exposed to weather and traffic.

(o) Permeability or conductivity of concrete at mid-depth of a core or cylinder versus values at the surface.

12.15 Keys to the concept of performance specification and the challenges they present

12.15.1 *Definitions*

There are multiple definitions of the term 'performance specification'. The Cement and Concrete Association of New Zealand clearly states 'A performance-based specification prescribes the required properties of the concrete but does not say how they are to be achieved'. The NRMCA defines a performance specification as 'a set of instructions that outlines the functional requirements for hardened concrete depending on the application. The instructions should be clear, achievable, measurable and enforceable'. In either of these definitions and in many others, there is room for interpretation as to whether the terms 'required properties' and 'hardened concrete' refer to hardened properties in the structure, hardened properties as sampled from the point of delivery, hardened properties as cast from prequalification testing or perhaps some combination. (Added complication comes from the likelihood that a contractor setting performance requirements for a concrete producer would include a number of fresh concrete requirements in addition to any hardened concrete requirements.) In the minds of many, the term 'performance specification' automatically conjures in-place properties of concrete as influenced by materials, proportions, construction operations and control of temperature and moisture. For others, the term 'performance specification' implies merely the freedom to supply concrete that meets point-of-discharge requirements without the customary qualifications or limitations on ingredients and proportions and without the need to submit documentation concerning materials or proportions. For still another group, the term 'performance specification' implies that concrete is supplied on the basis of historical record or pre-construction test results, with only spot checking to verify that the mix delivered remains substantially the same as originally approved. In reality, a performance specification can be any or all of these, and users of the convenient catch phrase must carefully define it to avoid miscommunication. It may not be helpful to secure agreement to use 'performance specifications' if the agreeing parties hold differing concepts. In all cases, however, there is a need to define the responsibility for product control and to allocate the authority to make the decisions about how to carry out that product responsibility.

12.15.2 Keys to the concept of performance specifications

(a) The ability of the specifications writer to discern the performance characteristics appropriate to the owner's intended use of the concrete.
(b) The ability of the specifications writer to describe these performance characteristics clearly, unambiguously and quantitatively so that performance can be evaluated.
(c) The availability of reliable, repeatable test methods that evaluate the required performance characteristics (along with performance compliance limits that take into account the inherent variability of each test method).
(d) The ability of the concrete producer–contractor team to choose combinations of materials, mixtures and construction techniques to meet the required characteristics so that projects can be planned and bid, risks and costs can be assessed, and materials and construction operations adjusted to comply with performance requirements.

These four keys present at least the following challenges:

(a) Under current, predominantly prescriptive specifications, end product performance is not always comprehensively spelled out at the specification stage. For example, prescriptive specifications may not explicitly include requirements for abrasion resistance, scaling resistance or limitations on concrete cracking. Nevertheless, unsatisfactory performance in any of these categories is often pointed out after the concrete has been installed. The rationale for finding the concrete unsatisfactory may be that these common end-result requirements are generally implied and that the concrete would have been satisfactory if the prescriptive requirements would have been met. In contrast, performance specifications require an 'up front' description of owner expectations. In most cases, this can take significant additional effort and expertise beyond that required for prescriptive specifications by design professionals working on the owner's behalf. In some cases the mechanisms involved are understood, but a reliable measure of end result performance is not easily obtained. In such cases it can be necessary to rely on an 'index test' or to retain a prescriptive option that has served well. For example, strength test results might be used to estimate abrasion resistance based on a correlation with a given set of materials.

(b) Some commonly expected (although uncommonly specified) performance characteristics are not readily clearly definable or readily quantified. In-place cracking, movements due to shrinkage, scaling, pop-outs, colour variations or local incidents of abrasion are easy to spot, but more difficult to describe in an unambiguous way.

(c) Despite an explosion of research and development into new concrete test methods, the industry does not yet have a comprehensive suite of test methods, or the predictive models to allow their use to reliably predict service life in general. As well, knowledge of each test is required in order to make safe decisions, given the variability inherent in the test results.

(d) Some contractors and concrete producers will need additional training to be able to select materials and construction operations that will produce the required concrete. Design professionals will also need additional training or special expertise (such as through certification) to be able to develop reliable performance requirements.

12.15.3 This is not new

The concept of performance specifications is not new. The ability to gain acceptance of concrete on the basis of proven performance requirements has been part of the ACI building code since its early days. Nevertheless, conventional concrete specifications include many prescriptive elements that also have their origins in the earliest code, perhaps reflecting an era when designers were masters of the selection and proportioning of concrete materials, and concrete producers purchased raw materials and batched in accordance with the specified instructions. Over the last several decades, however, concrete materials have become increasingly complex, with a growing number of combinations and permutations of cements, cementitious materials, admixtures, and aggregate types and grading. This not only means that there is a wider range of options for meeting any given concrete requirement, but also that it is more difficult for the design professional to stay current with concrete materials and construction technology as it has become a specialty field. Many producers have transitioned from being merely 'truckers' that deliver concrete mixed in accordance with a specified recipe to being well informed on concrete materials, including complex aggregate grading, chemical admixtures and a wide range of cementitious materials. More recently, fewer specifications require predetermined concrete recipes or materials and production inspections. Likewise, there has been a shift in the responsibility for concrete ingredients and mix proportions towards the concrete producer and away from the design professional. Today's review of concrete mixture submittals 'for general conformance with the contract documents', is a significant evolution from the fully specified mixture proportions of only a few years ago.

12.15.4   Advantages and disadvantages

There are advantages and disadvantages to the use of either performance or prescriptive specifications. The challenge is to find ways to combine the two types as appropriate for various applications to maximise the advantages and minimise the disadvantages. Performance may be called for when there are clear economic, logistical or scheduling benefits to be gained and those benefits can be shared with the designer and owner.

12.15.5   Combinations of specification methods

Arguments pitting performance against prescriptive specifications may not be productive. At issue is the most effective combination of specification requirements, with a sequence of prequalification, on-site testing at delivery and in-place evaluation of hardened concrete. One result of performance specifications is the ability to link payment to demonstrated quality. This results in penalties for quality that falls short of specified requirements but is acceptable (Concrete Institute of Australia, 2001;[10] Sprinkel, 2004;[11] Bognacki et al., 2002;[12] Bickley and Mitchell, 2001;[13] Ontario Ministry of Transportation Special Provision No. 904S13, December 2004).[14] In a few cases, bonuses are also paid for achieving or surpassing the specified minimum quality. Ideally, these bonuses or penalties could be connected to anticipated life-cycle

527

costs, but there are no clear examples of this to date. Where bonuses have been paid, concrete has been found to be either more consistent in quality or, in one documented case, cheaper (Sprinkel, 2004). Options on some projects would include separate sets or types of requirements for acceptance and for pay.

12.15.6 Prequalification

It makes good sense to qualify a mix for field use on the basis of pre-construction testing and/or historical records. Such prequalification demonstrates that the mixture and concrete producer have the potential to meet project requirements. However, the reality of batch-to-batch and day-to-day variability in some concrete production facilities (especially the variability in air and water content) make it necessary to demonstrate that the material delivered to the job (and in some cases as-placed, consolidated, finished and cured) lives up to its pre-qualification expectations. In some cases, the difference between prequalification results obtained on concrete cast and tested in the laboratory versus the performance of actual production concrete needs to be considered when evaluating laboratory test results.

12.15.7 Performance characteristics

On any given job, regardless of the type of specification, there may be a large number of concrete performance characteristics that are expected by the designer and owner. On that same job, some of those expectations will have been incorporated in the formal specifications, while others will have been assumed to develop if the few specified properties are achieved. Performance specifications focus on specific concrete properties, and hold the potential to clarify what is and what is not expected of the concrete.

12.15.8 Durability

The common concern worldwide is design for durable concrete. Common to most specifications is the use of exposure classes that clearly define expectations about the types of exposure that constitutes the service environment of the concrete. Even under current and predominantly prescriptive specifications, ambiguity is greatly diminished when the specifier clearly states the severity of the freeze–thaw environment, salt exposure, need for reduced permeability and sulfate exposure.

When this is not clear, it can be difficult for the producer–contractor to comply with building code requirements or to ensure designer–owner satisfaction. ACI 318-08 newly contains exposure classes within the tables of the code.

12.15.9 Code freedom for concrete in benign environments

The ACI 318 building code does not inhibit performance specifications for concrete that does not have special durability requirements. This type of concrete represents a significant segment of the concrete construction market that could be pursued for an entrée to performance specifications. Specifiers should avoid imposing special durability requirements in situations where no harsh service environment exists, as they may thus invite other problems in addition to higher cost. An example was pointed out in Section 12.11.1.1, in which a specification that leads to high paste content as a result of minimum cement or maximum w/c requirements can result in increased shrinkage cracking.

12.15.10 Code limits for concrete in aggressive environments

For durable concrete, the current ACI 318 building code limitations on w/cm and the percentage of supplementary cementing materials (for de-icer salt exposure, for example) are restrictive and can lead to undesirable consequences such as increased shrinkage and cracking while nevertheless providing the desired durability. However, these requirements might evolve towards performance alternatives, as in the case for w/cm code requirements that used to be in place for concrete strength. The statistical quality control features of the code are reasonable, and protect the owner and producer, even though the ACI chose the unfortunately misleading term of 'overdesign' to describe the difference between specified strength and required average strength. It is likely that any meaningful performance criteria will have a similar statistical basis.

12.15.11 Share and transfer of responsibility

In making the transition to performance specifications, the concrete supply industry needs to be clear about the potential for a very significant transfer of responsibility. If true end-product performance is specified, requiring the in-place assessment of hardened concrete, the concrete producer and contractor become jointly responsible for the quality of the finished structure. However, if the 'point of performance' is specified as the point of discharge of the concrete, the concrete producer's share of responsibility with the contractor is not significantly different than with a prescriptive specification. The owner's risk is likely to be minimised by an in-place, hardened concrete point of performance, with its joint responsibility of the producer–contractor team. The concrete producer's risk is likely to be minimised if acceptance were based on the delivery of a prequalified product. Decisions about the point in the construction sequence at which the concrete properties are to be evaluated thus have a critical impact on risk and responsibility. Further, taking advantage of performance specifications mandates greater dialogue between contractors and their concrete suppliers. Concrete suppliers will need to be proactive in determining the concrete mix characteristics required by the contractor in placing, compacting and finishing as well as meeting the performance criteria for the hardened concrete. Contractors cannot be relied upon entirely to clarify these needs when asking for prices.

12.16   International perspective and progress

12.16.1   *Overall assessment*
A study of the current world literature on specifications makes it clear that performance is a hot topic and that performance specifications are the 'Holy Grail' that many desire and are seeking. There is a worldwide concern about the durability of concrete and the effect of quality of installation on that final, in-place quality. In reviewing a large number of documents from around the world, however, few true performance specifications were found, and these only contained pure performance criteria for some properties of concrete. To paraphrase a popular quotation, 'When all is said and done, there is a great deal more said than done on the topic of performance specifications'.

  At the moment, most specifications that address performance delineate exposure conditions that affect the service life of concrete and provide parameters to be verified that are assumed to be indicative of the concrete's ability to achieve the desired service life. These parameters are based on experience and/or durability indices derived from research studies and, in some cases, field experience. In many cases, specifications provide tables of mandatory limits to w/cm, minimum cement contents and air entrainment. There is almost universal use of supplementary cementitious materials such as fly ash, ground granulated blast furnace slag and silica fume. Without exception, the levels of quality control required to guarantee a consistent and economic product are based on statistical analyses of test results, at least for strength.

The literature search uncovered few examples of performance specifications where the quality of the end-product was stated and the product details left entirely to the concrete supplier. In all instances where this issue was discussed, the consensus was that the test procedures needed before this approach can be extensively implemented have not yet been proven to be reliable or repeatable to the degree that would make them viable from a practical and legal point of view. On this basis, it would seem that at the moment any performance criteria in specifications for general use that are to be made the responsibility of the concrete supplier are limited to slump, air and strength.

Among the most highly developed new codes and specifications that have appeared on the international scene in recent years, the EU's Eurocode EN standards with their nationalised amendments may be the most comprehensive and the most complex. Cultural and language differences make the European EN standards difficult to navigate, however. BS 8500 is particularly complex, and it is considered that the approach to determining and ordering a concrete mixture in these standards, and in BS 8500 particularly, appear to be unnecessarily and unattractively complicated (Harrison 1997,[15] 2003a,[16] 2003b).[17] Some of the complication has resulted from having to accommodate the needs of the 27 EU countries in one standard. Australian and New Zealand documents are progressive, enlightened and clear, but are not truly performance based. The Canadian CSA A23.1 specification is also not strictly performance based, but contains many elements that are adaptable with little or no modification to a performance-oriented specification, particularly in regard to durable concrete.

12.16.2 US initiatives

Within the USA, many states have taken important steps towards performance specifications, both independently and under the leadership of the FHWA's Performance Specifications Strategic Roadmap. Virginia specifies shrinkage and rapid chloride permeability. Kansas has been exploring air void analysis, and Minnesota has been among the states that have encouraged 'contractor-based mix design' and has been controlling water content in certain applications. The Port Authority of New York and New Jersey has had good experience by focusing on measuring water content, shrinkage and rapid chloride permeability (Bognacki *et al.*, 2002). The Florida Department of Transportation requires concrete mixtures for exposure to marine environments to be prequalified using ASTM C1202, and recently has started to test surface resistivity on site-cast cylindrical specimens (Kessler *et al.*, 2008).[18]

12.17 Testing and quality management

12.17.1 Testing

To provide confidence for all parties under the adoption of performance specifications, there is a need for quick, reliable performance tests for concrete properties, including durability, which have to go far beyond current reliance on the 28 day compressive strength as the sole arbiter of concrete quality. The lack of adequate performance-related test methods certainly hinders the move from prescriptive to performance specifications, and there is no question that new developments in testing will facilitate the use of performance specifications.

12.17.2 Multi-stage testing

A transition to performance specifications literally means a transition to performance testing. Prequalification testing would be step one, followed by screening tests of fresh concrete at delivery and a programme of subsequent tests of hardened concrete, possibly of samples that had been taken from the truck chute, but more comprehensively from the structure itself. The Duracrete report (1997),[19] in establishing levels of quality assurance, rated in-place testing as the highest level. There is already a developing trend to make tests on the concrete in the finished structure, since the owner's primary concern is concrete that in the finished structure (Concrete Institute of Australia, 2001; Sprinkel, 2004).

12.17.3 Screening tests

Screening tests of fresh concrete are likely to continue to be based on slump, temperature and total air content, although slump may not be required by the specifier in lieu of more meaningful tests of the hardened concrete properties. Performance-oriented contractors may have slump requirements, however. Fresh concrete tests such as density (fresh density) may become more important as ways to identifying mixes, quantifying batch-to-batch uniformity and for providing additional data concerning air.

12.17.4   Other tests

Other tests that can be performed on fresh concrete, but cannot be used as a screening test due to the time required to obtain a result (prior to discharging the concrete), include the air void analyser and the microwave test for estimating water content. Even though a w/cm value may not be specified under a pure performance specification, once a proposed mixture has been pre-approved for use via a prequalification process, the water content may be an effective identity test to validate that the pre-approved mix has in fact been delivered to the job, within the limits of the precision of the test method.

12.17.5   Tests of mechanical properties

Standard cylinder tests are likely to play a role in performance requirements, but it is also likely that focus will shift towards in-place testing and/or to accelerated strength tests to provide early confidence (or early warning) that the desired outcomes will be achieved. (Given the need for early prediction of later-age performance, accelerated tests such as the four test methods described in ASTM C684 give an estimate of 28 day compressive strength between 5.5 and 49 hours, depending on which procedure is chosen.). Cylinder mass may become more commonly used as an early indicator of low-strength cylinders. Abrasion tests will be more common when abrasion has been defined as an in-service exposure. Given the strong association between cracking, functionality and owner–designer satisfaction, early and later age volume-change tests will become more common.

12.17.6   Durability concerns

The prime concern of all that the authorities reviewed is durability. Without exception, all include the use of supplementary cementitious materials, either as additions at the concrete plant or in blended cements as an aid to achieving durability. When it comes to testing hardened concrete for durability in regard to the ingress of aggressive substances, many tests have been developed (Baroghel-Bouny, 2002;[20] Cabrera *et al.*, 1988;[21] Claisse and Elsayad, 1997;[22] Dhir *et al.*,

1995;[23] Dinku and Reinhardt, 1997;[24] El-Dieb and Hooton, 1995;[25] Figg, 1973;[26] Hope and Malhotra, 1984;[27] Stanish *et al.*, 2000;[28] Streicker and Alexander, 1995),[29] but few are standardised, and none are considered ideal (Hooton, 2001).[30] The NORDTEST NT Build 492 rapid migration test (Tang and Nilsson, 1991;[31] Stanish *et al.*, 2001)[32] has achieved wide use in the Nordic countries, and can be adapted to evaluate curing effects (Hooton *et al.*, 2002).[33] Both bulk and surface resistivity tests have merit in simplicity (Andrade *et al.*, 1993;[34] Millard *et al.*, 1989;[35] Monfore, 1968;[36] Morris *et al.*, 1996).[37] Nevertheless, it is likely that AASHTO T277/ASTM C1202 rapid chloride permeability testing will remain a significant tool for evaluating concrete performance in the foreseeable future (Goodspeed *et al.*, 1996;[38] Hooton, 2001). The ASTM C1202 test has been used extensively in highway specifications in Virginia, in Australia and by highway departments in Canada. The Canadian CSA A23.1-04 specification uses this test to qualify concrete to be used in chloride and aggressive chemical exposures. Ontario specifications for the use of this test base acceptance on these results of tests on cores taken from the finished structure. Rate of absorption (sorptivity) tests are also quite useful (Balayssac *et al.*, 1993;[39] Hall, 1989;[40] Price and Bamforth, 1991;[41] McCarter *et al.*, 1996).[42] The ASTM C1585 sorptivity test has not had extensive use in North America, but similar tests have been used for acceptance in Australia (Ho and Chirgwin, 1996),[43] and a field version has been used in Canada (DeSouza *et al.*, 1997,[44] 1998,[45] 2000).[46]

12.17.7   ASTM C457

For evaluating the freeze–thaw durability of concrete in place, actual freezing and thawing tests are available, but are difficult to employ, expensive and time-consuming. ASTM C457 microscopic analysis of air void system parameters is likely to be important, and has been extensively used by Canadian highway departments. In Ontario, the test is made on core samples taken from the structure.

12.17.8   ASR and sulfate reactions

For the specific durability issues of ASR or sulfate reactions, a number of specialty tests are available. Given the role of flow through porous media as a factor in either of these deteriorative mechanisms, specific ASR or sulfate testing should be coupled with permeability and transport-type tests. ASTM C1567 is probably most appropriate for job mixture ingredient qualification, after aggregates are determined to be potentially reactive by ACI C1293 or prior field service.

12.17.9 Need for rapid approval

A practical difficulty in implementing performance specifications is the short bidding period followed by a short time before work commences. This is common in North American contracts. Tests to determine whether concrete mixtures meet ASR, diffusion and creep requirements need very long lead times, while freeze–thaw, shrinkage and scaling take at least 3 months. Further, the longer it takes to perform a test, the more likely it is that raw materials will have changed by the time the actual production concrete is produced. An example of a supplier's responsibility is given in the Australian specification (2001) that requires ready mixed concrete producers to determine the chloride and sulfate contents and the shrinkage of their most commonly supplied mixture every 6 months.

12.17.10 Quality management

In-place concrete properties depend on both the quality of the concrete delivered to the site and the quality and workmanship of construction operations such as placing, consolidating, finishing, curing and protection from the weather. Prior to approving concrete and allowing it to be placed, a design professional may require evidence from the concrete supplier that the mixture has the potential to meet the specified criteria, and may also require evidence from the contractor of the ability to install the concrete.

A performance specification will therefore require a quality management programme from the contractor (party holding contractual responsibility to the owner), who will in turn require such a programme from the concrete supplier. In view of the clarification of responsibilities inherent in any specification, but particularly in a performance specification, it must be clear that when the terms 'quality assurance' and 'quality control' are used, the ACI definitions in ACI 116, *Cement and Concrete Terminology*, apply as follows:

Quality assurance – actions taken by an owner or representative to provide and document assurance that what is being done and what is being provided are in accordance with the applicable standards of good practice and following the contract documents for the work.

535

Quality control – actions taken by a producer or contractor to provide and document control over what is being done and what is being provided so that the applicable standards of good practice and the contract documents for the work are followed.

It has been suggested in the EU specifications that not only independent entities such as the North American Cement and Concrete Reference Laboratory (CCRL inspects but does not certify) would be able to certify compliance of concrete mixtures, but that this could also be done by national organisations such as the NRMCA or regional concrete associations. (The NRMCA has a Quality Plan Guideline and Process of Certification for quality control projects close to completion, and has a similarly long-established programme that addresses the certification of plants and trucks, plant managers, delivery and sales professionals, and concrete technologists.) The Precast/Prestressed Concrete Institute has been operating a successful plant certification programme for over 30 years (Wilson, 1968;[47] PCI 1999,[48] PCI 2001[49]). However, owner acceptance of certification by producer associations will require demonstration of a clear lack of bias. Along these lines, the EU programme requires that concrete suppliers report deficiencies in their operation to owners and contractors in a timely fashion. For example, in the experience of one of the authors, this was done effectively for the new parking garage at Toronto's Pearson airport: the concrete supplier reported their errors immediately, and provided the corrective actions being taken. (While there was some initial resistance, in the end it did work, and everyone liked it, including the supplier.)

Concerns about suppliers who may not be technically capable of producing performance concrete will be allayed by maintaining and expanding the NRMCA's certification programmes. Such a programme must be, and must be perceived as, rigorously applied and maintained. This would also imply that only certified suppliers be allowed to supply to performance-based specifications. (Similar requirements exist in Quebec, Ontario, New Zealand, the UK, and in other locations.)

It has been suggested in Europe that where a supplier and its concrete mixtures are certified by an acceptable third party, quality assurance testing is not necessary unless a problem is seen. This will be a hard sell in North America, at least until the use of performance specifications is a mature practice or until enough on-site tests have proven the uniformly satisfactory behaviour of the concrete.

It is to be expected that concrete suppliers will have reservations about the validity of tests, particularly when a failure to meet specification is reported. Some jurisdictions deal with this by having a referee system that can carry out retests to satisfy all parties with regard to the disputed test result (e.g. CSA A23.1-04; Ontario Ministry of Transportation Special Provision No. 904S11, December 2004).[50] While questions may remain about how to select an impartial referee and who would pay for such a service, the specifications might state that referee testing is not only at the request of the contractor (or the contractor's supplier) but that the contractor pays for it.

12.17.11 Mixes

Typically, a concrete supplier will have a portfolio of many mixes that have been developed to meet price inquiries. Also typical of most contracts is a short bid time period and a rapid construction start after a bid award. Pre-testing all, or even a significant number, of these mixes to meet anticipated performance demands would be horrendously expensive and take up to several years to complete. How concrete suppliers will cope with this aspect of performance specifications is not clear.

One possible approach is the 'family of mixes' concept put forward in the UK. These are groups of related concretes for which a reliable relationship has been established and documented. Criteria are given in BS 8500 for confirming that a mix truly belongs to a particular family. A recognised authority can provide certificates of conformity to confirm membership of a family of mixes. In the UK, this certification can be provided either by a totally independent authority such as the British Standards Institute or the concrete industry's Quality Scheme for Ready Mixed Concrete.

While not detailed in this chapter, there are now many properties of concrete that can be included in a performance specification. Many properties are measurable by an existing ASTM standard test method, and can be specified as a minimum or maximum value. The more intractable problem now is the increasing number of durability criteria that can be specified for concrete. As new tests and criteria are developed, it will become necessary to demonstrate that mixtures in the current inventory can meet the new requirements. In some cases, current mixes will pass the new tests, and in others adjustments followed by retesting will be required. In such cases, suppliers will develop the required data and make adjustments as necessary. In a few cases, such as with the Ministry of Transportation of Ontario specifications, new criteria have been included for 1 or 2 years without being included as pay items, to allow suppliers to become familiar with the new tests and to develop suitable mixes.

12.18   Conclusions: how do we get there from here?

12.18.1   *Introduction*

The FHWA has charted a roadmap to performance-based specifications, and was working on arriving at the destination by 2008.[51] This mirrors similar programmes by the UK Highway Agency, and these planned approaches obviously involve many people for many hours over extended periods. The EU surely did not produce its documents without an enormous expenditure of time by representatives from 27 countries. All three must have and are expending huge resources on the development of their performance specifications. It would be naïve to assume that substantive changes to the philosophy for specifying and ordering concrete would occur without a well-planned and protracted programme. But there are other examples of the successful introduction of new and different technology in the concrete industry. When high-strength concrete at or above the 70 MPa (10 000 psi) level was introduced, significant changes were needed in the design and production of concrete mixtures. In Chicago, Seattle, Toronto and other cities, the consulting, testing and ready mixed concrete industries cooperated to develop concrete mixtures and quality assurance/control procedures. Ultimately, these became established practice, and were enshrined in an ACI guideline document. Similarly, at the beginning of the interest in high-performance concrete, the FHWA in the USA and Concrete Canada in Canada provided the impetus, encouragement and expertise to the industry that has led to the widespread adoption of this material. As would be expected, practical difficulties with such aspects as finishing and curing occurred, and original specifications were successively modified as experience was gained. The evolution of the ACI 318 building code itself shows clear evidence of the ability of the industry to adapt and to prove the validity of good ideas. With this background and the clear examples of important changes happening around the world and closer to home in Canada and various US state departments of transportation, we can expect progress on how we specify concrete. But how do we get there?

12.18.2 A multi-step plan

Taylor (2004)[52] has postulated nine steps in the transition from P2P. The following list of considerations and actions needed to achieve this transition builds on the 'Taylor' plan and on NRMCA intentions.

(*a*) Develop or modify an example performance specification (or minimally prescriptive specification) that takes advantage of the freedoms and opportunities afforded by current ACI 318 building code. Carefully describe the conditions under which this can be recommended for use.

(*b*) In the process of developing this model, identify those code provisions or other aspects of standard practice that represent a barrier to taking advantage of the performance specification philosophy.

(*c*) Draft specific change proposals for consideration by ACI 318 that would allow for advantageous use of performance specifications. Given that the cooperation and support of recognised authorities is a prerequisite to substantive changes in how concrete is specified, evaluated and accepted, it is essential that the P2P initiative be coordinated with groups such as ACI Committees 318 and 301. ACI 318, *Building Code Requirements for Structural Concrete*, is the 'Bible' for design professionals in the USA, much of South America and elsewhere, and ACI 301, *Specifications for Structural Concrete*, is also widely used. It will be necessary to work with these ACI committees to assist in code, commentary and specification amendments that can expand performance-oriented options and provisions. In addition to working with performance versus prescriptive requirements, attention would be given to establishing a clearer system of exposure classifications or requirements. In common with worldwide practice, these prescriptive tables would, for specific exposures, include restrictions on the type of cement or cementitious combination, the minimum cementitious content, w/c or w/cm and air content. Once a set of exposure classes has been agreed on, the next step is to produce criteria in tabular form that are essential requirements (initially some may be prescriptive) for concrete mixtures in order to assure the potential ability of concrete mixtures that incorporate these criteria's ability to meet durability requirements. Initially, the critical properties of concrete needed to meet the requirements of each exposure class may well be prescriptive, but in time would become performance criteria. Later, as practice changes and suitable test methods become available, any prescriptive requirements in the tables can be progressively deleted in favour of performance criteria.

(d) A more comprehensive model performance specification can be made available as code provisions permit.

(e) Industry acceptance is expected to be incremental as opportunities arise for project teams to capitalise on advantages of performance specifications. In the first such iteration, it is probable that only a few properties of concrete will be included as performance require-ments. These would be those for which proven tests with known and acceptable variability are available to check compliance. These tests would also be in common and extensive use, and already consid-ered useful by the design professionals. Candidate tests and properties could include the density of normal, heavy and light-weight con-cretes, early age compressive strength, flexural and splitting tensile strength, permeable voids, chloride permeability, freeze–thaw resis-tance, de-icer scaling, the modulus of elasticity, shrinkage and creep. For all these properties, only a minimum or maximum average value needs to be specified, along with a minimum or maximum value for a single test result (based on the precision of the test). Although early experience may only incorporate a few performance require-ments, this experience would inform developing code modifications.

(f) There are also tests already in somewhat specialised use that have the potential for wider application in the near future. These are the rapid chloride permeability test (ASTM C1202), the determination of air void system quality by ASTM C457, and sorptivity using the procedure in ASTM C1585, modified for in-place use.

(g) Use of performance specifications would expand along with demon-strated success and benefits, and with the incorporation of addi-tional properties as additional tests are proven to meet accuracy and variability. The procedure for bringing these tests into wider use can most easily be accomplished by an authority such as a department of transportation. A new test can be specified for use on contracts for one or more years, but for information rather than as a contractual issue. By these means, extensive field experi-ence can be gained, and the utility of the test determined. Emphasis on certification programmes is expected to increase, with attention to not only concrete production and construction but also for test-ing facilities and testing technicians.

(h) An essential part of effecting a change in mindset for stakeholders to adopt performance specifications will be education to not only encourage the appropriate use of performance specifications but also to encourage thinking the problem all the way through. (For example, similar to several 100 year service life projects in Europe, on two concrete tunnel liner contracts in Ontario (Canada), a diffusion coefficient was specified (Hart *et al.*, 1997).[53] However, the requisite test required 120 days per specimen, and test and tunnel liner production on the second contract was 1000 units per week. Clearly, the use of a 120 day test was technically justified from the perspective of performance, but logistically unmanageable.) It will be necessary to provide clear and detailed recommendations in the future based on the findings of field experience as it accumulates. The literature review identified several helpful and effective examples of articles, websites, papers, reports and brochures for paving the way for new specifications (Bickley *et al.*, 2006). For example, the current edition of the Canadian CSA A23.1, Annex J, 'Guide for selecting alternatives using Table 5 when ordering concrete', is a concise and clear aid. A similar annex should be part of the revision of ACI 318, or similar material should become part of the commentary. Such guidance will play a part in the gradual education and transitional thinking of those involved in choosing between prescription and performance or in deciding how to make the most effective blend of the two. As already initiated at the NRMCA, the education programme should include all stakeholders. Effective use of performance specifications requires that all parties understand basic concrete technology in all phases from material acquisition to in-place service. Through education and certification that qualifies companies to bid and participate on performance projects, there is a genuine opportunity to 'raise the bar.'

Acknowledgements

This chapter has been largely abstracted, with permission, from a more extensive report written for the RMC Research Foundation in the USA (Bickley, Hooton and Hover, 2006). The RMC Research Foundation is a non-profit organisation dedicated to continuous research and enhancing the educational opportunities of the ready mixed concrete industry.

References

1. Bickley, J.A., Hooton, R.D. and Hover, K.C. (2006) *Preparation of a performance-based specification for cast-in-place concrete.* National Ready Mixed Concrete Association, Silver Springs, MD.
2. Bickley, J.A., Hooton, R.D. and Hover, K.C. (2008) *Guide to specifying concrete performance: phase II report of preparation of a performance-based*

541

specification for cast-in-place concrete. National Ready Mixed Concrete Association, Silver Springs, MD.

3. Lord, A.R. (1928) *Handbook of reinforced concrete building design.* American Concrete Institute, Detroit (reprinted by Portland Cement Association, Concrete Reinforcing Steel Institute, and Rail Steel Bar Association).

4. Seelye, E.E. (1966) *Data book for civil engineers, specifications and cost.* John Wiley, New York, 3rd edn.

5. Hover, K.C. (2003) *Highway materials engineering module VI Portland cement and concrete.* NHI Course No. 131023, Publication No. HNI 04-112. Department of Transportation, Federal Highway Administration, Washington, DC.

6. Ruettgers, A., Vidal, E.N. and Wing, S.P. (1935) An investigation of the permeability of mass concrete with particular reference to Boulder Dam. *Proceedings of the American Concrete Institute*, 31, 382–416.

7. Hover, K.C. (1994) Air content and unit weight of concrete. In: P. Klieger and J.F. Lamond (eds), *Significance of tests and properties of concrete and concrete making materials*, 169C. American Society for Testing and Materials, West Conshohocken, PA, 282–295.

8. Boyd, A.J. and Hooton, R.D. (2007) Long-term scaling performance of concretes containing supplementary cementing materials. *Journal of Materials in Civil Engineering*, 16(10), 820–825.

9. ACI Committee on Responsibility in concrete construction (RCC) (2005) Guidelines for authorities and responsibilities in concrete design and construction. *Concrete International*, 27(4), 33–39.

10. Concrete Institute of Australia (2001) *Performance criteria for concrete in marine environments.* Concrete Institute of Australia, Crow's Nest.

11. Sprinkel, M.M. (2004) *Performance specification for high performance concrete overlays on bridges*, VTRC 05-R2. Virginia Transportation Research Council, Department of Transportation, Federal Highway Administration, Charlottesville, VA.

12. Bognacki, C., Marsano, J. and Wierciszewski, M. (2002) High performance aeronautical pavements. In: *2002 Federal Aviation Administration Airport Technology Transfer Conference*. Atlantic City.

13. Bickley, J.A. and Mitchell, D.A. (2001) *State-of-the-art review of high-performance concrete structures built in Canada: 1990–2000.* Cement Association of Canada, Ottawa.

14. Ontario Ministry of Transportation (2004) OPSS904 Special Provision No. 904S13 on High Performance Concrete.

15. Harrison, T.A. (1997) Performance testing of concrete for durability. *Concrete (London)*, 31(10), 14–15.

16. Harrison, T. (2003) BS EN 206-1/BS 8500, basics: conformity and identity testing. *Concrete*, 37(9), 50–51.

17. Harrison, T. (2003) *The new standards – getting started: an introductory guide to the new standards for concrete BS EN 206-1 and BS 8500.* Concrete Society and Quarry Products Association, Crowthorne.

542

18. Kessler, R.J., Powers, R.G., Vivas, E., Paredes, M.A. and Virmani, Y.P. (2008) Surface resistivity as an indicator of concrete chloride penetration resistance. In: *Proceedings of the Concrete Bridge Conference*. St Louis.
19. Duracrete (1997) *Compliance tests state-of the-art, Duracrete*, BE95-1347/ R6. CUR, Gouda.
20. Baroghel-Bouny, V. (2002) Which toolkit for durability evaluation as regards chloride ingress into concrete? Part II: development of a performance specification approach based on durability indicators and monitoring parameters. In: C. Andrade and J. Kropp (eds), *Proceedings of the 3rd International Workshop, Testing and Modelling Chloride Ingress into Concrete*, Madrid. RILEM, Bagneux.
21. Cabrera, J.G., Gowriplan, N. and Wainwright, P.J. (1989) An assessment of concrete curing efficiency using gas permeability. *Magazine of Concrete Research*, 41(149), 193–198.
22. Claisse, P.A., Elsayad, H.I. and Shaaban, I.G. (1997) Absorption and sorptivity of cover concrete. *Journal of Materials in Civil Engineering*, 9(3), 105–110.
23. Dhir, R.K., Hewlett, P.C., Byars, E.A. and Shaaban, I.G. (1995) A new technique for measuring the air permeability of near-surface concrete. *Magazine of Concrete Research*, 47(171), 167–176.
24. Dinku, A. and Reinhardt, H.W. (1997) Gas Permeability coefficient of cover concrete as a performance control. *Materials and Structures*, 30, 387–393.
25. El-Dieb, A.S. and Hooton, R.D. (1995) Water permeability measurement of high performance concrete using a high pressure triaxial cell. *Cement and Concrete Research*, 25(6), 1199–1208.
26. Figg, J.W. (1973) Methods of measuring the air and water permeability of concrete. *Magazine of Concrete Research*, 25(85), 213–219.
27. Hope, B.B. and Malhotra, V.M. (1984) The measurement of concrete permeability. *Canadian Journal of Civil Engineering*, 11, 287–292.
28. Stanish, K., Hooton, R.D. and Thomas, M.D.A. (2000) Evaluation of four short-term methods for determining the chloride penetrability in concrete. In: M.S. Khan (ed.) *Water–cement ratio and other durability parameters – techniques for determination*, ACI SP-191. American Concrete Institute, Farmington Hills, MI, 81–98.
29. Streicker, P.E. and Alexander, M.G. (1995) A chloride conduction test for concrete. *Cement and Concrete Research*, 25(6), 1284–1294.
30. Hooton, R.D. (2001) Development of standard test methods for measuring fluid penetration and ion transport rates. *Materials Science of Concrete: Fluid and Ion Transport Rates in Concrete*. Nov., 1–12.
31. Tang, L. and Nilsson, L.-O. (1991) Rapid Determination of the chloride diffusivity in concrete by applying an electrical field. *ACI Materials Journal*, 89(1), 49–53.
32. Stanish, K.D., Hooton, R.D. and Thomas, M.D.A. (2001) *Prediction of chloride penetration in concrete*, FHWA-RD-00-142. Department of Transportation, Federal Highway Administration, Washington, DC.

33. Hooton, R.D., Geiker, M.R. and Bentz, E.C. (2002) Effects of curing on chloride ingress and implications of service life. *ACI Materials Journal*, 99(2), 201–206.

34. Andrade, C., Alonso, C. and Goni, S. (1993) Possibilities for electrical resistivity to universally characterise mass transport processes in concrete. In: R.K. Dhir and M.R. Jones (eds), *Concrete 2000*. Spon, Dundee, vol. 2, 1640–1652.

35. Millard, S.G., Harrison, J.A. and Edwards, J.A. (1989) Measurements of the electrical resistivity of reinforced concrete structures for the assessment of corrosion risk. *Journal of Non-destructive Testing*, 31, 11.

36. Monfore, G.E. (1968) The electrical resistivity of concrete. *Journal of the PCA Research and Development Laboratories*, 10(2), 35–48.

37. Morris, W., Moreno, E.I. and Sagues, A.A. (1996) Practical evaluation of resistivity of concrete in concrete cylinders using a Wenner array probe. *Cement and Concrete Research*, 26(12), 1779–1788.

38. Goodspeed, C.M., Vanikar, S. and Cook, R.A. (1996) High-performance concrete defined for highway structures. *Concrete International*, 11, 62–67.

39. Balayssac, J.P., Detriche, C.H. and Grandet, J. (1993) Validity of the water absorption test for characterising cover concrete. *Materials and Structures*, 26, 226–230.

40. Hall, C. (1989) Water sorptivity of mortars and concretes: a review. *Magazine of Concrete Research*, 41(147), 51–61.

41. Price, W.F. and Bamforth, P.B. (1991) Initial surface absorption of concrete: examination of modified test apparatus for obtaining uniaxial flow. *Magazine of Concrete Research*, 43(155), 93–104.

42. McCarter, W.J., Ezirim, H. and Emerson, M. (1996) Properties of concrete in the cover zone: water penetration, sorptivity and ionic ingress. *Magazine of Concrete Research*, 48(176), 149–156.

43. Ho, D.W.S. and Chirgwin, G.J. (1996) A performance specification for durable concrete. *Construction and Building Materials*, 10(5), 375–379.

44. Desouza, S.J., Hooton, R.D. and Bickley, J.A. (1997) evaluation of laboratory drying procedures relevant to field conditions for concrete sorptivity measurements. *Cement, Concrete, and Aggregates*, 19(2), 92–96.

45. Desouza, S.J., Hooton, R.D. and Bickley, J.A. (1998) A field test for evaluating high performance concrete Covercrete quality. *Canadian Journal of Civil Engineering*, 25, 551–556.

46. Desouza, S.J., Hooton, R.D. and Bickley, J.A. (2000) *A practical QC test programme for HPC in precast concrete tunnel liners*, SP-191. American Concrete Institute, Farmington Hills, MI.

47. Wilson, C.W. (1968) Plant-certification – a program of merit in the prestressing industry. *PCI Journal*, 13(2), 12–18.

48. PCI (1999) *Manual for quality control for plants and production of structural precast concrete products*, MNL-116-99. Precast/Prestressed Concrete Institute, Chicago, IL, 4th edn.

49. PCI (2001) *Certified quality for people, products and performance*, PC-4. Precast/Prestressed Concrete Institute, Chicago, IL.

50. Ontario Ministry of Transportation (2004) OPSS904 Special Provision No. 904S11 on Air Void System in Hardened Concrete.
51. FHWA (2008) Concrete Pavement (CP) Roadmap http://www.cproadmap. org/index.cfm.
52. Taylor, P. (2004) Performance based specifications for concrete. *Concrete International*, 26(8), 91–93.
53. Hart, A.J., Ryell, J. and Thomas, M.D.A. (1997) High performance concrete in precast tunnel linings; meeting chloride diffusion and permeability requirements. In: *Proceedings of the PCI/FHWA International Symposium on High Performance Concrete*. New Orleans, 294–308.

Further reading

CEB (1997) *New approach to durability – an example for carbonation induced corrosion*. Bulletin 238. Comité Euro-International du Béton, Lausanne.

CEB (1997) *Quality management: guidelines for the implementation of the ISO standards of the 9000 series in the construction industry*. Bulletin 234. Comité Euro-International du Béton, Lausanne.

CEB (1998) *Quality management*, Thomas Telford, London.

Day, K.W. (2005) Perspective on prescriptions. *Concrete International*, 27(7), 27–30.

Duracrete (2000). *Probabilistic performance based durability design of concrete structures. Final technical report*. CUR, Gouda.

Highways Agency (2003) *Developing performance specifications: issues*. Highways Agency, London.

Highways Agency (2004) *Developing performance specifications: questions*. Highways Agency, London.

Highways Agency (2003) *Developing performance specifications: consultation document*. Highways Agency, London.

Highways Agency (2004) *Developing performance specifications: consultation response analysis report*. Highways Agency, London.

NMRA (2004) *Experimental case study demonstrating advantages of performance specifications*. Summary of proposal. National Ready Mixed Concrete Association, Silver Spring, MD. www.nrmca.org/P2P/Summary%20of% 20P2P%20Lab%20Study.pdf.

NMRA (2004) *Prescription to performance*. National Ready Mixed Concrete Association, Silver Spring, MD. www.nrmca.org/P2P/P2P%20Brochure. pdf.

NMRA (2004) *The P2P Initiative: a shift to performance-based specifications for concrete focuses on innovation, quality and customer satisfaction*. National Ready Mixed Concrete Association, Silver Spring, MD. www.nrmca.org/ P2P/P2P%20Article.pdf.

NMRA (2008) The P2P Initiative – continued prescriptive specification versus performance specifications: www.nrmca.org/P2P/p2p_cont.asp.

13

Advances in durability design and performance-based specification

Odd E. Gjorv, Norwegian University of Science and Technology, Norway

13.1   Introduction

Although most codes and specifications for concrete durability have been upgraded a number of times in recent years, current code specifications for concrete durability are still almost exclusively based on prescriptive requirements for concrete composition and execution of concrete work, the results of which are neither unique nor easy to verify and control during concrete construction. For many years, when concrete was mostly based on pure Portland cements and simple procedures for concrete production, the concept of water/cement ratio was the fundamental basis both for characterising and specifying concrete quality. Since a number of different cementitious materials and reactive fillers are now being applied for concrete production, however, the concrete properties are more and more being controlled by the various combinations of such materials. In addition, the concrete properties are also more and more being controlled by the use of various types of processed concrete aggregate, new concrete admixtures and sophisticated production equipment. As a result, the old and very simple terms 'water/cement ratio' and 'water/binder ratio' for characterising and specifying concrete quality have successively lost their meaning. As a consequence, there is a great need for performance-based definition and specification for concrete quality. In particular, this is true for characterising and specifying concrete durability.

In recent years, much research has been carried out in order to develop new procedures and recommendations for performance-based specification of concrete durability. Of all this research, very encouraging results based on the development of durability design have been obtained. Since all factors affecting the durability also show a large scatter and variability, the durability design has also mostly taken a probabilistic approach. As a result, it is now possible to specify performance-based requirements for the concrete durability, which also provide a basis both for performance-based concrete quality control during concrete construction and documentation of compliance with the specified durability.

Of the various deteriorating processes which may cause durability problems to concrete structures, current procedures for durability design is only available for corrosion of reinforcing steel either due to chloride penetration or carbonation. However, it is not the disintegration of the concrete itself, but rather electrochemical corrosion of the embedded steel, which poses the most critical and greatest threat to the durability and service life of concrete structures in severe environments.[1,2] In particular, this is true for concrete structures in chloride-containing environments.

For most types of deteriorating processes such as freezing and thawing and alkali–aggregate reactions as well as other types of chemical processes, much experience and many practical guidelines exist for how to reduce or avoid such durability problems for new concrete structures.[3] For new concrete structures to be produced for chloride-containing environments, however, it still appears to be difficult to avoid any steel corrosion within reasonable service periods of 20–30 years.[1,2] Therefore, in order to obtain a more controlled and improved durability of such concrete structures, probability-based durability design is increasingly being applied.

13.2   Durability design

13.2.1   General

In recent years, a rapid development of models and procedures for probability-based durability design of concrete structures has taken place,[3–7] and in many countries, such durability design has been applied to a number of important concrete structures.[8–11] Also, in Norway such durability design has been applied to a number of concrete structures, where safety, durability and service life have been of special importance.[3,12] In the beginning, this design was primarily based on the results and guidelines from the European research project Duracrete,[13] but successively as practical experience with such design was gained, the basis for the design was simplified and further developed for more practical applications. Thus, in 2004, this design was adopted by the Norwegian Association for Harbour Engineers as general recommendations and guidelines for durability design and performance-based concrete quality control of new concrete structures in Norwegian harbours,[14] a brief outline of which is given in the following.

13.2.2   Theoretical basis

13.2.2.1   Chloride penetration

As thoroughly discussed and presented by Poulsen and Mejlbro,[15] the penetration of chlorides into concrete is based on rather complex transport mechanisms. In a very simplified form, however, it is possible to estimate the rates of chloride penetration by use of Fick's second law of diffusion according to Collepardi *et al.*,[16,17] in combination with a time-dependent chloride diffusion coefficient according to Takewaka and Mastumoto,[18] Polder and deRooij[19] and Tang and Gulikers,[20] as shown in equations (13.1) and (13.2).

$$C(x,t) = C_S\left[1 - \mathrm{erf}\left(\frac{x}{2\sqrt{D(t)t}}\right)\right] \tag{13.1}$$

where $C(x,t)$ is the chloride concentration at depth x after time t, C_S is the chloride concentration at the concrete surface and D is the concrete chloride diffusion coefficient.

$$D(t) = \frac{D_0}{1-\alpha}\left[\left(1+\frac{t'}{t}\right)^{1-\alpha} - \left(\frac{t'}{t}\right)^{1-\alpha}\right]\left(\frac{t_0}{t}\right)^{\alpha} k_e \tag{13.2}$$

where D_0 is the diffusion coefficient after the reference time t_0, and t' is the age of the concrete at the time of chloride exposure. The parameter α represents the time dependence of the diffusion coefficient, while k_e is a parameter which takes the effect of temperature into account:

$$k_e = \exp\left[b_e\left(\frac{1}{293} - \frac{1}{T+273}\right)\right] \tag{13.3}$$

where b_e is a regression parameter and T is the temperature.

The criterion for steel corrosion then becomes

$$C(x) = C_{CR} \tag{13.4}$$

where $C(x)$ is the chloride concentration at the depth of the embedded steel, and C_{CR} is the critical chloride concentration in the concrete necessary for the onset of corrosion.

13.2.2.2 Calculation of probability

For the structural design of concrete structures, the main objective is always to establish the combined effects of loads (S) and the resistance to withstand the loads (R) in such a way that the design criterion becomes

$$R \geq S \quad \text{or} \quad R - S \geq 0 \tag{13.5}$$

When $R < S$, failure will occur, and since all the factors affecting R and S also show a high scatter and variability, all established design procedures have properly taken this into account.

In principle, the durability design takes the same approach as that of the structural design. In this case, the effect of both loads (S), which is the combined effect of both chloride loads and temperature conditions and the resistance to withstand the loads (R), which is the resistance against chloride penetration, must be established. Although neither S nor R is comparable to that of the structural design, the acceptance criterion for having the probability for 'failure' less than a given value is the same.

In Fig. 13.1, the scatter and variability of both R and S are demonstrated in the form of two distribution curves along the y axis. At an early stage, there is no overlapping between these two distribution curves, but over time, a gradual overlapping from time t_1 to t_2 takes place. This increasing overlapping will at any time reflect the probability of 'failure' or the probability for corrosion to occur, and, gradually, the upper acceptable level for the probability of 'failure' (t_{SLS}) is reached and exceeded.

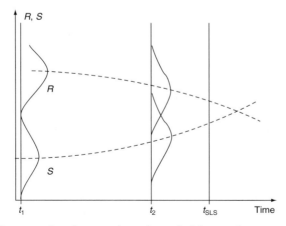

Fig. 13.1 The principles of a time-dependent reliability analysis

549

In principle, the probability of failure can be written as

$$P(\text{failure}) = P_f = P(R - S < 0) < P_0 \qquad (13.6)$$

where P_0 is a measure of the probability of failure.

In current codes for reliability of structures, an upper level for the probability of failure of 10% in the serviceability limit state is often specified.[21] Therefore, an upper probability level of 10% for the onset of corrosion has also been adopted as a basis for the current durability design.

Normally, the failure function includes a number of variables, all of which have their own statistical parameters. Therefore, the use of such a failure function requires numerical calculations and the application of special software. Currently, there are several mathematical methods available for the evaluation of the failure function,[22–25] such as

- FORM (first order reliability method)
- SORM (second order reliability method)
- Monte Carlo simulation (MCS).

13.2.2.3 Calculation of corrosion probability

In principle, the calculation of corrosion probability can be carried out by use of any of the above mathematical methods in combination with the use of appropriate software. Based on current experience with the durability design of concrete structures in chloride-containing environments, however, calculation of chloride penetration by use of equation (13.1) in combination with a Monte Carlo simulation has proved to give a very simple and appropriate basis for calculating the probability of corrosion.[12] Although such a combined calculation also can be carried out in different ways, special software[26] for this calculation has been developed.[27]

Primarily, the above calculation of corrosion probability provides the basis for the durability design of new concrete structures. As a result, it is possible to specify a certain service period before a probability of 10% for corrosion is reached. For the given environmental exposure, requirements for both concrete quality and concrete cover can then be specified. For the later operation of the concrete structure, however, the above calculation also provides a basis for condition assessment and preventive maintenance. Based on the observed rates of the real chloride penetration taking place during operation of the structure, calculations of the future probability of corrosion then provide a basis for the preventive maintenance of the structure. For both types of probability calculation, certain input parameters to the analyses are needed.

13.2.3 Input parameters

In general, the durability design should always be an integral part of the structural design for the given concrete structure. At an early stage of the design, therefore, the overall durability requirement of the structure should be based on the specification of a certain service period before 10% probability of corrosion is reached. For the given environmental exposure, the durability analysis then provides the basis for specification of a proper combination of concrete quality and concrete cover. Before the final requirements for concrete quality and concrete cover are given, however, it may be necessary to carry out several calculations for various combinations of possible concrete qualities and concrete covers. For all of these calculations, proper information about the following input parameters is needed:

- environmental loading:
  - chloride loads, C_S
  - temperature, T
- concrete quality:
  - chloride diffusivity, D_0
  - time dependence factor, α
  - critical chloride content, C_{CR}
- concrete cover, X.

All the above parameters may have different distribution characteristics. If nothing else is known, however, a statistical normal distribution may be assumed. For each parameter, proper information on both average value and standard deviation is then needed.

It is not possible to go into any details in this overview on how to determine or select the various input parameters to the above calculations: some general guidelines for the determination and selection can be found elsewhere.[14] For concrete structures in chloride-containing environments, however, the environmental loads are primarily based on current experience with observed surface chloride concentrations (C_S) on similar concrete structures in similar environments.

In the literature, there are several types and definitions of the chloride diffusivity of a given concrete as well as several methods for the testing of the chloride diffusivity.[28] As shown in Chapter 6,

NORDTEST has standardised three different types of test method, including the steady state migration method NT Build 355,[29] the immersion test method NT Build 443[30] and the non-steady state migration method NT Build 492.[31] Although all of these test methods show different values for the chloride diffusivity, they show a strong correlation.[28,32,33] Therefore, any of the established test methods can be used both for quantifying and comparing the resistivity against chloride penetration of various types of concrete (Chapter 8).

Although all of the above types of test method are accelerated test methods, the duration of the testing is very different. For the non-steady state migration method, there is no requirement for pre-curing of the concrete, and the testing mostly takes 24 hours, while both the steady state migration method and the immersion test method are based on well-cured concrete specimens, and the testing takes a very long time. Therefore, in order to also be used as a simple and rapid test method for quality control during concrete construction, extensive experience has shown that the chloride diffusivity based on the non-steady state migration method or the so-called rapid chloride migration (RCM) method proves to be a very appropriate test method. In particular, this is true if this test method is also combined with a corresponding testing of the electrical resistivity of the concrete.[34]

In order to meet the overall durability requirement of a 10% probability for steel corrosion during a certain service period, requirements for both the chloride diffusivity (D_0) and the concrete cover (X) for the given concrete structure must be specified. Then, a performance-based concrete quality control during concrete construction can be carried out.

13.3   Concrete quality control

13.3.1   *General*

Extensive experience demonstrates that the durability of concrete structures is closely related not only to design and material but also to construction issues. Although a probability-based approach to the durability design to a certain extent takes the great variability of construction quality into account, a numerical approach to the durability design alone is not sufficient for ensuring proper durability. For concrete structures in severe environments, construction quality and variability is a key issue which must be firmly grasped before a more rational approach to a more controlled durability can be achieved.[35] In order to provide some information about achieved construction quality, however, proper quality control of both the specified chloride diffusivity and the concrete cover must be carried out.

13.3.2   Chloride diffusivity

Although the RCM method has proved to be a simple, rapid and precise test method for determination of the chloride diffusivity (D_0),[36] the testing may still take a couple of days, which is not good enough for regular quality control during concrete construction. For all porous materials, however, the Nernst–Einstein equation expresses the following relationship between the ion diffusivity and the electrical resistivity of the material:[37]

$$D_i = \frac{RT}{Z^2F^2} \frac{t_i}{\gamma_i c_i \rho} \qquad (13.7)$$

where:

D_i = diffusivity for ion i
R = gas constant
T = absolute temperature
Z = ionic valence
F = Faraday constant
t_i = transfer number of ion i
γ_i = activity coefficient for ion i
c_i = concentration of ion i in the pore water
ρ = electrical resistivity

Since most of the factors in equation (13.7) are physical constants, the relationship can, for a given concrete with given temperature and moisture conditions, be simplified to

$$D = k\frac{1}{\rho} \qquad (13.8)$$

where D is the chloride diffusivity, k is a constant and ρ is the electrical resistivity of the concrete. Since the electrical resistivity of the concrete can be measured in a more rapid and simple way than the chloride diffusivity, it is primarily regular quality control of the electrical resistivity of the concrete which provides the basis for indirect quality control of the chloride diffusivity during concrete construction.[34] Therefore, as soon as the type of concrete is known, the above relationship between chloride diffusivity and electrical resistivity must be established. This is done by producing a certain number of concrete

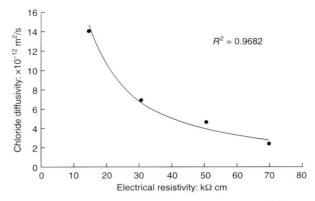

Fig. 13.2 A typical calibration curve for control of chloride diffusivity based on measurements of electrical resistivity[14]

specimens, on which parallel testing of both chloride diffusivity and electrical resistivity at different periods of water curing are carried out. After the relationship between the chloride diffusivity and the electrical resistivity has been established, this relationship is later on used as a calibration curve for indirect control of the chloride diffusivity based on regular measurements of the electrical resistivity during concrete construction (Fig. 13.2). Since the testing of electrical resistivity is a non-destructive type of test, these measurements are carried out on the same concrete specimens as those being used for the regular quality control of the 28 day compressive strength during concrete construction.

13.3.3   Concrete cover
Before any placing of the concrete takes place, regular control of the positioning of the reinforcing steel is always carried out. However, after placing of the concrete, it is very important to control whether the specified concrete cover has been achieved. It is the average value and standard deviation of the achieved concrete cover during concrete construction which are needed as the input into the durability analysis and for documentation of the achieved construction quality.

13.4   Achieved construction quality

13.4.1   General
From performance-based concrete quality control, average values and the standard deviation of both chloride diffusivity and concrete cover

are obtained. Upon completion of the concrete construction, therefore, these data are used as the input into a new durability analysis which provides documentation of compliance with the specified durability.

Since the specified chloride diffusivity is only based on the testing of small, separately produced concrete specimens water cured in the laboratory for 28 days, such a chloride diffusivity may be quite different from that achieved on the construction site. During concrete construction, therefore, some additional documentation on the achieved chloride diffusivity on the construction site must also be provided. At the end of concrete construction, such a chloride diffusivity in combination with the achieved concrete cover provide the basis for documentation of achieved in-situ durability during concrete construction.

Since neither the 28 day chloride diffusivity from small concrete specimens nor the achieved chloride diffusivity on the construction site during concrete construction reflects the potential chloride diffusivity of the given concrete, further documentation on the long-term chloride diffusivity of the given concrete must also be provided. Such a chloride diffusivity in combination with the achieved concrete cover provide the basis for documentation of the potential durability of the given structure.

Although the 28 day chloride diffusivity is normally used as the parameter for characterising the resistance of the given concrete against chloride penetration, this parameter is also being used as a durability parameter for characterising the general durability properties of a given concrete, not only for chloride-containing environments but also for other severe and aggressive environments. Thus, regular testing of the chloride diffusivity in the same way as outlined above was also used as part of the regular concrete quality control and documentation of the achieved construction quality of several concrete structures produced for a new sewage and water treatment plant in Trondheim some years ago.[38]

13.4.2 *Compliance with specified durability*
As a result of the durability design, an overall durability requirement based on a given service period with a probability for corrosion of less than 10% has been specified. In order to show compliance with such a durability requirement, a new durability analysis has to be carried out based on the average value and standard deviation of both the chloride diffusivity and the concrete cover obtained from the quality control during concrete construction. Although it may have been

difficult to select proper data for several of the input parameters to the original durability analysis, these input parameters are now the same for the new durability analysis. Therefore, this new durability analysis primarily reflects the achieved values for chloride diffusivity and concrete cover during concrete construction, including the observed scatter and variability. Hence, the new durability analysis provides a basis for the documentation of compliance with the specified durability.

13.4.3 In-situ durability

In principle, any documentation of the achieved chloride diffusivity on the construction site should be based on the testing of a number of concrete cores removed from the concrete structure under construction. In order not to weaken the structure, however, one or more unreinforced concrete elements should be separately produced on the construction site, from which much of the concrete coring can take place during the construction period. In addition, a certain extent of coring from the real concrete structure must also be carried out, but only from locations where the coring will not weaken the concrete structure.

The separately produced concrete elements, which can either be a wall or slab type of element or both, should always be produced and cured as representative as possible of the real concrete structure or various parts of the concrete structure. From these elements, which are produced at an early stage of the concrete construction work, a number of 100 mm-diameter concrete cores are later on removed at various ages, and immediately upon removal sent to the laboratory for testing of the achieved chloride diffusivity. In order to obtain a proper curve for the development of chloride diffusivity on the construction site, the cores should be removed and tested after periods of approximately 14, 28, 60, 90, 180 and up to, at least, 365 days. In addition, supplemental data on the achieved chloride diffusivity are also obtained from the testing of a certain number of concrete cores removed from the real concrete structure.

Depending somewhat on the type of binder system, the obtained development of chloride diffusivity on the construction site often tends to level out after a period of approximately 1 year. As an example, the observed development of the achieved chloride diffusivity from one particular construction site is shown in Fig. 13.3. In this figure, the development of chloride diffusivity for the same concrete based on separately cast and cured concrete specimens in the laboratory is also

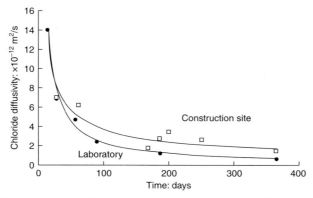

Fig. 13.3 *Development of the achieved chloride diffusivity on the construction site and in the laboratory*[14]

shown. Based on the achieved chloride diffusivity on the construction site after approximately 1 year, a new durability analysis is carried out. In combination with the achieved data on concrete cover, this new durability analysis provides the basis for the documentation of the achieved in-situ durability during the construction period.

13.4.4   Potential durability

In order to establish the calibration curve, the chloride diffusivity was determined on separately cast concrete specimens after curing periods in the laboratory of up to approximately 60 days. By continued testing of the chloride diffusivity on a few additional specimens after curing periods of up to approximately 1 year, a further development of chloride diffusivity is obtained, as shown in Fig. 13.3. Although it may take a long time before a final value for the chloride diffusivity is reached, such a development curve tends to level out after approximately 1 year for most types of concrete. Hence, the observed chloride diffusivity after approximately 1 year of water curing is used as the input into a new durability analysis. In combination with the achieved data on concrete cover, this analysis provides a basis for the documentation of the potential durability of the given concrete structure.

13.5   Evaluation and discussion of the obtained results

In order to obtain the best possible durability for a given concrete structure in a given environment, experience from a number of concrete structures produced in recent years has shown that durability analyses provide a proper basis for comparing and selecting one of several technical solutions.[14] For the evaluation of the obtained results, however, it is important to be aware of all the simplifications and assumptions made in the current procedures for the calculation of the corrosion probability. Although diffusion is a dominating transport mechanism through thick concrete covers in chloride-containing environments, only a very simple diffusion model for the calculation of chloride penetration rates is being used. Under more realistic conditions in the field, other transport mechanisms for chloride penetration than pure diffusion also exist. The characterisation of the resistance of the concrete against chloride penetration is further based on a rapid migration type of testing, where the chloride penetration is very different from what takes place under more realistic conditions in the field. This type of testing primarily distinguishes the differences in the chloride mobility from one type of concrete to another, and does not properly reflect the differences in the ability of the various binder systems to bind chlorides.[32] The durability analyses are also based on a number of other input parameters, for which there is a lack of reliable data and information. In particular, this is true for the input parameters such as the chloride loads (C_S) and the ageing factor for the chloride diffusivities (α). Although a selection of these parameters should preferably be based on current experience from other similar concrete structures in similar environments, such information is not necessarily available. Therefore, the selection of these parameters is normally based on general experience. The temperature is also another important factor, a proper value for which may also be difficult to select.

Based on the above simplifications and assumptions, therefore, the obtained 'service periods' with a probability for corrosion of less than 10% should not be taken as actual service periods for the given concrete structure. However, the durability analyses provide a basis for an engineering judgement of the most important factors which are considered relevant for durability, including the scatter and variability of all factors involved. Hence, a proper basis for comparing and selecting one of several technical solutions in order to obtain a best possible durability is obtained. As a result, a durability requirement can also be specified which is possible to verify and control in such a way that documentation of the achieved construction quality and compliance with the specified durability can be provided.

Even if the most significant requirements for both concrete quality and concrete cover have been specified and achieved during concrete construction, extensive experience demonstrates that for all concrete structures in severe chloride-containing environments, a certain rate of chloride penetration will always take place during the operation of the structure. As part of the durability design, therefore, it is also very important to provide the owner with a service manual for regular condition assessment and preventive maintenance of the structure.[14] For concrete structures in severe chloride-containing environments, it is only regular monitoring of the actual chloride penetration during the operation of the structure and evaluation of the future corrosion probability in combination with proper protective measures which provide the ultimate basis for achieving a more controlled durability and service life of the given concrete structure in the given environment.

References

1. Gjørv, O.E. (1994) Steel corrosion in concrete structures exposed to Norwegian marine environment. *Concrete International*, 16(4) 35–39.
2. Gjørv, O.E. (1996) Performance and serviceability of concrete structures in the marine environment. In: P.K. Mehta (ed.), *Proceedings of the Odd E. Gjørv Symposium on Concrete for Marine Structures*. Ottawa, 259–279.
3. Gjørv O.E. (2002) Durability and service life of concrete structures. In: *Proceedings of the 1st Fib Congress 2002*. Tokyo, session 8, vol. 6, 1–16.
4. Siemes, A.J.M. and Rostam, S. (1996) Durability safety and serviceability – a performance based design. In: *Proceedings of the IABSE Colloquium on Basis of Design and Actions on Structures*. Delft.
5. Engelund, S. and Sørensen, J.D. (1998) A probabilistic model for chloride-ingress and inititation of corrosion in reinforced concrete structures. *Structural Safety*, 20, 69–89.
6. Gehlen, C. (2000) *Probability-based service life calculations of concrete structures – reliability evaluation of reinforcement corrosion*. Dissertation, RWTH-Aachen (in German).
7. Fib (2006) *Model code for service life design*. Fib Bulletin 34. Federation International du Beton, Lausanne.
8. Stewart, M.G. and Rosowsky, D.V. (1998) Structural Safety and serviceability of concrete bridges subject to corrosion. *Journal of Infrastructure Systems*, 4(4), 146–155.
9. McGee, R. (1999) Modelling of durability performance of Tasmanian bridges. In: *Proceedings of the 8th International Conference on the Application of Statistics and Probability*. Sidney.
10. Gehlen, C. and Schiessl, P. (1999) Probability-based durability design for the Western Scheldt Tunnel. *Structural Concrete*, P.1(2), 1–7.
11. Gehlen, C. (2007) Durability design according to the new model code for service life design. In: F. Toutlemonde, K. Sakai, O.E. Gjørv and

N. Banthia (eds), *Proceedings of the 5th International Conference on Concrete Under Severe Conditions – Environment and Loading*. Paris, vol. 1.

12. Gjørv, O.E. (2004) Durability design and construction quality of concrete structures, In: B.H. Oh, K. Sakai, O.E. Gjørv and N. Banthia (eds), *Proceedings of the 4th International Conference on Concrete Under Severe Conditions – Environment and Loading*. Seoul, vol. 1.

13. Duracrete (2000) *General guidelines for durability design and redesign. Probabilistic performance based durability design of concrete structures*. Document R 15. EU – Brite EuRam III, Research Project No. BE95-1347. EU, Brussels.

14. Gjørv, O.E. (2009) *Durability design of concrete structures in severe environments*. Taylor and Francis, London.

15. Poulsen, E. and Mejlbro, L. (2006) *Diffusion of chlorides in concrete – theory and application*, Taylor and Francis, London.

16. Collepardi, M., Marcialis, A. and Turriziani, R. (1970) Kinetics of penetration of chloride ions in concrete. *L'Industria Italiana del Cemento*, 4, 157–164.

17. Collepardi, M., Marcialis, A. and Turriziani, R. (1972) Penetration of chloride ions into cement pastes and concretes. *Journal, American Ceramic Society*, 55(10), 534–535.

18. Takewaka, K. and Mastumoto, S. (1988) Quality and cover thickness of concrete based on the estimation of chloride penetration in marine environments. In: V.M. Malhotra (ed.), *Proceedings of the 2nd International Conference on Concrete in Marine Environment*. St-Andrews-by-the-Sea, ACI SP 109, 381–400.

19. Polder, R.B. and deRooij, M.R. (2005) Durability of marine concrete structures – field investigations and modelling. *HERON*, 50(3) 133–153.

20. Tang, L. and Gulikers, J. (2007) On the mathematics of time-dependent apparent chloride diffusion coefficient in concrete. *Cement and Concrete Research*, 37(4), 589–595.

21. Standard Norway (2004) *Design of structures – requirements to reliability*, NS 3490. Standard Norway, Oslo (in Norwegian).

22. Hammersley, J.M. and Handscomb, D.C. (1964) *Monte Carlo methods*. Methuen, London.

23. Walpole, R.E., Myers, R.H. and Meyers, S.L. (1972) *Probability and statistics for engineers and scientists*. Prentice-Hall, Englewood Cliffs, NJ.

24. Haugen, E.B. (1980) *Probabilistic mechanical design*. John Wiley, New York.

25. Elishakoff, I. (1983) *Probabilitic methods in the theory of structures*, John Wiley, New York.

26. DURACON (2004) *Probability-based durability analysis of concrete structures – software manual*. University do Minho, Department of Civil Engineering, Guimaraes. www.durabilityofconcrete.com.

27. Ferreira, M., Årskog, V., Jalali, S. and Gjørv, O.E. (2004) Software for probability-based durability analysis of concrete structures. In: B.H. Oh,

K. Sakai, O.E. Gjørv and N. Banthia, (eds), *Proceedings of the 4th International Conference on Concrete Under Severe Conditions – Environment and Loading*. Seoul, vol. 1.

28. Schiessl, P. and Lay, S. (2005) Influence of concrete composition. In: H. Böhni (ed.), *Corrosion in reinforced concrete structures*. Woodhead, Cambridge, 91–134.

29. NORDTEST (1989) *NT Build 355: concrete, repairing materials and protective coating: diffusion cell method, chloride permeability*. NORDTEST, Espoo.

30. NORDTEST (1995) *NT Build 443: concrete, hardened: accelerated chloride penetration*. NORDTEST, Espoo.

31. NORDTEST (1999) *NT Build 492: concrete, mortar and cement based repair materials: chloride migration coefficient from non-steady state migration experiments*. NORDTEST, Espoo.

32. Tong, L. and Gjørv, O.E. (2001) Chloride diffusivity based on migration testing. *Cement and Concrete Research*, 31, 973–982.

33. CHLORTEST (2005) *WP 5 report – final evaluation of test methods. Resistance of concrete to chloride ingress – from laboratory tests to in-field performance*. EU 5th Framework Program, Growth Project G6RD-CT-2002-00855. CORDIS, Brussels.

34. Gjørv, O.E. (2003) Durability of concrete structures and performance-based quality control. In: A.S. El-Dieb, M.M.R. Taha and S.L. Lissel (eds), *Proceedings of the International Conference on Performance of Construction Materials in The New Millenium*. Cairo.

35. Sommerville, G. (2000) A hollistic approach to structural durability design. In: O.E. Gjørv and K. Sakai (eds), *Concrete technology for a sustainable development in the 21st century* ed. Spon, London, 41–56.

36. Tang, L. (2008) *The RCM test (NT build 492) for evaluating the resistance of concrete to chloride ingress*. Tang's Cl Tech (tang.luping@bredband.net).

37. Atkins, P.W. and De Paula, J. (2006) *Physical chemistry*. Oxford University Press, Oxford, 8th edn.

38. Rindal, J., Gussiås, A. and Gjørv, O.E. (2000) *Høvringen Sewage Plant – documentation of achieved concrete construction quality*. BML report 200101. Department of Building Materials, Norwegian University of Science and Technology, Trondheim (in Norwegian).

Index

Page numbers in *italics* denote figures.

563